Lecture Notes in Physics

The Editorial Policy for Edited Volumes

The series Lecture Notes in Physics reports new developments in physical research and teaching - quickly, informally, and at a high level. The type of material considered for publication includes monographs presenting original research or new angles in a classical field. The timeliness of a manuscript is more important than its form, which may be preliminary or tentative. Manuscripts should be reasonably self-contained. They will often present not only results of the author(s) but also related work by other people and will provide sufficient motivation, examples, and applications.

Acceptance

The manuscripts or a detailed description thereof should be submitted either to one of the series editors or to the managing editor. The proposal is then carefully refereed. A final decision concerning publication can often only be made on the basis of the complete manuscript, but otherwise the editors will try to make a preliminary decision as definite as they can on the basis of the available information.

Contractual Aspects

Authors receive jointly 30 complimentary copies of their book. No royalty is paid on Lecture Notes in Physics volumes. But authors are entitled to purchase directly from Springer other books from Springer (excluding Hager and Landolt-Börnstein) at a $33\frac{1}{3}$% discount off the list price. Resale of such copies or of free copies is not permitted. Commitment to publish is made by a letter of interest rather than by signing a formal contract. Springer secures the copyright for each volume.

Manuscript Submission

Manuscripts should be no less than 100 and preferably no more than 400 pages in length. Final manuscripts should be in English. They should include a table of contents and an informative introduction accessible also to readers not particularly familiar with the topic treated. Authors are free to use the material in other publications. However, if extensive use is made elsewhere, the publisher should be informed. As a special service, we offer free of charge LaTeX macro packages to format the text according to Springer's quality requirements. We strongly recommend authors to make use of this offer, as the result will be a book of considerably improved technical quality. The books are hardbound, and quality paper appropriate to the needs of the author(s) is used. Publication time is about ten weeks. More than twenty years of experience guarantee authors the best possible service.

LNP Homepage (springerlink.com)

On the LNP homepage you will find:
— The LNP online archive. It contains the full texts (PDF) of all volumes published since 2000. Abstracts, table of contents and prefaces are accessible free of charge to everyone. Information about the availability of printed volumes can be obtained.
— The subscription information. The online archive is free of charge to all subscribers of the printed volumes.
— The editorial contacts, with respect to both scientific and technical matters.
— The author's / editor's instructions.

E. Bick F. D. Steffen (Eds.)

Topology and Geometry in Physics

 Springer

Editors

Eike Bick
d-fine GmbH
Opernplatz 2
60313 Frankfurt
Germany

Frank Daniel Steffen
DESY Theory Group
Notkestraße 85
22603 Hamburg
Germany

E. Bick, F.D. Steffen (Eds.), *Topology and Geometry in Physics*, Lect. Notes Phys. **659** (Springer, Berlin Heidelberg 2005), DOI 10.1007/b100632

ISBN 978-3-642-06209-4   e-ISBN 978-3-540-31532-2
ISSN 0075-8450

Springer is a part of Springer Science+Business Media

springeronline.com

The use of general descriptive names, registered names, trademarks, etc. in this publication does not imply, even in the absence of a specific statement, that such names are exempt from the relevant protective laws and regulations and therefore free for general use.

Cover design: *design & production*, Heidelberg

Printed on acid-free paper
54/3141/ts - 5 4 3 2 1 0

Preface

The concepts and methods of topology and geometry are an indispensable part of theoretical physics today. They have led to a deeper understanding of many crucial aspects in condensed matter physics, cosmology, gravity, and particle physics. Moreover, several intriguing connections between only apparently disconnected phenomena have been revealed based on these mathematical tools. Topological and geometrical considerations will continue to play a central role in theoretical physics. We have high hopes and expect new insights ranging from an understanding of high-temperature superconductivity up to future progress in the construction of quantum gravity.

This book can be considered an advanced textbook on modern applications of topology and geometry in physics. With emphasis on a pedagogical treatment also of recent developments, it is meant to bring graduate and postgraduate students familiar with quantum field theory (and general relativity) to the frontier of active research in theoretical physics.

The book consists of five lectures written by internationally well known experts with outstanding pedagogical skills. It is based on lectures delivered by these authors at the autumn school "Topology and Geometry in Physics" held at the beautiful baroque monastery in Rot an der Rot, Germany, in the year 2001. This school was organized by the graduate students of the Graduiertenkolleg "Physical Systems with Many Degrees of Freedom" of the Institute for Theoretical Physics at the University of Heidelberg. As this Graduiertenkolleg supports graduate students working in various areas of theoretical physics, the topics were chosen in order to optimize overlap with condensed matter physics, particle physics, and cosmology. In the introduction we give a brief overview on the relevance of topology and geometry in physics, describe the outline of the book, and recommend complementary literature.

We are extremely thankful to Frieder Lenz, Thomas Schücker, Misha Shifman, Jan-Willem van Holten, and Jean Zinn-Justin for making our autumn school a very special event, for vivid discussions that helped us to formulate the introduction, and, of course, for writing the lecture notes for this book. For the invaluable help in the proofreading of the lecture notes, we would like to thank Tobias Baier, Kurush Ebrahimi-Fard, Björn Feuerbacher, Jörg Jäckel, Filipe Paccetti, Volker Schatz, and Kai Schwenzer.

The organization of the autumn school would not have been possible without our team. We would like to thank Lala Adueva for designing the poster and the web page, Tobial Baier for proposing the topic, Michael Doran and Volker

Schatz for organizing the transport of the blackboard, Jörg Jäckel for financial management, Annabella Rauscher for recommending the monastery in Rot an der Rot, and Steffen Weinstock for building and maintaining the web page. Christian Nowak and Kai Schwenzer deserve a special thank for the organization of the magnificent excursion to Lindau and the boat trip on the Lake of Constance. The timing in coordination with the weather was remarkable. We are very thankful for the financial support from the Graduiertenkolleg "Physical Systems with Many Degrees of Freedom" and the funds from the Daimler-Benz Stiftung provided through Dieter Gromes. Finally, we want to thank Franz Wegner, the spokesperson of the Graduiertenkolleg, for help in financial issues and his trust in our organization.

We hope that this book has captured some of the spirit of the autumn school on which it is based.

Heidelberg   *Eike Bick*
July, 2004   *Frank Daniel Steffen*

Contents

List of Contributors

Jan-Willem van Holten
National Institute for Nuclear and High-Energy Physics
(NIKHEF)
P.O. Box 41882
1009 DB Amsterdam, the Netherlands
and
Department of Physics and Astronomy
Faculty of Science
Vrije Universiteit Amsterdam
t32@nikhef.nl

Frieder Lenz
Institute for Theoretical Physics III
University of Erlangen-Nürnberg
Staudstrasse 7
91058 Erlangen, Germany
flenz@theorie3.physik.uni-erlangen.de

Thomas Schücker
Centre de Physique Théorique
CNRS - Luminy, Case 907
13288 Marseille Cedex 9, France
Thomas.Schucker@cpt.univ-mrs.fr

Mikhail Shifman
William I. Fine Theoretical Physics Institute
University of Minnesota
116 Church Street SE
Minneapolis MN 55455, USA
shifman@umn.edu

Jean Zinn-Justin
Dapnia
CEA/Saclay
91191 Gif-sur-Yvette Cedex, France
jean.zinn-justin@cea.fr

Introduction and Overview

E. Bick[1] and F.D. Steffen[2]

[1] d-fine GmbH, Opernplatz 2, 60313 Frankfurt, Germany
[2] DESY Theory Group, Notkestr. 85, 22603 Hamburg, Germany

1 Topology and Geometry in Physics

The first part of the 20th century saw the most revolutionary breakthroughs in the history of theoretical physics, the birth of general relativity and quantum field theory. The seemingly nearly completed description of our world by means of classical field theories in a simple Euclidean geometrical setting experienced major modifications: Euclidean geometry was abandoned in favor of Riemannian geometry, and the classical field theories had to be quantized. These ideas gave rise to today's theory of gravitation and the standard model of elementary particles, which describe nature better than anything physicists ever had at hand. The dramatically large number of successful predictions of both theories is accompanied by an equally dramatically large number of problems.

The standard model of elementary particles is described in the framework of quantum field theory. To construct a quantum field theory, we first have to quantize some classical field theory. Since calculations in the quantized theory are plagued by divergencies, we have to impose a regularization scheme and prove renormalizability before calculating the physical properties of the theory. Not even one of these steps may be carried out without care, and, of course, they are not at all independent. Furthermore, it is far from clear how to reconcile general relativity with the standard model of elementary particles. This task is extremely hard to attack since both theories are formulated in a completely different mathematical language.

Since the 1970's, a lot of progress has been made in clearing up these difficulties. Interestingly, many of the key ingredients of these contributions are related to topological structures so that nowadays topology is an indispensable part of theoretical physics.

Consider, for example, the quantization of a gauge field theory. To quantize such a theory one chooses some particular gauge to get rid of redundant degrees of freedom. Gauge invariance as a symmetry property is lost during this process. This is devastating for the proof of renormalizability since gauge invariance is needed to constrain the terms appearing in the renormalized theory. *BRST quantization* solves this problem using concepts transferred from algebraic geometry. More generally, the BRST formalism provides an elegant framework for dealing with constrained systems, for example, in general relativity or string theories.

Once we have quantized the theory, we may ask for properties of the classical theory, especially symmetries, which are inherited by the quantum field theory. Somewhat surprisingly, one finds obstructions to the construction of quantized

E. Bick and F.D. Steffen, Introduction and Overview, Lect. Notes Phys. **659**, 1–5 (2005)
http://www.springerlink.com/

gauge theories when gauge fields couple differently to the two fermion chiral components, the so-called *chiral anomalies*. This puzzle is connected to the difficulties in regularizing such chiral gauge theories without breaking chiral symmetry. Physical theories are required to be anomaly-free with respect to local symmetries. This is of fundamental significance as it constrains the couplings and the particle content of the standard model, whose electroweak sector is a chiral gauge theory.

Until recently, because exact chiral symmetry could not be implemented on the lattice, the discussion of anomalies was only perturbative, and one could have feared problems with anomaly cancelations beyond perturbation theory. Furthermore, this difficulty prevented a numerical study of relevant quantum field theories. In recent years new lattice regularization schemes have been discovered (domain wall, overlap, and perfect action fermions or, more generally, Ginsparg–Wilson fermions) that are compatible with a generalized form of chiral symmetry. They seem to solve both problems. Moreover, these lattice constructions provide new insights into the topological properties of anomalies.

The questions of quantizing and regularizing settled, we want to calculate the physical properties of the quantum field theory. The spectacular success of the standard model is mainly founded on perturbative calculations. However, as we know today, the spectrum of effects in the standard model is much richer than perturbation theory would let us suspect. *Instantons, monopoles,* and *solitons* are examples of topological objects in quantum field theories that cannot be understood by means of perturbation theory. The implications of this subject are far reaching and go beyond the standard model: From new aspects of the confinement problem to the understanding of superconductors, from the motivation for cosmic inflation to intriguing phenomena in supersymmetric models.

Accompanying the progress in quantum field theory, attempts have been made to merge the standard model and general relativity. In the setting of *noncommutative geometry*, it is possible to formulate the standard model in geometrical terms. This allows us to discuss both the standard model and general relativity in the same mathematical language, a necessary prerequisite to reconcile them.

2   An Outline of the Book

This book consists of five separate lectures, which are to a large extend self-contained. Of course, there are cross relations, which are taken into account by the outline.

In the first lecture, "Topological Concepts in Gauge Theories," Frieder Lenz presents an introduction to topological methods in studies of gauge theories. He discusses the three paradigms of topological objects: the Nielsen–Olesen vortex of the abelian Higgs model, the 't Hooft–Polyakov monopole of the non-abelian Higgs model, and the instanton of Yang–Mills theory. The presentation emphasizes the common formal properties of these objects and their relevance in physics. For example, our understanding of superconductivity based on the abelian Higgs model, or Ginzburg–Landau model, is described. A compact review of Yang–Mills theory and the Faddeev–Popov quantization procedure of gauge theories is given, which addresses also the topological obstructions that arise when global gauge conditions are implemented. Our understanding of confinement, the key puzzle in quantum chromodynamics, is discussed in light of topological insights. This lecture also contains an introduction to the concept of homotopy with many illustrating examples and applications from various areas of physics.

The quantization of Yang–Mills theory is revisited as a specific example in the lecture "Aspects of BRST Quantization" by Jan-Willem van Holten. His lecture presents an elegant and powerful framework for dealing with quite general classes of constrained systems using ideas borrowed from algebraic geometry. In a very systematic way, the general formulation is always described first, which is then illustrated explicitly for the relativistic particle, the classical electro-magnetic field, Yang–Mills theory, and the relativistic bosonic string. Beyond the perturbative quantization of gauge theories, the lecture describes the construction of BRST-field theories and the derivation of the Wess–Zumino consistency condition relevant for the study of anomalies in chiral gauge theories.

The study of anomalies in gauge theories with chiral fermions is a key to most fascinating topological aspects of quantum field theory. Jean Zinn-Justin describes these aspects in his lecture "Chiral Anomalies and Topology." He reviews various perturbative and non-perturbative regularization schemes emphasizing possible anomalies in the presence of both gauge fields and chiral fermions. In simple examples the form of the anomalies is determined. In the non-abelian case it is shown to be compatible with the Wess–Zumino consistency conditions. The relation of anomalies to the index of the Dirac operator in a gauge background is discussed. Instantons are shown to contribute to the anomaly in CP(N-1) models and SU(2) gauge theories. The implications on the strong CP problem and the U(1) problem are mentioned. While the study of anomalies has been limited to the framework of perturbation theory for years, the lecture addresses also recent breakthroughs in lattice field theory that allow non-perturbative investigations of chiral anomalies. In particular, the overlap and domain wall fermion formulations are described in detail, where lessons on supersymmetric quantum mechanics and a two-dimensional model of a Dirac fermion in the background of a static soliton help to illustrate the general idea behind domain wall fermions.

The lecture of Misha Shifman is devoted to "Supersymmetric Solitons and Topology" and, in particular, on critical or BPS-saturated kinks and domain walls. His discussion includes minimal $\mathcal{N} = 1$ supersymmetric models of the Landau–Ginzburg type in 1+1 dimensions, the minimal Wess–Zumino model in 3+1 dimensions, and the supersymmetric CP(1) model in 1+1 dimensions, which is a hybrid model (Landau–Ginzburg model on curved target space) that possesses extended $\mathcal{N} = 2$ supersymmetry. One of the main subjects of this lecture is the variety of novel physical phenomena inherent to BPS-saturated solitons in the presence of fermions. For example, the phenomenon of multiplet shortening is described together with its implications on quantum corrections to the mass (or wall tension) of the soliton. Moreover, irrationalization of the

U(1) charge of the soliton is derived as an intriguing dynamical phenomena of the $\mathcal{N} = 2$ supersymmetric model with a topological term. The appendix of this lecture presents an elementary introduction to supersymmetry, which emphasizes its promises with respect to the problem of the cosmological constant and the hierarchy problem.

The high hopes that supersymmetry, as a crucial basis of string theory, is a key to a quantum theory of gravity and, thus, to the theory of everything must be confronted with still missing experimental evidence for such a boson–fermion symmetry. This demonstrates the importance of alternative approaches not relying on supersymmetry. A non-supersymmetric approach based on Connes' noncommutative geometry is presented by Thomas Schücker in his lecture "Forces from Connes' geometry." This lecture starts with a brief review of Einstein's derivation of general relativity from Riemannian geometry. Also the standard model of particle physics is carefully reviewed with emphasis on its mathematical structure. Connes' noncommutative geometry is illustrated by introducing the reader step by step to Connes' spectral triple. Einstein's derivation of general relativity is paralled in Connes' language of spectral triples as a commutative example. Here the Dirac operator defines both the dynamics of matter and the kinematics of gravity. A noncommutative example shows explicitly how a Yang–Mills–Higgs model arises from gravity on a noncommutative geometry. The noncommutative formulation of the standard model of particle physics is presented and consequences for physics beyond the standard model are addressed. The present status of this approach is described with a look at its promises towards a unification of gravity with quantum field theory and at its open questions concerning, for example, the construction of quantum fields in noncommutative space or spectral triples with Lorentzian signature. The appendix of this lecture provides the reader with a compact review of the crucial mathematical basics and definitions used in this lecture.

3   Complementary Literature

Let us conclude this introduction with a brief guide to complementary literature the reader might find useful. Further recommendations will be given in the lectures. For quantum field theory, we appreciate very much the books of Peskin and Schröder [1], Weinberg [2], and Zinn-Justin [3]. For general relativity, the books of Wald [4] and Weinberg [5] can be recommended. More specific texts we found helpful in the study of topological aspects of quantum field theory are the ones by Bertlmann [6], Coleman [7], Forkel [8], and Rajaraman [9]. For elaborate treatments of the mathematical concepts, we refer the reader to the texts of Göckeler and Schücker [10], Nakahara [11], Nash and Sen [12], and Schutz [13].

References

1. M. E. Peskin and D. V. Schroeder, *An Introduction to Quantum Field Theory* (Westview Press, Boulder 1995)

2. S. Weinberg, *The Quantum Theory Of Fields*, Vols. I, II, and III, (Cambridge University Press, Cambridge 1995, 1996, and 2000)
3. J. Zinn-Justin, *Quantum Field Theory and Critical Phenomena*, 4th edn. (Carendon Press, Oxford 2002)
4. R. Wald, *General Relativity* (The University of Chicago Press, Chicago 1984)
5. S. Weinberg, *Gravitation and Cosmology* (Wiley, New York 1972)
6. R. A. Bertlmann, *Anomalies in Quantum Field Theory* (Oxford University Press, Oxford 1996)
7. S. Coleman, *Aspects of Symmetry* (Cambridge University Press, Cambridge 1985)
8. H. Forkel, *A Primer on Instantons in QCD*, arXiv:hep-ph/0009136
9. R. Rajaraman, *Solitons and Instantons* (North-Holland, Amsterdam 1982)
10. M. Göckeler and T. Schücker, *Differential Geometry, Gauge Theories, and Gravity* (Cambridge University Press, Cambridge 1987)
11. M. Nakahara, *Geometry, Topology and Physics*, 2nd ed. (IOP Publishing, Bristol 2003)
12. C. Nash and S. Sen, *Topology and Geometry for Physicists* (Academic Press, London 1983)
13. B. F. Schutz, *Geometrical Methods of Mathematical Physics* (Cambridge University Press, Cambridge 1980)

Topological Concepts in Gauge Theories

F. Lenz

Institute for Theoretical Physics III, University of Erlangen-Nürnberg,
Staudstrasse 7, 91058 Erlangen, Germany

Abstract. In these lecture notes, an introduction to topological concepts and methods in studies of gauge field theories is presented. The three paradigms of topological
objects, the Nielsen–Olesen vortex of the abelian Higgs model, the 't Hooft–Polyakov
monopole of the non-abelian Higgs model and the instanton of Yang–Mills theory,
are discussed. The common formal elements in their construction are emphasized and
their different dynamical roles are exposed. The discussion of applications of topological
methods to Quantum Chromodynamics focuses on confinement. An account is given
of various attempts to relate this phenomenon to topological properties of Yang–Mills
theory. The lecture notes also include an introduction to the underlying concept of
homotopy with applications from various areas of physics.

1 Introduction

In a fragment [1] written in the year 1833, C. F. Gauß describes a profound
topological result which he derived from the analysis of a physical problem. He
considers the work W_m done by transporting a magnetic monopole (ein Element des "positiven nördlichen magnetischen Fluidums") with magnetic charge
g along a closed path C_1 in the magnetic field $\mathbf{B}$ generated by a current I flowing
along a closed loop C_2. According to the law of Biot–Savart, W_m is given by

$$W_m = g \oint_{C_1} \mathbf{B}(\mathbf{s}_1)\, d\mathbf{s}_1 = \frac{4\pi g}{c}\, I\, lk\{C_1, C_2\}.$$

Gauß recognized that W_m neither depends on the geometrical details of the
those of the closed path C_1.

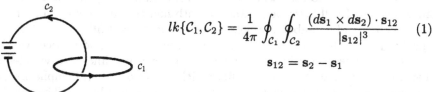

$$lk\{C_1, C_2\} = \frac{1}{4\pi} \oint_{C_1} \oint_{C_2} \frac{(d\mathbf{s}_1 \times d\mathbf{s}_2) \cdot \mathbf{s}_{12}}{|\mathbf{s}_{12}|^3} \qquad (1)$$

$$\mathbf{s}_{12} = \mathbf{s}_2 - \mathbf{s}_1$$

Fig. 1. Transport of a magnetic charge along C_1 in the magnetic field generated by a
current flowing along C_2

Under continuous deformations of these curves, the value of $lk\{C_1, C_2\}$, the *Linking Number ("Anzahl der Umschlingungen")*, remains unchanged. This quantity
is a topological invariant. It is an integer which counts the (signed) number of

F. Lenz, Topological Concepts in Gauge Theories, Lect. Notes Phys. **659**, 7–98 (2005)
http://www.springerlink.com/   © Springer-Verlag Berlin Heidelberg 2005

intersections of the loop $\mathcal{C}_1$ with an arbitrary (oriented) surface in $\mathbb{R}^3$ whose boundary is the loop $\mathcal{C}_2$ (cf. [2,3]). In the same note, Gauß deplores the little progress in topology ("Geometria Situs") since Leibniz's times who in 1679 postulated "another analysis, purely geometric or linear which also defines the position (situs), as algebra defines magnitude". Leibniz also had in mind applications of this new branch of mathematics to physics. His attempt to interest a physicist (Christiaan Huygens) in his ideas about topology however was unsuccessful. Topological arguments made their entrance in physics with the formulation of the Helmholtz laws of vortex motion (1858) and the circulation theorem by Kelvin (1869) and until today hydrodynamics continues to be a fertile field for the development and applications of topological methods in physics. The success of the topological arguments led Kelvin to seek for a description of the constituents of matter, the atoms at that time in terms of vortices and thereby explain topologically their stability. Although this attempt of a topological explanation of the laws of fundamental physics, the first of many to come, had to fail, a classification of knots and links by P. Tait derived from these efforts [4].

Today, the use of topological methods in the analysis of properties of systems is widespread in physics. Quantum mechanical phenomena such as the Aharonov–Bohm effect or Berry's phase are of topological origin, as is the stability of defects in condensed matter systems, quantum liquids or in cosmology. By their very nature, topological methods are insensitive to details of the systems in question. Their application therefore often reveals unexpected links between seemingly very different phenomena. This common basis in the theoretical description not only refers to obvious topological objects like vortices, which are encountered on almost all scales in physics, it applies also to more abstract concepts. "Helicity", for instance, a topological invariant in inviscid fluids, discovered in 1969 [5], is closely related to the topological charge in gauge theories. Defects in nematic liquid crystals are close relatives to defects in certain gauge theories. Dirac's work on magnetic monopoles [6] heralded in 1931 the relevance of topology for field theoretic studies in physics, but it was not until the formulation of non-abelian gauge theories [7] with their wealth of non-perturbative phenomena that topological methods became a common tool in field theoretic investigations.

In these lecture notes, I will give an introduction to topological methods in gauge theories. I will describe excitations with non-trivial topological properties in the abelian and non-abelian Higgs model and in Yang–Mills theory. The topological objects to be discussed are instantons, monopoles, and vortices which in space-time are respectively singular on a point, a world-line, or a world-sheet. They are solutions to classical non-linear field equations. I will emphasize both their common formal properties and their relevance in physics. The topological investigations of these field theoretic models is based on the mathematical concept of homotopy. These lecture notes include an introductory section on homotopy with emphasis on applications. In general, proofs are omitted or replaced by plausibility arguments or illustrative examples from physics or geometry. To emphasize the universal character in the topological analysis of physical systems, I will at various instances display the often amazing connections between very different physical phenomena which emerge from such analyses. Beyond the description of the paradigms of topological objects in gauge theories, these lecture notes contain an introduction to recent applications of topological methods to Quantum Chromodynamics with emphasis on the confinement issue. Confinement of the elementary degrees of freedom is the trademark of Yang–Mills theories. It is a non-perturbative phenomenon, i.e. the non-linearity of the theory is as crucial here as in the formation of topologically non-trivial excitations. I will describe various ideas and ongoing attempts towards a topological characterization of this peculiar property.

2   Nielsen–Olesen Vortex

The Nielsen–Olesen vortex [8] is a topological excitation in the abelian Higgs model. With topological excitation I will denote in the following a solution to the field equations with non-trivial topological properties. As in all the subsequent examples, the Nielsen–Olesen vortex owes its existence to vacuum degeneracy, i.e. to the presence of multiple, energetically degenerate solutions of minimal energy. I will start with a brief discussion of the abelian Higgs model and its (classical) "ground states", i.e. the field configurations with minimal energy.

2.1   Abelian Higgs Model

The abelian Higgs Model is a field theoretic model with important applications in particle and condensed matter physics. It constitutes an appropriate field theoretic framework for the description of phenomena related to superconductivity (cf. [9,10]) ("Ginzburg–Landau Model") and its topological excitations ("Abrikosov-Vortices"). At the same time, it provides the simplest setting for the mechanism of mass generation operative in the electro-weak interaction.

The abelian Higgs model is a gauge theory. Besides the electromagnetic field it contains a self-interacting scalar field (Higgs field) minimally coupled to electromagnetism. From the conceptual point of view, it is advantageous to consider this field theory in $2+1$ dimensional space-time and to extend it subsequently to $3+1$ dimensions for applications.

The abelian Higgs model Lagrangian

$$\mathcal{L} = -\frac{1}{4}F_{\mu\nu}F^{\mu\nu} + (D_\mu\phi)^*(D^\mu\phi) - V(\phi) \tag{2}$$

contains the complex (charged), self-interacting scalar field ϕ. The Higgs potential

$$V(\phi) = \frac{1}{4}\lambda(|\phi|^2 - a^2)^2. \tag{3}$$

as a function of the real and imaginary part of the Higgs field is shown in Fig. 2. By construction, this Higgs potential is minimal along a circle $|\phi| = a$ in the complex ϕ plane. The constant λ controls the strength of the self-interaction of the Higgs field and, for stability reasons, is assumed to be positive

$$\lambda \geq 0. \tag{4}$$

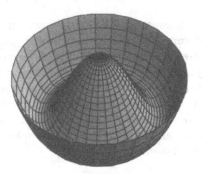

Fig. 2. Higgs Potential $V(\phi)$

The Higgs field is minimally coupled to the radiation field A_μ, i.e. the partial derivative ∂_μ is replaced by the covariant derivative

$$D_\mu = \partial_\mu + ieA_\mu. \tag{5}$$

Gauge fields and field strengths are related by

$$F_{\mu\nu} = \partial_\mu A_\nu - \partial_\nu A_\mu = \frac{1}{ie}\left[D_\mu, D_\nu\right].$$

Equations of Motion

- The (inhomogeneous) Maxwell equations are obtained from the principle of least action,

$$\delta S = \delta \int d^4x \mathcal{L} = 0,$$

by variation of S with respect to the gauge fields. With

$$\frac{\delta \mathcal{L}}{\delta \partial_\mu A_\nu} = -F^{\mu\nu}, \quad \frac{\delta \mathcal{L}}{\delta A_\nu} = -j^\nu,$$

we obtain

$$\partial_\mu F^{\mu\nu} = j^\nu, \quad j_\nu = ie(\phi^* \partial_\nu \phi - \phi \partial_\nu \phi^*) - 2e^2 \phi^* \phi A_\nu.$$

- The homogeneous Maxwell equations are not dynamical equations of motion – they are integrability conditions and guarantee that the field strength can be expressed in terms of the gauge fields. The homogeneous equations follow from the Jacobi identity of the covariant derivative

$$[D_\mu, [D_\nu, D_\sigma]] + [D_\sigma, [D_\mu, D_\nu]] + [D_\nu, [D_\sigma, D_\mu]] = 0.$$

Multiplication with the totally antisymmetric tensor, $\epsilon^{\mu\nu\rho\sigma}$, yields the homogeneous equations for the dual field strength $\tilde{F}^{\mu\nu}$

$$\left[D_\mu, \tilde{F}^{\mu\nu}\right] = 0 \quad , \quad \tilde{F}^{\mu\nu} = \frac{1}{2}\epsilon^{\mu\nu\rho\sigma} F_{\rho\sigma}.$$

The transition

$$F \to \tilde{F}$$

corresponds to the following duality relation of electric and magnetic fields

$$\mathbf{E} \to \mathbf{B} \,, \quad \mathbf{B} \to -\mathbf{E}.$$

- Variation with respect to the charged matter field yields the equation of motion

$$D_\mu D^\mu \phi + \frac{\delta V}{\delta \phi^*} = 0.$$

Gauge theories contain redundant variables. This redundancy manifests itself in the presence of local symmetry transformations; these "gauge transformations"

$$U(x) = e^{ie\alpha(x)} \tag{6}$$

rotate the phase of the matter field and shift the value of the gauge field in a space-time dependent manner

$$\phi \to \phi^{[U]} = U(x)\phi(x)\,, \quad A_\mu \to A_\mu^{[U]} = A_\mu + U(x)\frac{1}{ie}\partial_\mu U^\dagger(x)\,. \tag{7}$$

The covariant derivative D_μ has been defined such that $D_\mu \phi$ transforms covariantly, i.e. like the matter field ϕ itself.

$$D_\mu \phi(x) \to U(x)\, D_\mu \phi(x).$$

This transformation property together with the invariance of $F_{\mu\nu}$ guarantees invariance of $\mathcal{L}$ and of the equations of motion. A gauge field which is gauge equivalent to $A_\mu = 0$ is called a pure gauge. According to (7) a pure gauge satisfies

$$A_\mu^{pg}(x) = U(x)\frac{1}{ie}\partial_\mu U^\dagger(x) = -\partial_\mu \alpha(x)\,, \tag{8}$$

and the corresponding field strength vanishes.

Canonical Formalism. In the canonical formalism, electric and magnetic fields play distinctive dynamical roles. They are given in terms of the field strength tensor by

$$E^i = -F^{0i}\,, \quad B^i = -\frac{1}{2}\epsilon^{ijk}F_{jk} = (\text{rot}A)^i.$$

Accordingly,

$$-\frac{1}{4}F_{\mu\nu}F^{\mu\nu} = \frac{1}{2}\left(\mathbf{E}^2 - \mathbf{B}^2\right).$$

The presence of redundant variables complicates the formulation of the canonical formalism and the quantization. Only for independent dynamical degrees of freedom canonically conjugate variables may be defined and corresponding commutation relations may be associated. In a first step, one has to choose by a "gauge condition" a set of variables which are independent. For the development of the canonical formalism there is a particularly suited gauge, the "Weyl" – or "temporal" gauge

$$A_0 = 0. \tag{9}$$

We observe, that the time derivative of A_0 does not appear in $\mathcal{L}$, a property which follows from the antisymmetry of the field strength tensor and is shared by all gauge theories. Therefore in the canonical formalism A_0 is a constrained variable and its elimination greatly simplifies the formulation. It is easily seen that (9) is a legitimate gauge condition, i.e. that for an arbitrary gauge field a gauge transformation (7) with gauge function

$$\partial_0 \alpha(x) = A_0(x)$$

indeed eliminates A_0. With this gauge choice one proceeds straightforwardly with the definition of the canonically conjugate momenta

$$\frac{\delta \mathcal{L}}{\delta \partial_0 A_i} = -E^i , \quad \frac{\delta \mathcal{L}}{\delta \partial_0 \phi} = \pi ,$$

and constructs via Legendre transformation the Hamiltonian density

$$\mathcal{H} = \frac{1}{2}(\boldsymbol{E}^2 + \boldsymbol{B}^2) + \pi^*\pi + (\boldsymbol{D}\phi)^*(\boldsymbol{D}\phi) + V(\phi) , \quad H = \int d^3x \mathcal{H}(\mathbf{x}) . \tag{10}$$

With the Hamiltonian density given by a sum of positive definite terms (cf.(4)), the energy density of the fields of lowest energy must vanish identically. Therefore, such fields are static

$$\mathbf{E} = 0 , \quad \pi = 0 , \tag{11}$$

with vanishing magnetic field

$$\mathbf{B} = 0 . \tag{12}$$

The following choice of the Higgs field

$$|\phi| = a, \quad \text{i.e.} \quad \phi = ae^{i\beta} \tag{13}$$

renders the potential energy minimal. The ground state is not unique. Rather the system exhibits a "vacuum degeneracy", i.e. it possesses a continuum of field configurations of minimal energy. It is important to characterize the degree of this degeneracy. We read off from (13) that the manifold of field configurations of minimal energy is given by the manifold of zeroes of the potential energy. It is characterized by β and thus this manifold has the topological properties of a circle S^1. As in other examples to be discussed, this vacuum degeneracy is the source of the non-trivial topological properties of the abelian Higgs model.

To exhibit the physical properties of the system and to study the consequences of the vacuum degeneracy, we simplify the description by performing a time independent gauge transformation. Time independent gauge transformations do not alter the gauge condition (9). In the Hamiltonian formalism, these gauge transformations are implemented as canonical (unitary) transformations which can be regarded as symmetry transformations. We introduce the modulus and phase of the static Higgs field

$$\phi(\mathbf{x}) = \rho(\mathbf{x})e^{i\theta(\mathbf{x})},$$

and choose the gauge function

$$\alpha(\mathbf{x}) = -\theta(\mathbf{x}) \tag{14}$$

so that in the transformation (7) to the "unitary gauge" the phase of the matter field vanishes

$$\phi^{[U]}(\mathbf{x}) = \rho(\mathbf{x}), \quad \mathbf{A}^{[U]} = \mathbf{A} - \frac{1}{e}\nabla\theta(\mathbf{x}), \quad (\mathbf{D}\phi)^{[U]} = \nabla\rho(\mathbf{x}) - ie\mathbf{A}^{[U]}\rho(\mathbf{x}).$$

This results in the following expression for the energy density of the static fields

$$\epsilon(\mathbf{x}) = (\nabla\rho)^2 + \frac{1}{2}\mathbf{B}^2 + e^2\rho^2\mathbf{A}^2 + \frac{1}{4}\lambda(\rho^2 - a^2)^2. \tag{15}$$

In this unitary gauge, the residual gauge freedom in the vector potential has disappeared together with the phase of the matter field. In addition to condition (11), fields of vanishing energy must satisfy

$$\mathbf{A} = 0, \quad \rho = a. \tag{16}$$

In small oscillations of the gauge field around the ground state configurations (16) a restoring force appears as a consequence of the non-vanishing value a of the Higgs field ρ. Comparison with the energy density of a massive non-interacting scalar field φ

$$\epsilon_\varphi(\mathbf{x}) = \frac{1}{2}(\nabla\varphi)^2 + \frac{1}{2}M^2\varphi^2$$

shows that the term quadratic in the gauge field $\mathbf{A}$ in (15) has to be interpreted as a mass term of the vector field $\mathbf{A}$. In this Higgs mechanism, the photon has acquired the mass

$$M_\gamma = \sqrt{2}ea, \tag{17}$$

which is determined by the value of the Higgs field. For non-vanishing Higgs field, the zero energy configuration and the associated small amplitude oscillations describe electrodynamics in the so called Higgs phase, which differs significantly from the familiar Coulomb phase of electrodynamics. In particular, with photons becoming massive, the system does not exhibit long range forces. This is most directly illustrated by application of the abelian Higgs model to the phenomenon of superconductivity.

Meissner Effect. In this application to condensed matter physics, one identifies the energy density (15) with the free-energy density of a superconductor. This is called the Ginzburg–Landau model. In this model $|\phi|^2$ is identified with the density of the superconducting Cooper pairs (also the electric charge should be replaced $e \to e^* = 2e$) and serves as the order parameter to distinguish normal $a = 0$ and superconducting $a \neq 0$ phases.

Static solutions (11) satisfy the Hamilton equation (cf. (10), (15))

$$\frac{\delta H}{\delta \mathbf{A}(\mathbf{x})} = 0 \,,$$

which for a spatially constant scalar field becomes the Maxwell–London equation

$$\operatorname{rot} \mathbf{B} = \operatorname{rot} \operatorname{rot} \mathbf{A} = \mathbf{j} = 2e^2 a^2 \mathbf{A} \,.$$

The solution to this equation for a magnetic field in the normal conducting phase $(a = 0$ for $x < 0)$

$$\mathbf{B}(x) = \mathbf{B}_0 e^{-x/\lambda_L} \tag{18}$$

decays when penetrating into the superconducting region $(a \neq 0 \quad$ for $x > 0)$ within the penetration or London depth

$$\lambda_L = \frac{1}{M_\gamma} \tag{19}$$

determined by the photon mass. The expulsion of the magnetic field from the superconducting region is called Meissner effect.

Application of the gauge transformation ((7), (14)) has been essential for displaying the physics content of the abelian Higgs model. Its definition requires a well defined phase $\theta(x)$ of the matter field which in turn requires $\phi(x) \neq 0$. At points where the matter field vanishes, the transformed gauge fields $\mathbf{A}'$ are singular. When approaching the Coulomb phase $(a \to 0)$, the Higgs field oscillates around $\phi = 0$. In the unitary gauge, the transition from the Higgs to the Coulomb phase is therefore expected to be accompanied by the appearance of singular field configurations or equivalently by a "condensation" of singular points.

2.2 Topological Excitations

In the abelian Higgs model, the manifold of field configurations is a circle S^1 parameterized by the angle β in (13). The non-trivial topology of the manifold of vacuum field configurations is the origin of the topological excitations in the abelian Higgs model as well as in the other field theoretic models to be discussed later. We proceed as in the discussion of the ground state configurations and consider static fields (11) but allow for energy densities which do not vanish everywhere. As follows immediately from the expression (10) for the energy density, finite energy can result only if asymptotically $(|\mathbf{x}| \to \infty)$

$$\begin{aligned}
\phi(\mathbf{x}) &\to a e^{i\theta(\mathbf{x})} \\
\mathbf{B}(\mathbf{x}) &\to 0 \\
\mathbf{D}\phi(\mathbf{x}) = (\nabla - ie\mathbf{A}(\mathbf{x}))\,\phi(\mathbf{x}) &\to 0.
\end{aligned} \tag{20}$$

For these requirements to be satisfied, scalar and gauge fields have to be correlated asymptotically. According to the last equation, the gauge field is asymptotically given by the phase of the scalar field

$$A(x) = \frac{1}{ie} \nabla \ln \phi(x) = \frac{1}{e} \nabla \theta(x). \tag{21}$$

The vector potential is by construction asymptotically a "pure gauge" (8) and no magnetic field strength is associated with $A(x)$.

Quantization of Magnetic Flux. The structure (21) of the asymptotic gauge field implies that the magnetic flux of field configurations with finite energy is quantized. Applying Stokes' theorem to a surface Σ which is bounded by an asymptotic curve $\mathcal{C}$ yields

$$\Phi_B^n = \int_\Sigma B \, d^2x = \oint_\mathcal{C} A \cdot ds = \frac{1}{e} \oint_\mathcal{C} \nabla \theta(x) \cdot ds = n \frac{2\pi}{e}. \tag{22}$$

Being an integer multiple of the fundamental unit of magnetic flux, Φ_B^n cannot change as a function of time, it is a conserved quantity. The appearance of this conserved quantity does not have its origin in an underlying symmetry, rather it is of topological origin. Φ_B^n is also considered as a topological invariant since it cannot be changed in a continuous deformation of the asymptotic curve $\mathcal{C}$. In order to illustrate the topological meaning of this result, we assume the asymptotic curve $\mathcal{C}$ to be a circle. On this circle, $|\phi| = a$ (cf. (13)). Thus the scalar field $\phi(x)$ provides a mapping of the asymptotic circle $\mathcal{C}$ to the circle of zeroes of the Higgs potential ($V(a) = 0$). To study this mapping in detail, it is convenient to introduce polar coordinates

$$\phi(x) = \phi(r, \varphi) \underset{r \to \infty}{\longrightarrow} a e^{i\theta(\varphi)} \quad , \quad e^{i\theta(\varphi + 2\pi)} = e^{i\theta(\varphi)}.$$

The phase of the scalar field defines a non-trivial mapping of the asymptotic circle

$$\theta : S^1 \to S^1 \quad , \quad \theta(\varphi + 2\pi) = \theta(\varphi) + 2n\pi \tag{23}$$

to the circle $|\phi| = a$ in the complex plane. These mappings are naturally divided into (equivalence) classes which are characterized by their winding number n. This winding number counts how often the phase θ winds around the circle when the asymptotic circle (φ) is traversed once. A formal definition of the winding number is obtained by decomposing a continuous but otherwise arbitrary $\theta(\varphi)$ into a strictly periodic and a linear function

$$\theta_n(\varphi) = \theta^{period}(\varphi) + n\varphi \qquad n = 0, \pm 1, \dots$$

where

$$\theta^{period}(\varphi + 2\pi) = \theta^{period}(\varphi).$$

The linear functions can serve as representatives of the equivalence classes. Elements of an equivalence class can be obtained from each other by continuous

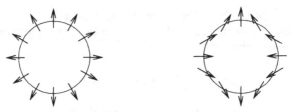

Fig. 3. Phase of a matter field with winding number $n = 1$ (left) and $n = -1$ (right)

deformations. The magnetic flux is according to (22) given by the phase of the Higgs field and is therefore quantized by the winding number n of the mapping (23). For instance, for field configurations carrying one unit of magnetic flux, the phase of the Higgs field belongs to the equivalence class θ_1. Figure 3 illustrates the complete turn in the phase when moving around the asymptotic circle. For $n = 1$, the phase $\theta(\mathbf{x})$ follows, up to continuous deformations, the polar angle φ, i.e. $\theta(\varphi) = \varphi$. Note that by continuous deformations the radial vector field can be turned into the velocity field of a vortex $\theta(\varphi) = \varphi + \pi/4$. Because of their shape, the $n = -1$ singularities, $\theta(\varphi) = \pi - \varphi$, are sometimes referred to as "hyperbolic" (right-hand side of Fig. 3). Field configurations $\mathbf{A}(\mathbf{x}), \phi(\mathbf{x})$ with $n \neq 0$ are called vortices and possess indeed properties familiar from hydrodynamics. The energy density of vortices cannot be zero everywhere with the magnetic flux $\Phi_B^n \neq 0$. Therefore in a finite region of space $\mathbf{B} \neq 0$. Furthermore, the scalar field must at least have one zero, otherwise a singularity arises when contracting the asymptotic circle to a point. Around a zero of $|\phi|$, the Higgs field displays a rapidly varying phase $\theta(\mathbf{x})$ similar to the rapid change in direction of the velocity field close to the center of a vortex in a fluid. However, with the modulus of the Higgs field approaching zero, no infinite energy density is associated with this infinite variation in the phase. In the Ginzburg–Landau theory, the core of the vortex contains no Cooper pairs ($\phi = 0$), the system is locally in the ordinary conducting phase containing a magnetic field.

The Structure of Vortices. The structure of the vortices can be studied in detail by solving the Euler–Lagrange equations of the abelian Higgs model (2). To this end, it is convenient to change to dimensionless variables (note that in 2+1 dimensions ϕ, A_μ, and e are of dimension length$^{-1/2}$)

$$\mathbf{x} \to \frac{1}{ea}\mathbf{x}, \quad \mathbf{A} \to \frac{1}{a}\mathbf{A}, \quad \phi \to \frac{1}{a}\phi, \quad \beta = \frac{\lambda}{2e^2}. \tag{24}$$

Accordingly, the energy of the static solutions becomes

$$\frac{E}{a^2} = \int d^2x \left\{ |(\mathbf{\nabla} - i\mathbf{A})\phi|^2 + \frac{1}{2}(\mathbf{\nabla} \times \mathbf{A})^2 + \frac{\beta}{2}(\phi\phi^* - 1)^2 \right\}. \tag{25}$$

The static spherically symmetric Ansatz

$$\phi = |\phi(r)|e^{in\varphi}, \quad \mathbf{A} = n\frac{\alpha(r)}{r}\mathbf{e}_\varphi,$$

converts the equations of motion into a system of (ordinary) differential equations coupling gauge and Higgs fields

$$\left(-\frac{d^2}{dr^2} - \frac{1}{r}\frac{d}{dr}\right)|\phi| + \frac{n^2}{r^2}(1-\alpha)^2|\phi| + \beta(|\phi|^2 - 1)|\phi| = 0, \tag{26}$$

$$\frac{d^2\alpha}{dr^2} - \frac{1}{r}\frac{d\alpha}{dr} - 2(\alpha - 1)|\phi|^2 = 0. \tag{27}$$

The requirement of finite energy asymptotically and in the core of the vortex leads to the following boundary conditions

$$r \to \infty: \ \alpha \to 1, \ |\phi| \to 1, \quad \alpha(0) = |\phi(0)| = 0. \tag{28}$$

From the boundary conditions and the differential equations, the behavior of Higgs and gauge fields is obtained in the core of the vortex

$$\alpha \sim -2r^2, \quad |\phi| \sim r^n,$$

and asymptotically

$$\alpha - 1 \sim \sqrt{r}e^{-\sqrt{2}r}, \quad |\phi| - 1 \sim \sqrt{r}e^{-\sqrt{2\beta}r}.$$

The transition from the core of the vortex to the asymptotics occurs on different scales for gauge and Higgs fields. The scale of the variations in the gauge field is the penetration depth λ_L determined by the photon mass (cf. (18) and (19)). It controls the exponential decay of the magnetic field when reaching into the superconducting phase. The coherence length

$$\xi = \frac{1}{ea\sqrt{2\beta}} = \frac{1}{a\sqrt{\lambda}} \tag{29}$$

controls the size of the region of the "false" Higgs vacuum ($\phi = 0$). In superconductivity, ξ sets the scale for the change in the density of Cooper pairs. The Ginzburg–Landau parameter

$$\kappa = \frac{\lambda_L}{\xi} = \sqrt{\beta} \tag{30}$$

varies with the substance and distinguishes Type I ($\kappa < 1$) from Type II ($\kappa > 1$) superconductors. When applying the abelian Higgs model to superconductivity, one simply reinterprets the vortices in 2 dimensional space as 3 dimensional objects by assuming independence of the third coordinate. Often the experimental setting singles out one of the 3 space dimensions. In such a 3 dimensional interpretation, the requirement of finite vortex energy is replaced by the requirement of finite energy/length, i.e. finite tension. In Type II superconductors, if the strength of an applied external magnetic field exceeds a certain critical value, magnetic flux is not completely excluded from the superconducting region. It penetrates the superconducting region by exciting one or more vortices each of which carrying a single quantum of magnetic flux Φ_B^1 (22). In Type I superconductors, the large coherence length ξ prevents a sufficiently fast rise of the Cooper pair density. In turn the associated shielding currents are not sufficiently strong to contain the flux within the penetration length λ_L and therefore no vortex can form. Depending on the applied magnetic field and the temperature, the Type II superconductors exhibit a variety of phenomena related to the intricate dynamics of the vortex lines and display various phases such as vortex lattices, liquid or amorphous phases (cf. [11,12]). The formation of magnetic flux lines inside Type II superconductors by excitation of vortices can be viewed as mechanism for confining magnetic monopoles. In a Gedankenexperiment we may imagine to introduce a north and south magnetic monopole inside a type II superconductor separated by a distance d. Since the magnetic field will be concentrated in the core of the vortices and will not extend into the superconducting region, the field energy of this system becomes

$$V = \frac{1}{2} \int d^3x \, \mathbf{B}^2 \propto \frac{4\pi d}{e^2 \lambda_L^2}. \tag{31}$$

Thus, the interaction energy of the magnetic monopoles grows linearly with their separation. In Quantum Chromodynamics (QCD) one is looking for mechanisms of confinement of (chromo-) electric charges. Thus one attempts to transfer this mechanism by some "duality transformation" which interchanges the role of electric and magnetic fields and charges. In view of such applications to QCD, it should be emphasized that formation of vortices does not happen spontaneously. It requires a minimal value of the applied field which depends on the microscopic structure of the material and varies over three orders of magnitude [13].

The point $\kappa = \beta = 1$ in the parameter space of the abelian Higgs model is very special. It separates Type I from Type II superconductors. I will now show that at this point the energy of a vortex is determined by its charge. To this end, I first derive a bound on the energy of the topological excitations, the "Bogomol'nyi bound" [14]. Via an integration by parts, the energy (25) can be written in the following form

$$\frac{E}{a^2} = \int d^2x \left| [(\partial_x - iA_x) \pm i(\partial_y - iA_y)] \phi \right|^2 + \frac{1}{2} \int d^2x \left[B \pm (\phi\phi^* - 1) \right]^2$$
$$\pm \int d^2x \, B + \frac{1}{2}(\beta - 1) \int d^2x \left[\phi^*\phi - 1 \right]^2$$

with the sign chosen according to the sign of the winding number n (cf. (22)). For "critical coupling" $\beta = 1$ (cf. (24)), the energy is bounded by the third term on the right-hand side, which in turn is given by the winding number (22)

$$E \geq 2\pi |n| \,.$$

The Bogomol'nyi bound is saturated if the vortex satisfies the following first order differential equations

$$[(\partial_x - iA_x) \pm i(\partial_y - iA_y)] \phi = 0$$

$$B = \pm(\phi\phi^* - 1).$$

It can be shown that for $\beta = 1$ this coupled system of first order differential equations is equivalent to the Euler–Lagrange equations. The energy of these particular solutions to the classical field equations is given in terms of the magnetic charge. Neither the existence of solutions whose energy is determined by topological properties, nor the reduction of the equations of motion to a first order system of differential equations is a peculiar property of the Nielsen–Olesen vortices. We will encounter again the Bogomol'nyi bound and its saturation in our discussion of the 't Hooft monopole and of the instantons. Similar solutions with the energy determined by some charge play also an important role in supersymmetric theories and in string theory.

A wealth of further results concerning the topological excitations in the abelian Higgs model has been obtained. Multi-vortex solutions, fluctuations around spherically symmetric solutions, supersymmetric extensions, or extensions to non-commutative spaces have been studied. Finally, one can introduce fermions by a Yukawa coupling

$$\delta\mathcal{L} \sim g\phi\bar{\psi}\psi + e\bar{\psi}A\!\!\!/\psi$$

to the scalar and a minimal coupling to the Higgs field. Again one finds what will turn out to be a quite general property. Vortices induce fermionic zero modes [15,16]. We will discuss this phenomenon in the context of instantons.

3   Homotopy

3.1   The Fundamental Group

In this section I will describe extensions and generalizations of the rather intuitive concepts which have been used in the analysis of the abelian Higgs model. From the physics point of view, the vacuum degeneracy is the essential property of the abelian Higgs model which ultimately gives rise to the quantization of the magnetic flux and the emergence of topological excitations. More formally, one views fields like the Higgs field as providing a mapping of the asymptotic circle in configuration space to the space of zeroes of the Higgs potential. In this way, the quantization is a consequence of the presence of integer valued topological invariants associated with this mapping. While in the abelian Higgs model these properties are almost self-evident, in the forthcoming applications the structure of the spaces to be mapped is more complicated. In the non-abelian Higgs model, for instance, the space of zeroes of the Higgs potential will be a subset of a non-abelian group. In such situations, more advanced mathematical tools have proven to be helpful for carrying out the analysis. In our discussion and for later applications, the concept of homotopy will be central (cf. [17,18]). It is a concept which is relevant for the characterization of global rather than local properties of spaces and maps (i.e. fields). In the following we will assume that the spaces are "topological spaces", i.e. sets in which open subsets with certain properties are defined and thereby the concept of continuity ("smooth maps") can be introduced (cf. [19]). In physics, one often requires differentiability of functions. In this case, the topological spaces must possess additional properties (differentiable manifolds). We start with the formal definition of homotopy.

Definition: Let X, Y be smooth manifolds and $f : X \to Y$ a smooth map between them. A *homotopy* or *deformation* of the map f is a smooth map

$$F : X \times I \to Y \quad (I = [0, 1])$$

with the property

$$F(x, 0) = f(x)$$

Each of the maps $f_t(x) = F(x, t)$ is said to be homotopic to the initial map $f_0 = f$ and the map of the whole cylinder $X \times I$ is called a homotopy. The relation of homotopy between maps is an equivalence relation and therefore allows to divide the set of smooth maps $X \to Y$ into equivalence classes, *homotopy classes*.

Definition: Two maps f, g are called *homotopic*, $f \sim g$, if they can be deformed continuously into each other.
The mappings

$$\mathbb{R}^n \to \mathbb{R}^n : f(x) = x, \ g(x) = x_0 = \text{const.}$$

are homotopic with the homotopy given by

$$F(x, t) = (1 - t)x + tx_0. \tag{32}$$

Spaces X in which the identity mapping 1_X and the constant mapping are homotopic, are homotopically equivalent to a point. They are called *contractible*.

Definition: Spaces X and Y are defined to be *homotopically equivalent* if continuous mappings exist

$$f : X \to Y \quad , \quad g : Y \to X$$

such that

$$g \circ f \sim 1_X \quad , \quad f \circ g \sim 1_Y$$

An important example is the equivalence of the $n-$sphere and the punctured $\mathbb{R}^{n+1}$ (one point removed)

$$S^n = \{\mathbf{x} \in \mathbb{R}^{n+1} | x_1^2 + x_2^2 + \ldots + x_{n+1}^2 = 1\} \sim \mathbb{R}^{n+1} \backslash \{0\}. \tag{33}$$

which can be proved by stereographic projection. It shows that with regard to homotopy, the essential property of a circle is the hole inside. Topologically identical (*homeomorphic*) spaces, i.e. spaces which can be mapped continuously and bijectively onto each other, possess the same connectedness properties and are therefore homotopically equivalent. The converse is not true.

In physics, we often can identify the parameter t as time. Classical fields, evolving continuously in time are examples of homotopies. Here the restriction to

Fig. 4. Phase of matter field with winding number $n = 0$

continuous functions follows from energy considerations. Discontinuous changes of fields are in general connected with infinite energies or energy densities. For instance, a homotopy of the "spin system" shown in Fig. 4 is provided by a spin wave connecting some initial $F(x, 0)$ with some final configuration $F(x, 1)$. Homotopy theory classifies the different sectors (equivalence classes) of field configurations. Fields of a given sector can evolve into each other as a function of time. One might be interested, whether the configuration of spins in Fig. 3 can evolve with time from the ground state configuration shown in Fig. 4.

The Fundamental Group. The fundamental group characterizes connectedness properties of spaces related to properties of loops in these spaces. The basic idea is to detect defects – like a hole in the plane – by letting loops shrink to a point. Certain defects will provide a topological obstruction to such attempts. Here one considers arcwise (or path) connected spaces, i.e. spaces where any pair of points can be connected by some path.

A **loop** (closed path) through x_0 in M is formally defined as a map

$$\alpha : [0, 1] \to M \qquad \text{with} \qquad \alpha(0) = \alpha(1) = x_0 .$$

A product of two loops is defined by

$$\gamma = \alpha * \beta , \qquad \gamma(t) = \left\{ \begin{array}{ll} \alpha(2t) & , \quad 0 \leq t \leq \dfrac{1}{2} \\[2mm] \beta(2t - 1) & , \quad \dfrac{1}{2} \leq t \leq 1 \end{array} \right\} ,$$

and corresponds to traversing the loops consecutively. Inverse and constant loops are given by

$$\alpha^{-1}(t) = \alpha(1 - t), \qquad 0 \leq t \leq 1$$

and

$$c(t) = x_0$$

respectively. The inverse corresponds to traversing a given loop in the opposite direction.

Definition: Two loops through $x_0 \in M$ are said to be homotopic, $\alpha \sim \beta$, if they can be continuously deformed into each other, i.e. if a mapping H exists,

$$H : [0, 1] \times [0, 1] \to M ,$$

with the properties

$$H(s,0) = \alpha(s), \quad 0 \leq s \leq 1; \quad H(s,1) = \beta(s),$$
$$H(0,t) = H(1,t) = x_0, \quad 0 \leq t \leq 1. \tag{34}$$

Once more, we may interpret t as time and the homotopy H as a time-dependent evolution of loops into each other.

Definition: $\pi_1(M, x_0)$ denotes the set of equivalence classes (homotopy classes) of loops through $x_0 \in M$.

The product of equivalence classes is defined by the product of their representatives. It can be easily seen that this definition does not depend on the loop chosen to represent a certain class. In this way, $\pi_1(M, x_0)$ acquires a group structure with the constant loop representing the neutral element. Finally, in an arcwise connected space M, the equivalence classes $\pi_1(M, x_0)$ are independent of the base point x_0 and one therefore denotes with $\pi_1(M)$ the *fundamental group* of M.

For applications, it is important that the fundamental group (or more generally the homotopy groups) of homotopically equivalent spaces X, Y are identical

$$\pi_1(X) = \pi_1(Y).$$

Examples and Applications. Trivial topological spaces as far as their connectedness is concerned are *simply connected* spaces.

Definition: A topological space X is said to be *simply connected* if any loop in X can be continuously shrunk to a point.

The set of equivalence classes consists of one element, represented by the constant loop and one writes

$$\pi_1 = 0.$$

Obvious examples are the spaces $\mathbb{R}^n$.

Non-trivial connectedness properties are the source of the peculiar properties of the abelian Higgs model. The phase of the Higgs field θ defined on a loop at infinity, which can continuously be deformed into a circle at infinity, defines a mapping

$$\theta : S^1 \to S^1.$$

An arbitrary phase χ defined on S^1 has the properties

$$\chi(0) = 0, \quad \chi(2\pi) = 2\pi m. \tag{35}$$

It can be continuously deformed into the linear function $m\varphi$. The mapping

$$H(\varphi, t) = (1 - t)\chi(\varphi) + t\varphi \frac{\chi(2\pi)}{2\pi}$$

with the properties

$$H(0, t) = \chi(0) = 0 \quad , \quad H(2\pi, t) = \chi(2\pi),$$

is a homotopy and thus

$$\chi(\varphi) \sim m\varphi.$$

The equivalence classes are therefore characterized by integers m and since these winding numbers are additive when traversing two loops

$$\pi_1(S^1) \sim \mathbb{Z}. \tag{36}$$

Vortices are defined on $\mathbb{R}^2 \backslash \{0\}$ since the center of the vortex, where $\theta(\mathbf{x})$ is ill-defined, has to be removed. The homotopic equivalence of this space to S^1 (33) implies that a vortex with winding number $N \neq 0$ is stable; it cannot evolve with time into the homotopy class of the ground-state configuration where up to continuous deformations, the phase points everywhere into the same direction.

This argument also shows that the (abelian) vortex is not topologically stable in higher dimensions. In $\mathbb{R}^n \backslash \{0\}$ with $n \geq 3$, by continuous deformation, a loop can always avoid the origin and can therefore be shrunk to a point. Thus

$$\pi_1(S^n) = 0, \; n \geq 2, \tag{37}$$

i.e. n–spheres with $n > 1$ are simply connected. In particular, in 3 dimensions a "point defect" cannot be detected by the fundamental group. On the other hand, if we remove a line from the $\mathbb{R}^3$, the fundamental group is again characterized by the winding number and we have

$$\pi_1(\mathbb{R}^3 \backslash \mathbb{R}) \sim \mathbb{Z}. \tag{38}$$

This result can also be seen as a consequence of the general homotopic equivalence

$$\mathbb{R}^{n+1} \backslash \mathbb{R} \sim S^{n-1}. \tag{39}$$

The result (37) implies that stringlike objects in 3-dimensional spaces can be detected by loops and that their topological stability is determined by the nontriviality of the fundamental group. For constructing pointlike objects in higher dimensions, the fields must assume values in spaces with different connectedness properties.

The fundamental group of a product of spaces X, Y is isomorphic to the product of their fundamental groups

$$\pi_1(X \otimes Y) \sim \pi_1(X) \otimes \pi_1(Y). \tag{40}$$

For a torus T and a cylinder C we thus have

$$\pi_1(T) \sim \mathbb{Z} \otimes \mathbb{Z}, \quad \pi_1(C) = \mathbb{Z} \otimes \{0\}. \tag{41}$$

3.2 Higher Homotopy Groups

The fundamental group displays the properties of loops under continuous deformations and thereby characterizes topological properties of the space in which the loops are defined. With this tool only a certain class of non-trivial topological properties can be detected. We have already seen above that a point defect cannot be detected by loops in dimensions higher than two and therefore the concept of homotopy groups must be generalized to higher dimensions. Although in $\mathbb{R}^3$ a circle cannot enclose a pointlike defect, a 2-sphere can. The higher homotopy groups are obtained by suitably defining higher dimensional analogs of the (one dimensional) loops. For technical reasons, one does not choose directly spheres and starts with $n-$cubes which are defined as

$$I^n = \{(s_1, \ldots, s_n) \,|\, 0 \le s_i \le 1 \quad \text{all } i\}$$

whose boundary is given by

$$\partial I^n = \{(s_1, \ldots, s_n) \in I^n \,|\, s_i = 0 \quad \text{or } s_i = 1 \quad \text{for at least one } i\}.$$

Loops are curves with the initial and final points identified. Correspondingly, one considers continuous maps from the $n-$cube to the topological space X

$$\alpha : \; I^n \to X$$

with the properties that the image of the boundary is one point in X

$$\alpha : I^n \to X \quad , \quad \alpha(s) = x_0 \quad \text{for} \quad s \in \partial I^n.$$

$\alpha(I^n)$ is called an $n-$loop in X. Due to the identification of the points on the boundary these $n-$loops are topologically equivalent to $n-$spheres. One now proceeds as above and introduces a homotopy, i.e. continuous deformations of $n-$loops

$$F : I^n \times I \to X$$

and requires

$$F(s_1, s_2, \ldots, 0) = \alpha(s_1, \ldots, s_n)$$
$$F(s_1, s_2, \ldots, 1) = \beta(s_1, \ldots, s_n)$$
$$F(s_1, s_2, \ldots, t) = x_0 \quad , \quad \text{for} \quad (s_1, \ldots, s_n) \in \partial I^n \quad \Rightarrow \quad \alpha \sim \beta .$$

The homotopy establishes an equivalence relation between the $n-$loops. The space of $n-$loops is thereby partitioned into disjoint classes. The set of equivalence classes is, for arcwise connected spaces (independence of x_0), denoted by

$$\pi_n(X) = \{\alpha | \alpha : I^n \to X, \; \alpha(s \in \partial I^n) = x_0\} .$$

As π_1, also π_n can be equipped with an algebraic structure. To this end one defines a product of maps α, β by connecting them along a common part of the boundary, e.g. along the part given by $s_1=1$

$$\alpha \circ \beta(s_1, s_2, \ldots, s_n) = \begin{cases} \alpha(2s_1, s_2, \ldots, s_n) & , \quad 0 \le s_1 \le \dfrac{1}{2} \\ \beta(2s_1 - 1, s_2, \ldots, s_n) & , \quad \dfrac{1}{2} \le s_1 \le 1 \end{cases}$$

$$\alpha^{-1}(s_1, s_2, \ldots, s_n) = \alpha(1 - s_1, s_2, \ldots, s_n).$$

After definition of the unit element and the inverse respectively

$$e(s_1, s_2 \ldots s_n) = x_0, \qquad \alpha^{-1}(s_1, s_2 \ldots s_n) = \alpha(1 - s_1, s_2 \ldots s_n)$$

π_n is seen to be a group. The algebraic structure of the higher homotopy groups is simple

$$\pi_n(X) \text{ is abelian for } n > 1. \tag{42}$$

The fundamental group, on the other hand, may be non-abelian, although most of the applications in physics deal with abelian fundamental groups. An example of a non-abelian fundamental group will be discussed below (cf. (75)).

The mapping between spheres is of relevance for many applications of homotopy theory. The following result holds

$$\pi_n(S^n) \sim \mathbb{Z}. \tag{43}$$

In this case the integer n characterizing the mapping generalizes the winding number of mappings between circles. By introducing polar coordinates θ, φ and θ', φ' on two spheres, under the mapping

$$\theta' = \theta, \ \varphi' = \varphi,$$

the sphere $S^{2'}$ is covered once if θ and φ wrap the sphere S^2 once. This 2-loop belongs to the class $k = 1 \in \pi_2(S^2)$. Under the mapping

$$\theta' = \theta, \ \varphi' = 2\varphi$$

$S^{2'}$ is covered twice and the 2-loop belongs to the class $k = 2 \in \pi_2(S^2)$. Another important result is

$$\pi_m(S^n) = 0 \quad m < n, \tag{44}$$

a special case of which ($\pi_1(S^2)$) has been discussed above. There are no simple intuitive arguments concerning the homotopy groups $\pi_n(S^m)$ for $n > m$, which in general are non-trivial. A famous example (cf. [2]) is

$$\pi_3(S^2) \sim \mathbb{Z}, \tag{45}$$

a result which is useful in the study of Yang–Mills theories in a certain class of gauges (cf. [20]). The integer k labeling the equivalence classes has a geometric interpretation. Consider two points $y_1, y_2 \in S^2$, which are regular points in the (differentiable) mapping

$$f : S^3 \to S^2$$

i.e. the differential df is 2-dimensional in y_1 and y_2. The preimages of these points $M_{1,2} = f^{-1}(y_{1,2})$ are curves C_1, C_2 on S^3; the integer k is the linking number $lk\{C_1, C_2\}$ of these curves, cf. (1). It is called the *Hopf invariant*.

3.3 Quotient Spaces

Topological spaces arise in very different fields of physics and are frequently of complex structure. Most commonly, such non-trivial topological spaces are obtained by identification of certain points which are elements of simple topological spaces. The mathematical concept behind such identifications is that of a *quotient space*. The identification of points is formulated as an equivalence relation between them.

Definition: Let X be a topological space and $\sim$ an equivalence relation on X. Denote by

$$[x] = \{y \in X | y \sim x\}$$

the equivalence class of x and with $X/\sim$ the set of equivalence classes; the projection taking each $x \in X$ to its equivalence class be denoted by

$$\pi(x) = [x].$$

$X/\sim$ is then called quotient space of X relative to the relation $\sim$. The quotient space is a topological space with subsets $V \subset X/\sim$ defined to be open if $\pi^{-1}(V)$ is an open subset of X.

- An elementary example of a quotient space is a circle. It is obtained by an equivalence relation of points in $\mathbb{R}$ and therefore owes its non-trivial topological properties to this identification. Let the equivalence relation be defined by:

$$X = \mathbb{R}, \; x, y \in \mathbb{R}, \quad x \sim y \quad \text{if} \quad x - y \in \mathbb{Z}.$$

  $\mathbb{R}/\sim$ can be identified with

$$S^1 = \{z \in \mathbb{C} | |z| = 1\},$$

  the unit circle in the complex plane and the projection is given by

$$\pi(x) = e^{2i\pi x}.$$

  The circle is the topological space in which the phase of the Higgs field or of the wave function of a superconductor lives. Also the orientation of the spins of magnetic substances with restricted to a plane can be specified by points on a circle. In field theory such models are called $O(2)$ models. If the spins can have an arbitrary direction in 3-dimensions ($O(3)$ models), the relevant manifold representing such spins is the surface of a ball, i.e. S^2.

- Let us consider

$$X = \mathbb{R}^{n+1} \backslash \{0\},$$

  i.e. the set of all (n+1) tuples $x = (x^1, x^2, ..., x^{n+1})$ except $(0, 0, ..., 0)$, and define

$$x \sim y \quad \text{if for real } t \neq 0 \quad (y^1, y^2, ..., y^{n+1}) = (tx^1, tx^2, ..., tx^{n+1}).$$

The equivalence classes $[x]$ may be visualized as lines through the origin. The resulting quotient space is called the *real projective space* and denoted by $\mathbb{R}P^n$; it is a differentiable manifold of dimension n. Alternatively, the projective spaces can be viewed as spheres with antipodal points identified

$$\mathbb{R}P^n = \{x|x \in S^n, x \sim -x\}. \tag{46}$$

These topological spaces are important in condensed matter physics. These are the topological spaces of the degrees of freedom of (nematic) liquid crystals. Nematic liquid crystals consist of long rod-shaped molecules which spontaneously orient themselves like spins of a magnetic substances. Unlike spins, there is no distinction between head and tail. Thus, after identification of head and tail, the $n-$spheres relevant for the degrees of freedom of magnetic substances, the spins, turn into the projective spaces relevant for the degrees of freedom of liquid crystals, the *directors*.

- The $n-$spheres are the central objects of homotopy; physical systems in general are defined in the $\mathbb{R}^n$. In order to apply homotopy arguments, often the space $\mathbb{R}^n$ has to be replaced by S^n. Formally this is possible by adjoining the point $\{\infty\}$ to $\mathbb{R}^n$

$$\mathbb{R}^n \cup \{\infty\} = S^n. \tag{47}$$

This procedure is called the *one-point (or Alexandroff) compactification* of $\mathbb{R}^n$ ([21]). Geometrically this is achieved by the stereographic projection with the infinitely remote points being mapped to the north-pole of the sphere. For this to make sense, the fields which are defined in $\mathbb{R}^n$ have to approach a constant with $|\mathbf{x}| \to \infty$. Similarly the process of compactification of a disc D^2 or equivalently a square to S^2 as shown in Fig. 5 requires the field (phase and modulus of a complex field) to be constant along the boundary.

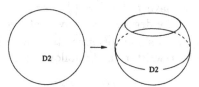

Fig. 5. Compactification of a disc D^2 to S^2 can be achieved by deforming the disc and finally adding a point, the north-pole

3.4   Degree of Maps

For mappings between closed oriented manifolds X and Y of equal dimension (n), a homotopy invariant, the *degree* can be introduced [2,3]. Unlike many other topological invariants, the degree possesses an integral representation, which is extremely useful for actually calculating the value of topological invariants. If $y_0 \in Y$ is a regular value of f, the set $f^{-1}(y_0)$ consists of only a finite number of points $x_1, ...x_m$. Denoting with x_i^β, y_0^α the local coordinates, the Jacobian defined by

$$J_i = \det \left(\frac{\partial y_0^\alpha}{\partial x_i^\beta} \right) \neq 0$$

is non-zero.

Definition: The *degree* of f with respect to $y_0 \in Y$ is defined as

$$\deg f = \sum_{x_i \in f^{-1}(y_0)} \mathrm{sgn}\,(J_i) \,. \tag{48}$$

The degree has the important property of being independent of the choice of the regular value y_0 and to be invariant under homotopies, i.e. the degree can be used to classify homotopic classes. In particular, it can be proven that a pair of smooth maps from a closed oriented n-dimensional manifold X^n to the n-sphere S^n, $f, g : X^n \to S^n$, are homotopic *iff* their degrees coincide.

For illustration, return to our introductory example and consider maps from the unit circle to the unit circle $S^1 \to S^1$. As we have seen above, we can picture the unit circle as arising from $\mathbb{R}^1$ by identification of the points $x + 2n\pi$ and $y + 2n\pi$ respectively. We consider a map with the property

$$f(x + 2\pi) = f(x) + 2k\pi \,,$$

i.e. if x moves around *once* the unit circle, its image $y = f(x)$ has turned around k times. In this case, every y_0 has at least k preimages with slopes (i.e. values of the Jacobian) of the same sign. For the representative of the k-th homotopy class, for instance,

$$f_k(x) = k \cdot x$$

and with the choice $y_0 = \pi$ we have $f^{-1}(y_0) = \{\frac{1}{k}\pi, \frac{2}{k}\pi, ...\pi\}$.

Since $\partial y_0 / \partial x |_{x=l/(k\pi)} = 1$, the degree is k. Any continuous deformation can only add pairs of pre-images with slopes of opposite signs which do not change the degree. The degree can be rewritten in the following integral form:

$$\deg f = k = \frac{1}{2\pi} \int_0^{2\pi} dx \left(\frac{df}{dx} \right) \,.$$

Many of the homotopy invariants appearing in our discussion can actually be calculated after identification with the degree of an appropriate map and its evaluation by the integral representation of the degree. In the Introduction we have seen that the work of transporting a magnetic monopole around a closed curve in the magnetic field generated by circular current is given by the linking number lk (1) of these two curves. The topological invariant lk can be identified with the degree of the following map [22]

$$T^2 \to S^2 : \quad (t_1, t_2) \to \hat{\mathbf{s}}_{12} = \frac{\mathbf{s}_1(t_1) - \mathbf{s}_2(t_2)}{|\mathbf{s}_1(t_1) - \mathbf{s}_2(t_2)|} \,.$$

The generalization of the above integral representation of the degree is usually formulated in terms of differential forms as

$$\int_X f^*\omega = \deg f \int_Y \omega \tag{49}$$

where f^* is the induced map (pull back) of differential forms of degree n defined on Y. In the $\mathbb{R}^n$ this reduces to the formula for changing the variables of integrations over some function χ

$$\int_{f^{-1}(U_i)} \chi(y(x)) \det \left(\frac{\partial y_0^\alpha}{\partial x_i^\beta} \right) dx_1...dx_n = \text{sgn} \det \left(\frac{\partial y_0^\alpha}{\partial x_i^\beta} \right) \int_{U_i} \chi(y) dy_1...dy_n$$

where the space is represented as a union of disjoint neighborhoods U_i with $y_0 \in U_i$ and non-vanishing Jacobian.

3.5   Topological Groups

In many application of topological methods to physical systems, the relevant degrees of freedom are described by fields which take values in topological groups like the Higgs field in the abelian or non-abelian Higgs model or link variables and Wilson loops in gauge theories. In condensed matter physics an important example is the order parameter in superfluid ^{3}He in the "A-phase" in which the pairing of the Helium atoms occurs in p-states with the spins coupled to 1. This pairing mechanism is the source of a variety of different phenomena and gives rise to the rather complicated manifold of the order parameter $SO(3) \otimes S^2/\mathbb{Z}_2$ (cf. [23]).

SU(2) as Topological Space. The group $SU(2)$ of unitary transformations is of fundamental importance for many applications in physics. It can be generated by the Pauli-matrices

$$\tau^1 = \begin{pmatrix} 0 & 1 \\ 1 & 0 \end{pmatrix}, \ \tau^2 = \begin{pmatrix} 0 & -i \\ i & 0 \end{pmatrix}, \ \tau^3 = \begin{pmatrix} 1 & 0 \\ 0 & -1. \end{pmatrix} \tag{50}$$

Every element of $SU(2)$ can be parameterized in the following way

$$U = e^{i\boldsymbol{\phi}\cdot\boldsymbol{\tau}} = \cos\phi + i\boldsymbol{\tau}\cdot\hat{\boldsymbol{\phi}}\sin\phi = a + i\boldsymbol{\tau}\cdot\boldsymbol{b}. \tag{51}$$

Here ϕ denotes an arbitrary vector in internal (e.g. isospin or color) space and we do not explicitly write the neutral element e. This vector is parameterized by the 4 (real) parameters $a, \boldsymbol{b}$ subject to the unitarity constraint

$$UU^\dagger = (a + i\boldsymbol{b}\cdot\boldsymbol{\tau})(a - i\boldsymbol{b}\cdot\boldsymbol{\tau}) = a^2 + b^2 = 1.$$

This parameterization establishes the topological equivalence (homeomorphism) of $SU(2)$ and S^3

$$SU(2) \sim S^3. \tag{52}$$

This homeomorphism together with the results (43) and (44) shows

$$\pi_{1,2}\Big(SU(2)\Big) = 0, \quad \pi_3\Big(SU(2)\Big) = \mathbb{Z}. \tag{53}$$

One can show more generally the following properties of homotopy groups

$$\pi_k\Big(SU(n)\Big) = 0 \ k < n.$$

The triviality of the fundamental group of $SU(2)$ (53) can be verified by constructing an explicit homotopy between the loop

$$u_{2n}(s) = \exp\{i2n\pi s\tau^3\} \tag{54}$$

and the constant map

$$u_c(s) = 1. \tag{55}$$

The mapping

$$H(s,t) = \exp\left\{ -i\frac{\pi}{2}t\tau^1 \right\} \exp\left\{ i\frac{\pi}{2}t(\tau^1 \cos 2\pi n s + \tau^2 \sin 2\pi n s) \right\}$$

has the desired properties (cf. (34))

$$H(s,0) = 1, \quad H(s,1) = u_{2n}(s), \quad H(0,t) = H(1,t) = 1,$$

as can be verified with the help of the identity (51). After continuous deformations and proper choice of the coordinates on the group manifold, any loop can be parameterized in the form 54.

Not only Lie groups but also quotient spaces formed from them appear in important physical applications. The presence of the group structure suggests the following construction of quotient spaces. Given any subgroup H of a group G, one defines an equivalence between two arbitrary elements $g_1, g_2 \in G$ if they are identical up to multiplication by elements of H

$$g_1 \sim g_2 \text{ iff } g_1^{-1}g_2 \in H. \tag{56}$$

The set of elements in G which are equivalent to $g \in G$ is called the left coset (modulo H) associated with g and is denoted by

$$g H = \{gh \,|\, h \in H\}. \tag{57}$$

The space of cosets is called the coset space and denoted by

$$G/H = \{gH \,|\, g \in G\}. \tag{58}$$

If N is an *invariant* or *normal* subgroup, i.e. if $gNg^{-1} = N$ for all $g \in G$, the coset space is actually a group with the product defined by $(g_1N) \cdot (g_2N) = g_1g_2N$. It is called the *quotient* or *factor* group of G by N.

As an example we consider the group of translations in $\mathbb{R}^3$. Since this is an abelian group, each subgroup is normal and can therefore be used to define factor groups. Consider $N = T_x$ the subgroup of translations in the $x-$direction. The cosets are translations in the y-z plane followed by an arbitrary translation in the $x-$direction. The factor group consists therefore of translations with unspecified parameter for the translation in the $x-$direction. As a further example consider rotations $R(\varphi)$ around a point in the $x - y$ plane. The two elements

$$e = R(0) \quad , \quad r = R(\pi)$$

form a normal subgroup N with the factor group given by

$$G/N = \{R(\varphi) N | 0 \leq \varphi < \pi\}.$$

Homotopy groups of coset spaces can be calculated with the help of the following two identities for connected and simply connected Lie-groups such as $SU(n)$. With H_0 we denote the component of H which is connected to the neutral element e. This component of H is an invariant subgroup of H. To verify this, denote with $\gamma(t)$ the continuous curve which connects the unity e at $t = 0$ with an arbitrary element h_0 of H_0. With $\gamma(t)$ also $h\gamma(t)h^{-1}$ is part of H_0 for arbitrary $h \in H$. Thus H_0 is a normal subgroup of H and the coset space H/H_0 is a group. One extends the definition of the homotopy groups and defines

$$\pi_0(H) = H/H_0. \tag{59}$$

The following identities hold (cf. [24,25])

$$\pi_1(G/H) = \pi_0(H), \tag{60}$$

and

$$\pi_2(G/H) = \pi_1(H_0). \tag{61}$$

Applications of these identities to coset spaces of $SU(2)$ will be important in the following. We first observe that, according to the parameterization (51), together with the neutral element e also $-e$ is an element of $SU(2)$ $(\phi = 0, \pi$ in (51)$)$. These 2 elements commute with all elements of SU(2) and form a subgroup, the *center* of $SU(2)$

$$Z\big(SU(2)\big) = \{e, -e\} \sim \mathbb{Z}_2. \tag{62}$$

According to the identity (60) the fundamental group of the factor group is non-trivial

$$\pi_1\big(SU(2)/Z(SU(2))\big) = \mathbb{Z}_2. \tag{63}$$

As one can see from the following argument, this result implies that the group of rotations in 3 dimensions $SO(3)$ is not simply connected. Every rotation matrix $R_{ij} \in SO(3)$ can be represented in terms of $SU(2)$ matrices (51)

$$R_{ij}[U] = \frac{1}{4} \operatorname{tr} \left\{ U\tau^i U^\dagger \tau^j \right\}.$$

The $SU(2)$ matrices U and $-U$ represent the same $SO(3)$ matrix. Therefore,

$$SO(3) \sim SU(2)/\mathbb{Z}_2 \tag{64}$$

and thus

$$\pi_1\Big(SO(3)\Big) = \mathbb{Z}_2, \tag{65}$$

i.e. $SO(3)$ is not simply connected.

We have verified above that the loops $u_{2n}(s)$ (54) can be shrunk in SU(2) to a point. They also can be shrunk to a point on $SU(2)/\mathbb{Z}_2$. The loop

$$u_1(s) = \exp\{i\pi s\tau^3\} \tag{66}$$

connecting antipodal points however is topologically stable on $SU(2)/\mathbb{Z}_2$, i.e. it cannot be deformed continuously to a point, while its square, $u_1^2(s) = u_2(s)$ can be.

The identity (61) is important for the spontaneous symmetry breakdown with a remaining $U(1)$ gauge symmetry. Since the groups $SU(n)$ are simply connected, one obtains

$$\pi_2\Big(SU(n)/U(1)\Big) = \mathbb{Z}. \tag{67}$$

3.6   Transformation Groups

Historically, groups arose as collections of permutations or one-to-one transformations of a set X onto itself with composition of mappings as the group product. If X contains just n elements, the collection $S(X)$ of all its permutations is the symmetric group with $n!$ elements. In F. Klein's approach, to each geometry is associated a group of transformations of the underlying space of the geometry. For example, the group $E(2)$ of Euclidean plane geometry is the subgroup of $S(E^2)$ which leaves the distance $d(x,y)$ between two arbitrary points in the plane (E^2) invariant, i.e. a transformation

$$T : E^2 \to E^2$$

is in the group iff

$$d(Tx, Ty) = d(x, y).$$

The group $E(2)$ is also called the group of *rigid motions*. It is generated by translations, rotations, and reflections. Similarly, the general Lorentz group is the group of Poincaré transformations which leave the (relativistic) distance between two space-time points invariant. The interpretation of groups as transformation groups is very important in physics. Mathematically, transformation groups are defined in the following way (cf.[26]):

Definition: A Lie group G is represented as a group of transformations of a manifold X (left action on X) if there is associated with each $g \in G$ a diffeomorphism of X to itself

$$x \to T_g(x), \quad x \in X \quad \text{with} \quad T_{g_1 g_2} = T_{g_1} T_{g_2}$$

("right action" $T_{g_1 g_2} = T_{g_2} T_{g_1}$) and if $T_g(x)$ depends smoothly on the arguments g, x.

If G is any of the Lie groups $GL(n, R), O(n, R), GL(n, C), U(n)$ then G acts in the obvious way on the manifold $\mathbb{R}^n$ or C^n.

The *orbit* of $x \in X$ is the set

$$Gx = \{T_g(x) \,|\, g \in G\} \subset X. \tag{68}$$

The action of a group G on a manifold X is said to be *transitive* if for every two points $x, y \in X$ there exists $g \in G$ such that $T_g(x) = y$, i.e. if the orbits satisfy $Gx = X$ for every $x \in X$. Such a manifold is called a *homogeneous* space of the Lie group. The prime example of a homogeneous space is $\mathbb{R}^3$ under translations; every two points can be connected by translations. Similarly, the group of translations acts transitively on the n-dimensional torus $T^n = \left(S^1\right)^n$ in the following way:

$$T_y(z) = \left(e^{2i\pi(\varphi_1 + t_1)}, ..., e^{2i\pi(\varphi_n + t_n)}\right)$$

with

$$y = (t_1, ..., t_n) \in \mathbb{R}^n, \quad z = \left(e^{2i\pi(\varphi_1)}, ..., e^{2i\pi(\varphi_n)}\right) \in T^n.$$

If the translations are given in terms of integers, $t_i = n_i$, we have $T_\mathbf{n}(z) = z$. This is a subgroup of the translations and is defined more generally:

Definition: The *isotropy group* H_x of the point $x \in X$ is the subgroup of all elements of G leaving x fixed and is defined by

$$H_x = \{g \in G \,|\, T_g(x) = x\}. \tag{69}$$

The group $O(n+1)$ acts transitively on the sphere S^n and thus S^n is a homogeneous space for the Lie group $O(n+1)$ of orthogonal transformations of $\mathbb{R}^{n+1}$. The isotropy group of the point $x = (1, 0, ...0) \in S^n$ is comprised of all matrices of the form

$$\begin{pmatrix} 1 & 0 \\ 0 & A \end{pmatrix}, \quad A \in O(n)$$

describing rotations around the x_1 axis.

Given a transformation group G acting on a manifold X, we define orbits as the equivalence classes, i.e.

$$x \sim y \quad \text{if for some } g \in G \quad y = g\,x.$$

For $X = \mathbb{R}^n$ and $G = O(n)$ the orbits are concentric spheres and thus in one to one correspondence with real numbers $r \geq 0$. This is a homeomorphism of $\mathbb{R}^n / O(n)$ on the ray $0 \leq r \leq \infty$ (which is almost a manifold).

If one defines points on S^2 to be equivalent if they are connected by a rotation around a fixed axis, the z axis, the resulting quotient space $S^2/O(2)$ consists of all the points on S^2 with fixed azimuthal angle, i.e. the quotient space is a segment

$$S^2/O(2) = \{\theta \,|\, 0 \leq \theta \leq \pi\}. \tag{70}$$

Note that in the integration over the coset spaces $\mathbb{R}^n/O(n)$ and $S^2/O(2)$ the radial volume element r^{n-1} and the volume element of the polar angle $\sin\theta$ appear respectively.

The quotient space X/G needs not be a manifold, it is then called an *orbifold*. If G is a discrete group, the fixed points in X under the action of G become singular points on X/G. For instance, by identifying the points $\mathbf{x}$ and $-\mathbf{x}$ of a plane, the fixed point $\mathbf{0} \in \mathbb{R}^2$ becomes the tip of the cone $\mathbb{R}^2/\mathbb{Z}_2$.

Similar concepts are used for a proper description of the topological space of the degrees of freedom in gauge theories. Gauge theories contain redundant variables, i.e. variables which are related to each other by gauge transformations. This suggests to define an equivalence relation in the space of gauge fields (cf. (7) and (90))

$$A_\mu \sim \tilde{A}_\mu \quad \text{if} \quad \tilde{A}_\mu = A_\mu^{[U]} \quad \text{for some } U, \tag{71}$$

i.e. elements of an equivalence class can be transformed into each other by gauge transformations U, they are *gauge copies* of a chosen representative. The equivalence classes

$$O = \left\{ A^{[U]} | U \in G \right\} \tag{72}$$

with A fixed and U running over the set of gauge transformations are called the *gauge orbits*. Their elements describe the same physics. Denoting with $\mathcal{A}$ the space of gauge configurations and with G the space of gauge transformations, the coset space of gauge orbits is denoted with $\mathcal{A}/\mathcal{G}$. It is this space rather than $\mathcal{A}$ which defines the physical configuration space of the gauge theory. As we will see later, under suitable assumptions concerning the asymptotic behavior of gauge fields, in Yang–Mills theories, each gauge orbit is labeled by a topological invariant, the *topological charge*.

3.7   Defects in Ordered Media

In condensed matter physics, topological methods find important applications in the investigations of properties of defects occurring in ordered media [27]. For applying topological arguments, one has to specify the topological space X in which the fields describing the degrees of freedom are defined and the topological space M (target space) of the values of the fields. In condensed matter physics the (classical) fields $\psi(\mathbf{x})$ are called the order parameter and M correspondingly the order parameter space. A system of spins or directors may be defined on lines, planes or in the whole space, i.e. $X = \mathbb{R}^n$ with $n = 1, 2$ or 3. The fields or order parameters describing spins are spatially varying unit vectors with arbitrary orientations: $M = S^2$ or if restricted to a plane $M = S^1$. The target spaces of directors are the corresponding projective spaces $\mathbb{R}P^n$. A defect is a point, a line or a surface on which the order parameter is ill-defined. The defects are defined accordingly as *point defects (monopoles), line defects (vortices, disclinations), or surface defects (domain walls)*. Such defects are topologically stable if they cannot be removed by a continuous change in the order parameter. Discontinuous changes require in physical systems of e.g. spin degrees of freedom substantial changes in a large number of the degrees of freedom and therefore large energies. The existence of singularities alter the topology of the space X. Point and line defects induce respectively the following changes in the topology: $X = \mathbb{R}^3 \to \mathbb{R}^3 \backslash \{0\} \sim S^2$ and $X = \mathbb{R}^3 \to \mathbb{R}^3 \backslash \mathbb{R}^1 \sim S^1$. Homotopy provides the appropriate tools to study the stability of defects. To this end, we proceed as in the abelian Higgs model and investigate the order parameter on a circle or a 2-sphere sufficiently far away from the defect. In this way, the order parameter defines a mapping $\psi : S^n \to M$ and the stability of the defects is guaranteed if the homotopy group $\pi_n(M)$ is non-trivial. Alternatively one may study the defects by removing from the space X the manifold on which the order parameter becomes singular. The structure of the homotopy group has important implications for the dynamics of the defects. If the asymptotic circle encloses two defects, and if the homotopy group is abelian, than in a merger of the two defects the resulting defect is specified by the sum of the two integers characterizing the individual defects. In particular, winding numbers $(\pi_1(S^1))$ and monopole charges $(\pi_2(S^2))$ (cf. (43)) are additive.

I conclude this discussion by illustrating some of the results using the examples of magnetic systems represented by spins and nematic liquid crystals represented by directors, i.e. spins with indistinguishable heads and tails (cf. (46) and the following discussion). If 2-dimensional spins ($M = S^1$) or directors ($M = \mathbb{R}P^1$) live on a plane ($X = \mathbb{R}^2$), a defect is topologically stable. The punctured plane obtained by the removal of the defect is homotopically equivalent to a circle (33) and the topological stability follows from the non-trivial homotopy group $\pi_1(S^1)$ for magnetic substances. The argument applies to nematic substances as well since identification of antipodal points of a circle yields again a circle

$$\mathbb{R}P^1 \sim S^1 .$$

On the other hand, a point defect in a system of 3-dimensional spins $M = S^2$ defined on a plane $X = \mathbb{R}^2$ – or equivalently a line defect in $X = \mathbb{R}^3$ – is not stable. Removal of the defect manifold generates once more a circle. The triviality of $\pi_1(S^2)$ (cf. (37)) shows that the defect can be continuously deformed into a configuration where all the spins point into the same direction. On S^2 a loop can always be shrunk to a point (cf. (37)). In nematic substances, there are stable point and line defects for $X = \mathbb{R}^2$ and $X = \mathbb{R}^3$, respectively, since

$$\pi_1(\mathbb{R}P^2) = \mathbb{Z}_2.$$

Non-shrinkable loops on RP^2 are obtained by connecting a given point on S^2 with its antipodal one. Because of the identification of antipodal points, the line connecting the two points cannot be contracted to a point. In the identification, this line on S^2 becomes a non-contractible loop on RP^2. Contractible and non-contractible loops on RP^2 are shown in Fig. 6 . This figure also demonstrates that connecting two antipodal points with two different lines produces a contractible loop. Therefore the space of loops contains only two inequivalent classes. More generally, one can show (cf. [19])

$$\pi_1(\mathbb{R}P^n) = \mathbb{Z}_2 , \qquad n \geq 2, \qquad (73)$$

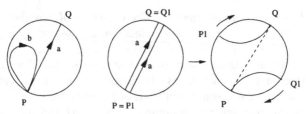

Fig. 6. The left figure shows loops a, b which on $\mathbb{R}P^2$ can (b) and cannot (a) be shrunk to a point. The two figures on the right demonstrate how two loops of the type a can be shrunk to one point. By moving the point $P1$ together with its antipodal point $Q1$ two shrinkable loops of the type b are generated

and (cf. [18])

$$\pi_n(\mathbb{R}P^m) = \pi_n(S^m)\,, \quad n \geq 2\,. \tag{74}$$

Thus, in 3-dimensional nematic substances point defects (monopoles), also present in magnetic substances, and line defects (disclinations), absent in magnetic substances, exist. In Fig. 7 the topologically stable line defect is shown. The circles around the defect are mapped by $\theta(\varphi) = \frac{\varphi}{2}$ into $\mathbb{R}P^2$. Only due to the identification of the directions $\theta \sim \theta + \pi$ this mapping is continuous. For magnetic substances, it would be discontinuous along the $\varphi = 0$ axis.

Liquid crystals can be considered with regard to their underlying dynamics as close relatives to some of the fields of particle physics. They exhibit spontaneous orientations, i.e. they form ordered media with respect to 'internal' degrees of freedom not joined by formation of a crystalline structure. Their topologically stable defects are also encountered in gauge theories as we will see later. Unlike the fields in particle physics, nematic substances can be manipulated and, by their birefringence property, allow for a beautiful visualization of the structure and dynamics of defects (for a thorough discussion of the physics of liquid crystals and their defects (cf. [29,30]). These substances offer the opportunity to study on a macroscopic level, emergence of monopoles and their dynamics. For instance, by enclosing a water droplet in a nematic liquid drop, the boundary conditions on the surface of the water droplet and on the surface of the nematic drop cooperate to generate a monopole (hedgehog) structure which, as Fig. 8C demonstrates,

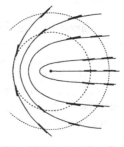

Fig. 7. Line defect in $\mathbb{R}P^2$. In addition to the directors also the integral curves are shown

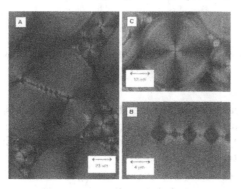

Fig. 8. Nematic drops (A) containing one (C) or more water droplets (B) (the figure is taken from [31]). The distance between the defects is about 5 μm

can be observed via its peculiar birefringence properties, as a four armed star of alternating bright and dark regions. If more water droplets are dispersed in a nematic drop, they form chains (Fig. 8A) which consist of the water droplets alternating with hyperbolic defects of the nematic liquid (Fig. 8B). The non-trivial topological properties stabilize these objects for as long as a couple of weeks [31]. In all the examples considered so far, the relevant fundamental groups were abelian. In nematic substances the "biaxial nematic phase" has been identified (cf. [29]) which is characterized by a non-abelian fundamental group. The elementary constituents of this phase can be thought of as rectangular boxes rather than rods which, in this phase are aligned. Up to 180° rotations around the 3 mutually perpendicular axes (R_i^π), the orientation of such a box is specified by an element of the rotation group $SO(3)$. The order parameter space of such a system is therefore given by

$$M = SO(3)/\mathbb{D}_2, \quad \mathbb{D}_2 = \{\mathbb{1}, R_1^\pi, R_2^\pi, R_3^\pi\}.$$

By representing the rotations by elements of $SU(2)$ (cf. (64)), the group $\mathbb{D}_2$ is extended to the group of 8 elements, containing the Pauli matrices (50),

$$\mathbb{Q} = \{\pm\mathbb{1}, \pm\tau^1, \pm\tau^2, \pm\tau^3\},$$

the group of *quaternions*. With the help of the identities (59) and (60), we derive

$$\pi_1(SO(3)/\mathbb{D}_2) \sim \pi_1(SU(2)/\mathbb{Q}) \sim \mathbb{Q}. \tag{75}$$

In the last step it has been used that in a discrete group the connected component of the identity contains the identity only.

The non-abelian nature of the fundamental group has been predicted to have important physical consequences for the behavior of defects in the nematic biaxial phase. This concerns in particular the coalescence of defects and the possibility of entanglement of disclination lines [29].

4   Yang–Mills Theory

In this introductory section I review concepts, definitions, and basic properties of gauge theories.

Gauge Fields. In non-abelian gauge theories, gauge fields are matrix-valued functions of space-time. In SU(N) gauge theories they can be represented by the generators of the corresponding Lie algebra, i.e. gauge fields and their color components are related by

$$A_\mu(x) = A_\mu^a(x)\frac{\lambda^a}{2}, \tag{76}$$

where the color sum runs over the $N^2 - 1$ generators. The generators are hermitian, traceless $N \times N$ matrices whose commutation relations are specified by the structure constants f^{abc}

$$\left[\frac{\lambda^a}{2}, \frac{\lambda^b}{2}\right] = i f^{abc}\frac{\lambda^c}{2}.$$

The normalization is chosen as

$$\mathrm{tr}\left(\frac{\lambda^a}{2} \cdot \frac{\lambda^b}{2}\right) = \frac{1}{2}\delta_{ab}.$$

Most of our applications will be concerned with $SU(2)$ gauge theories; in this case the generators are the Pauli matrices (50)

$$\lambda^a = \tau^a,$$

with structure constants

$$f^{abc} = \epsilon^{abc}.$$

Covariant derivative, field strength tensor, and its color components are respectively defined by

$$D_\mu = \partial_\mu + ig A_\mu, \tag{77}$$

$$F^{\mu\nu} = \frac{1}{ig}[D_\mu, D_\nu], \quad F_{\mu\nu}^a = \partial_\mu A_\nu^a - \partial_\nu A_\mu^a - g f^{abc} A_\mu^b A_\nu^c. \tag{78}$$

The definition of electric and magnetic fields in terms of the field strength tensor is the same as in electrodynamics

$$E^{ia}(x) = -F^{0ia}(x) \quad , \quad B^{ia}(x) = -\frac{1}{2}\epsilon^{ijk}F^{jka}(x) . \tag{79}$$

The dimensions of gauge field and field strength in 4 dimensional space-time are

$$[A] = \ell^{-1}, \quad [F] = \ell^{-2}$$

and therefore in absence of a scale

$$A_\mu^a \sim M_{\mu\nu}^a \frac{x^\nu}{x^2},$$

with arbitrary constants $M_{\mu\nu}^a$. In general, the action associated with these fields exhibits infrared and ultraviolet logarithmic divergencies. In the following we will discuss

- Yang–Mills Theories
  Only gauge fields are present. The Yang–Mills Lagrangian is

$$\mathcal{L}_{YM} = -\frac{1}{4} F^{\mu\nu a} F_{\mu\nu}^a = -\frac{1}{2} \text{tr} \{ F^{\mu\nu} F_{\mu\nu} \} = \frac{1}{2} (\mathbf{E}^2 - \mathbf{B}^2). \tag{80}$$

- Quantum Chromodynamics
  QCD contains besides the gauge fields (gluons), fermion fields (quarks). Quarks are in the fundamental representation, i.e. in SU(2) they are represented by 2-component color spinors. The QCD Lagrangian is (flavor dependences suppressed)

$$\mathcal{L}_{QCD} = \mathcal{L}_{YM} + \mathcal{L}_m, \quad \mathcal{L}_m = \bar{\psi} (i\gamma^\mu D_\mu - m) \psi, \tag{81}$$

  with the action of the covariant derivative on the quarks given by

$$(D_\mu \psi)^i = (\partial_\mu \delta^{ij} + ig A_\mu^{ij}) \psi^j \quad i,j = 1 \ldots N$$

- Georgi–Glashow Model
  In the Georgi–Glashow model [32] (non-abelian Higgs model), the gluons are coupled to a scalar, self-interacting $(V(\phi))$ (Higgs) field in the adjoint representation. The Higgs field has the same representation in terms of the generators as the gauge field (76) and can be thought of as a 3-component color vector in $SU(2)$. Lagrangian and action of the covariant derivative are respectively

$$\mathcal{L}_{GG} = \mathcal{L}_{YM} + \mathcal{L}_m, \quad \mathcal{L}_m = \frac{1}{2} D_\mu \phi D^\mu \phi - V(\phi), \tag{82}$$

$$(D_\mu \phi)^a = [D_\mu, \phi]^a = (\partial_\mu \delta^{ac} - g f^{abc} A_\mu^b) \phi^c. \tag{83}$$

Equations of Motion. The principle of least action

$$\delta S = 0, \quad S = \int d^4 x \mathcal{L}$$

yields when varying the gauge fields

$$\delta S_{YM} = -\int d^4 x \, \text{tr} \{ F_{\mu\nu} \delta F^{\mu\nu} \} = -\int d^4 x \, \text{tr} \left\{ F_{\mu\nu} \frac{2}{ig} [D^\mu, \delta A^\nu] \right\}$$

$$= 2 \int d^4 x \, \text{tr} \{ \delta A^\nu [D^\mu, F_{\mu\nu}] \}$$

the inhomogeneous field equations

$$[D_\mu, F^{\mu\nu}] = j^\nu \,, \tag{84}$$

with j^ν the color current associated with the matter fields

$$j^{a\nu} = \frac{\delta \mathcal{L}_m}{\delta A_\nu^a} \,. \tag{85}$$

For QCD and the Georgi–Glashow model, these currents are given respectively by

$$j^{a\nu} = g \bar{\psi} \gamma^\nu \frac{\tau^a}{2} \psi \,, \quad j^{a\nu} = g f^{abc} \phi^b (D^\nu \phi)^c \,. \tag{86}$$

As in electrodynamics, the homogeneous field equations for the Yang–Mills field strength

$$\left[D_\mu, \tilde{F}^{\mu\nu} \right] = 0 \,,$$

with the dual field strength tensor

$$\tilde{F}^{\mu\nu} = \frac{1}{2} \epsilon^{\mu\nu\sigma\rho} F_{\sigma\rho} \,, \tag{87}$$

are obtained as the Jacobi identities of the covariant derivative

$$[D_\mu, [D_\nu, D_\rho]] + [D_\nu, [D_\rho, D_\mu]] + [D_\rho, [D_\nu, D_\mu]] = 0 \,,$$

i.e. they follow from the mere fact that the field strength is represented in terms of gauge potentials.

Gauge Transformations. Gauge transformations change the color orientation of the matter fields locally, i.e. in a space-time dependent manner, and are defined as

$$U(x) = \exp \{ig\alpha(x)\} = \exp \left\{ ig\alpha^a(x) \frac{\tau^a}{2} \right\} \,, \tag{88}$$

with the arbitrary gauge function $\alpha^a(x)$. Matter fields transform covariantly with U

$$\psi \to U\psi \,, \quad \phi \to U\phi U^\dagger. \tag{89}$$

The transformation property of A is chosen such that the covariant derivatives of the matter fields $D_\mu \psi$ and $D_\mu \phi$ transform as the matter fields ψ and ϕ respectively. As in electrodynamics, this requirement makes the gauge fields transform inhomogeneously

$$A_\mu(x) \to U(x) \left(A_\mu(x) + \frac{1}{ig} \partial_\mu \right) U^\dagger(x) = A_\mu^{[U]}(x) \tag{90}$$

resulting in a covariant transformation law for the field strength

$$F_{\mu\nu} \to U F_{\mu\nu} U^\dagger. \tag{91}$$

Under infinitesimal gauge transformations $(|g\alpha^a(x)| \ll 1)$

$$A^a_\mu(x) \rightarrow A^a_\mu(x) - \partial_\mu \alpha^a(x) - g f^{abc} \alpha^b(x) A^c_\mu(x). \qquad (92)$$

As in electrodynamics, gauge fields which are gauge transforms of $A_\mu = 0$ are called pure gauges (cf. (8)) and are, according to (90), given by

$$A^{pg}_\mu(x) = U(x) \frac{1}{ig} \partial_\mu U^\dagger(x). \qquad (93)$$

Physical observables must be independent of the choice of gauge (coordinate system in color space). Local quantities such as the Yang–Mills action density $\mathrm{tr}\, F^{\mu\nu}(x)F_{\mu\nu}(x)$ or matter field bilinears like $\bar{\psi}(x)\psi(x)$, $\phi^a(x)\phi^a(x)$ are gauge invariant, i.e. their value does not change under local gauge transformations. One also introduces non-local quantities which, in generalization of the transformation law (91) for the field strength, change homogeneously under gauge transformations. In this construction a basic building block is the path ordered integral

$$\Omega(x,y,\mathcal{C}) = P \exp\left\{-ig \int_{s_0}^{s} d\sigma \frac{dx^\mu}{d\sigma} A_\mu\big(x(\sigma)\big)\right\} = P \exp\left\{-ig \int_\mathcal{C} dx^\mu A_\mu\right\}. \qquad (94)$$

It describes a gauge string between the space-time points $x = x(s_0)$ and $y = x(s)$. Ω satisfies the differential equation

$$\frac{d\Omega}{ds} = -ig \frac{dx^\mu}{ds} A_\mu \Omega. \qquad (95)$$

Gauge transforming this differential equation yields the transformation property of Ω

$$\Omega(x,y,\mathcal{C}) \rightarrow U(x)\,\Omega(x,y,\mathcal{C})\,U^\dagger(y). \qquad (96)$$

With the help of Ω, non-local, gauge invariant quantities like

$$\mathrm{tr}\Omega^\dagger(x.y,\mathcal{C})F^{\mu\nu}(x)\Omega(x,y,\mathcal{C})\,F_{\mu\nu}(y), \quad \bar{\psi}(x)\Omega(x,y,\mathcal{C})\,\psi(y),$$

or closed gauge strings – $(SU(N))$ Wilson loops

$$W_\mathcal{C} = \frac{1}{N}\mathrm{tr}\,\Omega(x,x,\mathcal{C}) \qquad (97)$$

can be constructed. For pure gauges (93), the differential equation (95) is solved by

$$\Omega^{pg}(x,y,\mathcal{C}) = U(x)\,U^\dagger(y). \qquad (98)$$

While $\bar{\psi}(x)\Omega(x,y,\mathcal{C})\,\psi(y)$ is an operator which connects the vacuum with meson states for $SU(2)$ and $SU(3)$, fermionic baryons appear only in $SU(3)$ in which gauge invariant states containing an odd number of fermions can be constructed.

In SU(3) a point-like gauge invariant baryonic state is obtained by creating three quarks in a color antisymmetric state at the same space-time point

$$\psi(x) \sim \epsilon^{abc}\psi^a(x)\psi^b(x)\psi^c(x).$$

Under gauge transformations,

$$\psi(x) \to \epsilon^{abc}U_{a\alpha}(x)\psi^\alpha(x)U_{b\beta}(x)\psi^\beta(x)U_{c\gamma}(x)\psi^\gamma(x)$$
$$= \det\left(U(x)\right)\epsilon^{abc}\psi^a(x)\psi^b(x)\psi^c(x).$$

Operators that create finite size baryonic states must contain appropriate gauge strings as given by the following expression

$$\psi(x,y,z) \sim \epsilon^{abc}[\Omega(u,x,\mathcal{C}_1)\psi(x)]^a \, [\Omega(u,y,\mathcal{C}_2)\psi(y)]^b \, [\Omega(u,z,\mathcal{C}_3)\psi(z)]^c .$$

The presence of these gauge strings makes ψ gauge invariant as is easily verified with the help of the transformation property (96). Thus, gauge invariance is enforced by color exchange processes taking place between the quarks.

Canonical Formalism. The canonical formalism is developed in the same way as in electrodynamics. Due to the antisymmetry of $F_{\mu\nu}$, the Lagrangian (80) does not contain the time derivative of A_0 which, in the canonical formalism, has to be treated as a constrained variable. In the Weyl gauge [33,34]

$$A_0^a = 0, \quad a = 1....N^2 - 1, \tag{99}$$

these constrained variables are eliminated and the standard procedure of canonical quantization can be employed. In a first step, the canonical momenta of gauge and matter fields (quarks and Higgs fields) are identified

$$\frac{\delta\mathcal{L}_{YM}}{\partial_0 A_i^a} = -E^{ai}, \quad \frac{\delta\mathcal{L}_{mq}}{\partial_0\psi^\alpha} = i\psi^{\alpha\dagger}, \quad \frac{\delta\mathcal{L}_{mH}}{\partial_0\phi^a} = \pi^a.$$

By Legendre transformation, one obtains the Hamiltonian density of the gauge fields

$$\mathcal{H}_{YM} = \frac{1}{2}(E^2 + B^2), \tag{100}$$

and of the matter fields

$$\text{QCD}: \quad \mathcal{H}_m = \psi^\dagger\left(\frac{1}{i}\gamma^0\gamma^i D_i + \gamma^0 m\right)\psi, \tag{101}$$

$$\text{Georgi–Glashow model}: \quad \mathcal{H}_m = \frac{1}{2}\pi^2 + \frac{1}{2}(D\phi)^2 + V(\phi). \tag{102}$$

The gauge condition (99) does not fix the gauge uniquely, it still allows for time-independent gauge transformations $U(\mathbf{x})$, i.e. gauge transformations which are generated by time-independent gauge functions $\alpha(\mathbf{x})$ (88). As a consequence the Hamiltonian exhibits a local symmetry

$$H = U(\mathbf{x})\,H\,U(\mathbf{x})^\dagger \tag{103}$$

This residual gauge symmetry is taken into account by requiring physical states $|\Phi\rangle$ to satisfy the Gauß law, i.e. the 0-component of the equation of motion (cf. (84))

$$([D_i, E^i] + j^0)|\Phi\rangle = 0.$$

In general, the non-abelian Gauß law cannot be implemented in closed form which severely limits the applicability of the canonical formalism. A complete canonical formulation has been given in axial gauge [35] as will be discussed below. The connection of canonical to path-integral quantization is discussed in detail in [36].

5   't Hooft–Polyakov Monopole

The t' Hooft–Polyakov monopole [37,38] is a topological excitation in the non-abelian Higgs or Georgi–Glashow model ($SU(2)$ color). We start with a brief discussion of the properties of this model with emphasis on ground state configurations and their topological properties.

5.1   Non-Abelian Higgs Model

The Lagrangian (82) and the equations of motion (84) and (85) of the non-abelian Higgs model have been discussed in the previous section. For the following discussion we specify the self-interaction, which as in the abelian Higgs model is assumed to be a fourth order polynomial in the fields with the normalization chosen such that its minimal value is zero

$$V(\phi) = \frac{1}{4}\lambda(\phi^2 - a^2)^2, \quad \lambda > 0. \tag{104}$$

Since ϕ is a vector in color space and gauge transformations rotate the color direction of the Higgs field (89), V is gauge invariant

$$V(g\phi) = V(\phi). \tag{105}$$

We have used the notation

$$g\phi = U\phi U^\dagger, \quad g \in G = SU(2).$$

The analysis of this model parallels that of the abelian Higgs model. Starting point is the energy density of static solutions, which in the Weyl gauge is given by ((100), (102))

$$\epsilon(\mathbf{x}) = \frac{1}{2}\mathbf{B}^2 + \frac{1}{2}(\mathbf{D}\phi)^2 + V(\phi). \tag{106}$$

The choice

$$\mathbf{A} = 0, \quad \phi = \phi_0 = \text{const.}, \quad V(\phi_0) = 0 \tag{107}$$

minimizes the energy density. Due to the presence of the local symmetry of the Hamiltonian (cf. (103)), this choice is not unique. Any field configuration connected to (107) by a time-independent gauge transformation will also have vanishing energy density. Gauge fixing conditions by which the Gauß law constraint is implemented remove these gauge ambiguities; in general a global gauge symmetry remains (cf. [39,35]). Under a space-time independent gauge transformation

$$g = \exp\left\{ig\alpha^a \frac{\tau^a}{2}\right\}, \quad \alpha = \text{const}, \tag{108}$$

applied to a configuration (107), the gauge field is unchanged as is the modulus of the Higgs field. The transformation rotates the spatially constant ϕ_0. In such a ground-state configuration, the Higgs field exhibits a spontaneous orientation analogous to the spontaneous magnetization of a ferromagnet,

$$\phi = \phi_0, \quad |\phi_0| = a.$$

This appearance of a phase with spontaneous orientation of the Higgs field is a consequence of a vacuum degeneracy completely analogous to the vacuum degeneracy of the abelian Higgs model with its spontaneous orientation of the phase of the Higgs field.

Related to the difference in the topological spaces of the abelian and non-abelian Higgs fields, significantly different phenomena occur in the spontaneous symmetry breakdown. In the Georgi–Glashow model, the loss of rotational symmetry in color space is not complete. While the configuration (107) changes under the (global) color rotations (108) and does therefore not reflect the invariance of the Lagrangian or Hamiltonian of the system, it remains invariant under rotations around the axis in the direction of the Higgs field $\alpha \sim \phi_0$. These transformations form a subgroup of the group of rotations (108), it is the isotropy group (little group, stability group) (for the definition cf. (69)) of transformations which leave ϕ_0 invariant

$$H_{\phi_0} = \{h \in SU(2)|h\phi_0 = \phi_0\}. \tag{109}$$

The space of the zeroes of V, i.e. the space of vectors ϕ of fixed length a, is S^2 which is a homogeneous space (cf. the discussion after (68)) with all elements being generated by application of arbitrary transformations $g \in G$ to a (fixed) ϕ_0. The space of zeroes of V and the coset space G/H_{ϕ_0} are mapped onto each other by

$$F_{\phi_0} : G/H_{\phi_0} \to \{\phi|V(\phi) = 0\}, \quad F_{\phi_0}(\tilde{g}) = g\phi_0 = \phi$$

with g denoting a representative of the coset $\tilde{g}$. This mapping is bijective. The space of zeroes is homogeneous and therefore all zeroes of V appear as an image of some $\tilde{g} \in G/H_{\phi_0}$. This mapping is injective since $\tilde{g}_1\phi_0 = \tilde{g}_2\phi_0$ implies $g_1^{-1}\phi_0 g_2 \in H_{\phi_0}$ with $g_{1,2}$ denoting representatives of the corresponding cosets $\tilde{g}_{1,2}$ and therefore the two group elements belong to the same equivalence class (cf. (56)) i.e. $\tilde{g}_1 = \tilde{g}_2$. Thus, these two spaces are homeomorphic

$$G/H_{\phi_0} \sim S^2. \tag{110}$$

It is instructive to compare the topological properties of the abelian and non-abelian Higgs model.

- In the abelian Higgs model, the gauge group is

$$G = U(1)$$

and by the requirement of gauge invariance, the self-interaction is of the form

$$V(\phi) = V(\phi^* \phi).$$

The vanishing of V determines the modulus of ϕ and leaves the phase undetermined

$$V = 0 \quad \Rightarrow |\phi_0| = a e^{i\beta}.$$

After choosing the phase β, no residual symmetry is left, only multiplication with 1 leaves ϕ_0 invariant, i.e.

$$H = \{e\}, \tag{111}$$

and thus

$$G/H = G \sim S^1. \tag{112}$$

- In the non-abelian Higgs model, the gauge group is

$$G = SU(2),$$

and by the requirement of gauge invariance, the self-interaction is of the form

$$V(\phi) = V(\phi^2), \quad \phi^2 = \sum_{a=1,3} \phi^{a\,2}.$$

The vanishing of V determines the modulus of ϕ and leaves the orientation undetermined

$$V = 0 \quad \Rightarrow \quad \phi_0 = a \hat{\phi}_0.$$

After choosing the orientation $\hat{\phi}_0$, a residual symmetry persists, the invariance of ϕ_0 under (true) rotations around the ϕ_0 axis and under multiplication with an element of the center of $SU(2)$ (cf. (62))

$$H = U(1) \otimes \mathbb{Z}_2, \tag{113}$$

and thus

$$G/H = SU(2)/(U(1) \otimes \mathbb{Z}_2) \sim S^2. \tag{114}$$

5.2   The Higgs Phase

To display the physical content of the Georgi–Glashow model we consider small oscillations around the ground-state configurations (107) – the normal modes of the classical system and the particles of the quantized system. The analysis of the normal modes simplifies greatly if the gauge theory is represented in the unitary gauge, the gauge which makes the particle content manifest. In this gauge, components of the Higgs field rather than those of the gauge field (like the longitudinal gauge field in Coulomb gauge) are eliminated as redundant variables. The Higgs field is used to define the coordinate system in internal space

$$\phi(x) = \phi^a(x)\frac{\tau^a}{2} = \rho(x)\frac{\tau^3}{2}. \tag{115}$$

Since this gauge condition does not affect the gauge fields, the Yang–Mills part of the Lagrangian (80) remains unchanged and the contribution of the Higgs field (82) simplifies

$$\mathcal{L} = -\frac{1}{4}F^{a\mu\nu}F^a_{\mu\nu} + \frac{1}{2}\partial_\mu\rho\partial^\mu\rho + g^2\rho^2 A^-_\mu A^{+\,\mu} - V(|\rho|), \tag{116}$$

with the "charged" components of the gauge fields defined by

$$A^\pm_\mu = \frac{1}{\sqrt{2}}(A^1_\mu \mp iA^2_\mu). \tag{117}$$

For small oscillations we expand the Higgs field $\rho(x)$ around the value in the zero-energy configuration (107)

$$\rho(x) = a + \sigma(x), \quad |\sigma| \ll a. \tag{118}$$

To leading order, the interaction with the Higgs field makes the charged components (117) of the gauge fields massive with the value of the mass given by the value of $\rho(x)$ in the zero-energy configuration

$$M^2 = g^2 a^2. \tag{119}$$

The fluctuating Higgs field $\sigma(x)$ acquires its mass through the self-interaction

$$m^2_\sigma = V''_{\rho=a} = 2\,a^2. \tag{120}$$

The neutral vector particles A^3_μ, i.e. the color component of the gauge field along the Higgs field, remains massless. This is a consequence of the survival of the non-trivial isotropy group $H_{\phi_0} \sim U(1)$ (cf. (109)) in the symmetry breakdown of the gauge group $SU(2)$. By coupling to a second Higgs field, with expectation value pointing in a color direction different from ϕ_0, a further symmetry breakdown can be achieved which is complete up to the discrete $\mathbb{Z}_2$ symmetry (cf. (114)). In such a system no massless vector particles can be present [8,40].

Superficially it may appear that the emergence of massive vector particles in the Georgi–Glashow model happens almost with necessity. The subtleties of the procedure are connected to the gauge choice (115). Definition of a coordinate system in the internal color space via the Higgs field requires

$$\phi \neq 0.$$

This requirement can be enforced by the choice of form (controlled by a) and strength λ of the Higgs potential V (104). Under appropriate circumstances, quantum or thermal fluctuations will only rarely give rise to configurations where

$\phi(x)$ vanishes at certain points and singular gauge fields (monopoles) are present. On the other hand, one expects at fixed a and λ with increasing temperature the occurrence of a phase transition to a gluon–Higgs field plasma. Similarly, at $T = 0$ a "quantum phase transition" ($T = 0$ phase transition induced by variation of external parameters, cf. [41]) to a confinement phase is expected to happen when decreasing a, λ . In the unitary gauge, these phase transitions should be accompanied by a condensation of singular fields. When approaching either the plasma or the confined phase, the dominance of the equilibrium positions $\phi = 0$ prohibits a proper definition of a coordinate system in color space based on the the color direction of the Higgs field.

The fate of the discrete $\mathbb{Z}_2$ symmetry is not understood in detail. As will be seen, realization of the center symmetry indicates confinement. Thus, the $\mathbb{Z}_2$ factor should not be part of the isotropy group (113) in the Higgs phase. The gauge choice (115) does not break this symmetry. Its breaking is a dynamical property of the symmetry. It must occur spontaneously. This $\mathbb{Z}_2$ symmetry must be restored in the quantum phase transition to the confinement phase and will remain broken in the transition to the high temperature plasma phase.

5.3 Topological Excitations

As in the abelian Higgs model, the non-trivial topology (S^2) of the manifold of vacuum field configurations of the Georgi–Glashow model is the origin of the topological excitations. We proceed as above and discuss field configurations of finite energy which differ in their topological properties from the ground-state configurations. As follows from the expression (106) for the energy density, finite energy can result only if asymptotically, $|\mathbf{x}| \to \infty$

$$\phi(\mathbf{x}) \to a\phi_0(\mathbf{x})$$

$$\mathbf{B} \to 0$$

$$[D_i\phi((\mathbf{x}))]^a = [\partial_i\delta^{ac} - g\epsilon^{abc}A_i^b(\mathbf{x})]\phi^c(\mathbf{x}) \to 0, \tag{121}$$

where $\phi_0(\mathbf{x})$ is a unit vector specifying the color direction of the Higgs field. The last equation correlates asymptotically the gauge and the Higgs field. In terms of the scalar field, the asymptotic gauge field is given by

$$A_i^a \to \frac{1}{ga^2}\epsilon^{abc}\phi^b\partial_i\phi^c + \frac{1}{a}\phi^a\mathcal{A}_i, \tag{122}$$

where $\mathcal{A}$ denotes the component of the gauge field along the Higgs field. It is arbitrary since (121) determines only the components perpendicular to ϕ. The asymptotic field strength associated with this gauge field (cf. (78)) has only a color component parallel to the Higgs field – the "neutral direction" (cf. the definition of the charged gauge fields in (117)) and we can write

$$F^{aij} = \frac{1}{a}\phi^a F^{ij}, \quad \text{with} \quad F^{ij} = \frac{1}{ga^3}\epsilon^{abc}\phi^a\partial^i\phi^b\partial^j\phi^c + \partial^i\mathcal{A}^j - \partial^j\mathcal{A}^i. \tag{123}$$

One easily verifies that the Maxwell equations

$$\partial_i F^{ij} = 0 \tag{124}$$

are satisfied. These results confirm the interpretation of $F_{\mu\nu}$ as a legitimate field strength related to the unbroken $U(1)$ part of the gauge symmetry. As the magnetic flux in the abelian Higgs model, the magnetic charge in the non-abelian Higgs model is quantized. Integrating over the asymptotic surface S^2 which encloses the system and using the integral form of the degree (49) of the map defined by the scalar field (cf. [42]) yields

$$m = \int_{S^2} \mathbf{B} \cdot d\boldsymbol{\sigma} = -\frac{1}{2ga^3} \int_{S^2} \epsilon^{ijk} \epsilon^{abc} \phi^a \partial^j \phi^b \partial^k \phi^c d\sigma^i = -\frac{4\pi N}{g} . \tag{125}$$

No contribution to the magnetic charge arises from $\nabla \times \mathcal{A}$ when integrated over a surface without boundary. The existence of a winding number associated with the Higgs field is a direct consequence of the topological properties discussed above. The Higgs field ϕ maps the asymptotic S^2 onto the space of zeroes of V which topologically is S^2 and has been shown (110) to be homeomorph to the coset space G/H_{ϕ_0}. Thus, asymptotically, the map

$$\phi : \quad S^2 \to S^2 \sim G/H_{\phi_0} \tag{126}$$

is characterized by the homotopy group $\pi_2(G/H_{\phi_0}) \sim \mathbb{Z}$. Our discussion provides an illustration of the general relation (61)

$$\pi_2\left(SU(2)/U(1) \otimes \mathbb{Z}_2\right) = \pi_1\left(U(1)\right) \sim \mathbb{Z} .$$

The non-triviality of the homotopy group guarantees the stability of topological excitations of finite energy.

An important example is the spherically symmetric hedgehog configuration

$$\phi^a(\mathbf{r}) \underset{r \to \infty}{\to} \phi_0^a(\mathbf{r}) = a \cdot \frac{x^a}{r}$$

which on the asymptotic sphere covers the space of zeroes of V exactly once. Therefore, it describes a monopole with the asymptotic field strength (apart from the $\mathcal{A}$ contribution) given, according to (123), by

$$F^{ij} = \epsilon^{ijk} \frac{x^k}{g r^3} , \quad \mathbf{B} = -\frac{\mathbf{r}}{g \, r^3} . \tag{127}$$

Monopole Solutions. The asymptotics of Higgs and gauge fields suggest the following spherically symmetric Ansatz for monopole solutions

$$\phi^a = a \frac{x^a}{r} H(agr) , \quad A_i^a = \epsilon^{aij} \frac{x^j}{gr^2}[1 - K(agr)] \tag{128}$$

with the boundary conditions at infinity

$$H(r) \underset{r \to \infty}{\to} 1, \quad K(r) \underset{r \to \infty}{\to} 0 .$$

As in the abelian Higgs model, topology forces the Higgs field to have a zero. Since the winding of the Higgs field ϕ cannot be removed by continuous deformations, ϕ has to have a zero. This defines the center of the monopole. The boundary condition

$$H(0) = 0, \quad K(0) = 1$$

in the core of the monopole guarantees continuity of the solution. As in the abelian Higgs model, the changes in the Higgs and gauge field are occurring on two different length scales. Unlike at asymptotic distances, in the core of the monopole also charged vector fields are present. The core of the monopole represents the perturbative phase of the Georgi–Glashow model, as the core of the vortex is made of normal conducting material and ordinary gauge fields.

With the Ansatz (128) the equations of motion are converted into a coupled system of ordinary differential equations for the unknown functions H and K which allows for analytical solutions only in certain limits. Such a limiting case is obtained by saturation of the Bogomol'nyi bound. As for the abelian Higgs model, this bound is obtained by rewriting the total energy of the static solutions

$$E = \int d^3x \left[\frac{1}{2}\mathbf{B}^2 + \frac{1}{2}(\mathbf{D}\phi)^2 + V(\phi) \right] = \int d^3x \left[\frac{1}{2}(\mathbf{B} \pm \mathbf{D}\phi)^2 + V(\phi) \mp \mathbf{B}\mathbf{D}\phi \right],$$

and by expressing the last term via an integration by parts (applicable for covariant derivatives due to antisymmetry of the structure constants in the definition of D in (83)) and with the help of the equation of motion $\mathbf{DB} = 0$ by the magnetic charge (125)

$$\int d^3x \mathbf{B}\,\mathbf{D}\phi = a \int_{S^2} \mathbf{B}\,d\sigma = a\,m.$$

The energy satisfies the Bogomol'nyi bound

$$E \geq |m|\,a.$$

For this bound to be saturated, the strength of the Higgs potential has to approach zero

$$V = 0, \quad \text{i.e.} \quad \lambda = 0,$$

and the fields have to satisfy the first order equation

$$\mathbf{B}^a \pm (\mathbf{D}\phi)^a = 0.$$

In the approach to vanishing λ, the asymptotics of the Higgs field $|\phi| \xrightarrow[r \to \infty]{} a$ must remain unchanged. The solution to this system of first order differential equations is known as the Prasad–Sommerfield monopole

$$H(agr) = \coth agr - \frac{1}{agr}, \quad K(agr) = \frac{\sinh agr}{agr}.$$

In this limiting case of saturation of the Bogomol'nyi bound, only one length scale exists $((ag)^{-1})$. The energy of the excitation, i.e. the mass of the monopole is given in terms of the mass of the charged vector particles (119) by

$$E = M \frac{4\pi}{g^2}.$$

As for the Nielsen–Olesen vortices, a wealth of further results have been obtained concerning properties and generalizations of the 't Hooft–Polyakov monopole solution. Among them I mention the "Julia–Zee" dyons [43]. These solutions of the field equations are obtained using the Ansatz (128) for the Higgs field and the spatial components of the gauge field but admitting a non-vanishing time component of the form

$$A_0^a = \frac{x^a}{r^2} J(agr).$$

This time component reflects the presence of a source of electric charge q. Classically the electric charge of the dyon can assume any value, semiclassical arguments suggest quantization of the charge in units of g [44].

As the vortices of the Abelian Higgs model, 't Hooft–Polyakov monopoles induce zero modes if massless fermions are coupled to the gauge and Higgs fields of the monopole

$$\mathcal{L}_\psi = i\bar{\psi}\gamma^\mu D_\mu \psi - g\phi^a \bar{\psi} \frac{\tau^a}{2} \psi. \tag{129}$$

The number of zero modes is given by the magnetic charge $|m|$ (125) [45]. Furthermore, the coupled system of a t' Hooft–Polyakov monopole and a fermionic zero mode behaves as a boson if the fermions belong to the fundamental representation of $SU(2)$ (as assumed in (129)) while for isovector fermions the coupled system behaves as a fermion. Even more puzzling, fermions can be generated by coupling bosons in the fundamental representation to the 't Hooft–Polyakov monopole. The origin of this conversion of isospin into spin [46–48] is the correlation between angular and isospin dependence of Higgs and gauge fields in solutions of the form (128). Such solutions do not transform covariantly under spatial rotations generated by **J**. Under combined spatial and isospin rotations (generated by **I**)

$$\mathbf{K} = \mathbf{J} + \mathbf{I}, \tag{130}$$

monopoles of the type (128) are invariant. **K** has to be identified with the angular momentum operator. If added to this invariant monopole, matter fields determine by their spin and isospin the angular momentum **K** of the coupled system.

Formation of monopoles is not restricted to the particular model. The Georgi–Glashow model is the simplest model in which this phenomenon occurs. With the topological arguments at hand, one can easily see the general condition for the existence of monopoles. If we assume electrodynamics to appear in the process of symmetry breakdown from a simply connected topological group G, the isotropy group H (69) must contain a $U(1)$ factor. According to the identities (61) and (40), the resulting non trivial second homotopy group of the coset

space
$$\pi_2(G/[\tilde{H} \otimes U(1)]) = \pi_1(\tilde{H}) \otimes \mathbf{Z} \tag{131}$$

guarantees the existence of monopoles. This prediction is independent of the group G, the details of the particular model, or of the process of the symmetry breakdown. It applies to Grand Unified Theories in which the structure of the standard model ($SU(3) \otimes SU(2) \otimes U(1)$) is assumed to originate from symmetry breakdown. The fact that monopoles cannot be avoided has posed a serious problem to the standard model of cosmology. The predicted abundance of monopoles created in the symmetry breakdown occurring in the early universe is in striking conflict with observations. Resolution of this problem is offered by the inflationary model of cosmology [49,50].

6 Quantization of Yang–Mills Theory

Gauge Copies. Gauge theories are formulated in terms of redundant variables. Only in this way, a covariant, local representation of the dynamics of gauge degrees of freedom is possible. For quantization of the theory both canonically or in the path integral, redundant variables have to be eliminated. This procedure is called gauge fixing. It is not unique and the implications of a particular choice are generally not well understood. In the path integral one performs a sum over all field configurations. In gauge theories this procedure has to be modified by making use of the decomposition of the space of gauge fields into equivalence classes, the gauge orbits (72). Instead of summing in the path integral over formally different but physically equivalent fields, the integration is performed over the equivalence classes of such fields, i.e. over the corresponding gauge orbits. The value of the action is gauge invariant, i.e. the same for all members of a given gauge orbit,

$$S\left[A^{[U]}\right] = S[A] .$$

Therefore, the action is seen to be a functional defined on classes (gauge orbits). Also the integration measure

$$d\left[A^{[U]}\right] = d[A] \quad , \quad d[A] = \prod_{x,\mu,a} dA_\mu^a(x) .$$

is gauge invariant since shifts and rotations of an integration variable do not change the value of an integral. Therefore, in the naive path integral

$$\tilde{Z} = \int d[A] \, e^{iS[A]} \propto \int \prod_x dU(x) .$$

a "volume" associated with the gauge transformations $\prod_x dU(x)$ can be factorized and thereby the integration be performed over the gauge orbits. To turn this property into a working algorithm, redundant variables are eliminated by imposing a gauge condition

$$f[A] = 0, \tag{132}$$

which is supposed to eliminate all gauge copies of a certain field configuration A. In other words, the functional f has to be chosen such that for arbitrary field configurations the equation

$$f[A^{[U]}] = 0$$

determines uniquely the gauge transformation U. If successful, the set of all gauge equivalent fields, the gauge orbit, is represented by exactly one representative. In order to write down an integral over gauge orbits, we insert into the integral the gauge-fixing δ-functional

$$\delta[f(A)] = \prod_x \prod_{a=1}^{N^2-1} \delta[f^a(A(x))].$$

This modification of the integral however changes the value depending on the representative chosen, as the following elementary identity shows

$$\delta(g(x)) = \frac{\delta(x-a)}{|g'(a)|} \quad , \quad g(a) = 0. \tag{133}$$

This difficulty is circumvented with the help of the Faddeev–Popov determinant $\Delta_f[A]$ defined implicitly by

$$\Delta_f[A] \int d[U] \, \delta\left[f\left(A^{[U]}\right)\right] = 1.$$

Multiplication of the path integral $\tilde{Z}$ with the above "1" and taking into account the gauge invariance of the various factors yields

$$\tilde{Z} = \int d[U] \int d[A] \, e^{iS[A]} \Delta_f[A] \, \delta\left[f\left(A^{[U]}\right)\right]$$

$$= \int d[U] \int d[A] \, e^{iS[A^{[U]}]} \Delta_f\left[A^{[U]}\right] \delta\left[f\left(A^{[U]}\right)\right] = \int d[U] \, Z.$$

The gauge volume has been factorized and, being independent of the dynamics, can be dropped. In summary, the final definition of the generating functional for gauge theories

$$Z[J] = \int d[A] \, \Delta_f[A] \, \delta\left(f[A]\right) e^{iS[A]+i\int d^4x \, J^\mu A_\mu} \tag{134}$$

is given in terms of a sum over gauge orbits.

Faddeev–Popov Determinant. For the calculation of $\Delta_f[A]$, we first consider the change of the gauge condition $f^a[A]$ under infinitesimal gauge transformations . Taylor expansion

$$f^a_x\left[A^{[U]}\right] \approx f^a_x[A] + \int d^4y \sum_{b,\mu} \frac{\delta f^a_x[A]}{\delta A^b_\mu(y)} \delta A^b_\mu(y)$$

$$= f^a_x[A] + \int d^4y \sum_b M(x,y;a,b) \, \alpha^b(y)$$

with δA_μ^a given by infinitesimal gauge transformations (92), yields

$$M\left(x,y;a,b\right) = \left(\partial_\mu \delta^{b,c} + g f^{bcd} A_\mu^d\left(y\right)\right) \frac{\delta f_x^a\left[A\right]}{\delta A_\mu^c\left(y\right)}. \tag{135}$$

In the second step, we compute the integral

$$\Delta_f^{-1}\left[A\right] = \int d\left[U\right] \delta\left[f\left(A^{[U]}\right)\right]$$

by expressing the integration $d\left[U\right]$ as an integration over the gauge functions α. We finally change to the variables $\beta = M\alpha$

$$\Delta_f^{-1}\left[A\right] = \left|\det M\right|^{-1} \int d\left[\beta\right] \delta\left[f\left(A\right) - \beta\right]$$

and arrive at the final expression for the Faddeev–Popov determinant

$$\Delta_f\left[A\right] = \left|\det M\right|. \tag{136}$$

Examples:

- Lorentz gauge

$$f_x^a\left(A\right) \qquad = \partial^\mu A_\mu^a\left(x\right) - \chi^a\left(x\right)$$
$$M\left(x,y;a,b\right) = -\left(\delta^{ab}\Box - g f^{abc} A_\mu^c\left(y\right) \partial_y^\mu\right) \delta^{(4)}\left(x - y\right) \tag{137}$$

- Coulomb gauge

$$f_x^a\left(A\right) \qquad = \mathrm{div}\mathbf{A}^a\left(x\right) - \chi^a\left(x\right)$$
$$M\left(x,y;a,b\right) = \left(\delta^{ab}\Delta + g f^{abc} \mathbf{A}^c\left(y\right) \boldsymbol{\nabla}_y\right) \delta^{(4)}\left(x - y\right) \tag{138}$$

- Axial gauge

$$f_x^a\left(A\right) \qquad = n^\mu A_\mu^a\left(x\right) - \chi^a\left(x\right)$$
$$M\left(x,y;a,b\right) = -\delta^{ab} n_\mu \partial_y^\mu \delta^{(4)}\left(x - y\right) \tag{139}$$

We note that in axial gauge, the Faddeev–Popov determinant does not depend on the gauge fields and therefore changes the generating functional only by an irrelevant factor.

Gribov Horizons. As the elementary example (133) shows, a vanishing Faddeev–Popov determinant $\left(g'(a) = 0\right)$ indicates the gauge condition to exhibit a quadratic or higher zero. This implies that at this point in function space, the gauge condition is satisfied by at least two gauge equivalent configurations, i.e. vanishing of $\Delta_f\left[A\right]$ implies the existence of zero modes associated with M (135)

$$M\chi_0 = 0$$

and therefore the gauge choice is ambiguous. The (connected) spaces of gauge fields which make the gauge choice ambiguous

$$\mathcal{M}_H = \big\{ A \big|\ \det M = 0 \big\}$$

are called Gribov horizons [51]. Around Gribov horizons, pairs of infinitesimally close gauge equivalent fields exist which satisfy the gauge condition. If on the other hand two gauge fields satisfy the gauge condition and are separated by an infinitesimal gauge transformation, these two fields are separated by a Gribov horizon. The region beyond the horizon thus contains gauge copies of fields inside the horizon. In general, one therefore needs additional conditions to select exactly one representative of the gauge orbits. The structure of Gribov horizons and of the space of fields which contain no Gribov copies depends on the choice of the gauge. Without specifying further the procedure, we associate an infinite potential energy $\mathcal{V}[A]$ with every gauge copy of a configuration which already has been taken into account, i.e. after gauge fixing, the action is supposed to contain implicitly this potential energy

$$S[A] \to S[A] - \int d^4x\, \mathcal{V}[A]. \qquad (140)$$

With the above expression, and given a reasonable gauge choice, the generating functional is written as an integral over gauge orbits and can serve as starting point for further formal developments such as the canonical formalism [36] or applications e.g. perturbation theory.

The occurrence of Gribov horizons points to a more general problem in the gauge fixing procedure. Unlike in electrodynamics, global gauge conditions may not exist in non-abelian gauge theories [52]. In other words, it may not be possible to formulate a condition which in the whole space of gauge fields selects exactly one representative. This difficulty of imposing a global gauge condition is similar to the problem of a global coordinate choice on e.g. S^2. In this case, one either has to resort to some patching procedure and use more than one set of coordinates (like for the Wu–Yang treatment of the Dirac Monopole [53]) or deal with singular fields arising from these gauge ambiguities (Dirac Monopole). Gauge singularities are analogous to the coordinate singularities on non-trivial manifolds (azimuthal angle on north pole).

The appearance of Gribov-horizons poses severe technical problems in analytical studies of non-abelian gauge theories. Elimination of redundant variables is necessary for proper definition of the path-integral of infinitely many variables. In the gauge fixing procedure it must be ascertained that every gauge orbit is represented by exactly one field-configuration. Gribov horizons may make this task impossible. On the other hand, one may regard the existence of global gauge conditions in QED and its non-existence in QCD as an expression of a fundamental difference in the structure of these two theories which ultimately could be responsible for their vastly different physical properties.

7   Instantons

7.1   Vacuum Degeneracy

Instantons are solutions of the classical Yang–Mills field equations with distinguished topological properties [54]. Our discussion of instantons follows the pattern of that of the Nielsen–Olesen vortex or the 't Hooft–Polyakov monopole and starts with a discussion of configurations of vanishing energy (cf. [55,34,57]). As follows from the Yang–Mills Hamiltonian (100) in the Weyl gauge (99), static zero-energy solutions of the equations of motion (84) satisfy

$$\mathbf{E} = 0, \quad \mathbf{B} = 0,$$

and therefore are pure gauges (93)

$$\mathbf{A} = \frac{1}{ig} U(\mathbf{x}) \mathbf{\nabla} U^\dagger(\mathbf{x}). \tag{141}$$

In the Weyl gauge, pure gauges in electrodynamics are gradients of time-independent scalar functions. In $SU(2)$ Yang–Mills theory, the manifold of zero-energy solutions is according to (141) given by the set of all $U(\mathbf{x}) \in SU(2)$. Since topologically $SU(2) \sim S^3$ (cf. (52)), each $U(\mathbf{x})$ defines a mapping from the base space $\mathbb{R}^3$ to S^3. We impose the requirement that at infinity, $U(\mathbf{x})$ approaches a unique value independent of the direction of $\mathbf{x}$

$$U(\mathbf{x}) \to \text{const.} \quad \text{for } |\mathbf{x}| \to \infty. \tag{142}$$

Thereby, the configuration space becomes compact $\mathbb{R}^3 \to S^3$ (cf. (47)) and pure gauges define a map

$$U(\mathbf{x}) : \ S^3 \longrightarrow S^3 \tag{143}$$

to which, according to (43), a winding number can be assigned. This winding number counts how many times the 3-sphere of gauge transformations $U(\mathbf{x})$ is covered if $\mathbf{x}$ covers once the 3-sphere of the compactified configuration space. Via the degree of the map (49) defined by $U(\mathbf{x})$, this winding number can be calculated [42,56] and expressed in terms of the gauge fields

$$n_w = \frac{g^2}{16\pi^2} \int d^3x \ \epsilon_{ijk} \left(A_i^a \, \partial_j A_k^a - \frac{g}{3} \, \epsilon^{abc} \, A_i^a \, A_j^b \, A_k^c \right). \tag{144}$$

The expression on the right hand side yields an integer only if A is a pure gauge. Examples of gauge transformations giving rise to non-trivial winding (hedgehog solution for $n = 1$) are

$$U_n(\mathbf{x}) = \exp\{i\pi n \, \frac{\mathbf{x}\boldsymbol{\tau}}{\sqrt{x^2 + p^2}}\} \tag{145}$$

with winding number $n_w = n$ (cf. (51) for verifying the asymptotic behavior (142)). Gauge transformations which change the winding number n_w are

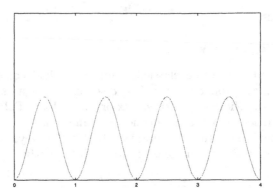

Fig. 9. Schematic plot of the potential energy $V[A] = \int d^3\, x\mathbf{B}[A]^2$ as a function of the winding number (144)

called *large* gauge transformations. Unlike *small* gauge transformations, they cannot be deformed continuously to $U = 1$.

These topological considerations show that Yang–Mills theory considered as a classical system possesses an infinity of different lowest energy ($E = 0$) solutions which can be labeled by an integer n. They are connected to each other by gauge fields which cannot be pure gauges and which therefore produce a finite value of the magnetic field, i.e. of the potential energy. The schematic plot of the potential energy in Fig. 9 shows that the ground state of QCD can be expected to exhibit similar properties as that of a particle moving in a periodic potential. In the quantum mechanical case too, an infinite degeneracy is present with the winding number in gauge theories replaced by the integer characterizing the equilibrium positions of the particle.

7.2 Tunneling

"Classical vacua" are states with values of the coordinate of a mechanical system $x = n$ given by the equilibrium positions. Correspondingly, in gauge theories the classical vacua, the "n-vacua" are given by the pure gauges ((141) and (145)). To proceed from here to a description of the quantum mechanical ground state, tunneling processes have to be included which, in such a semi-classical approximation, connect classical vacua with each other. Thereby the quantum mechanical ground state becomes a linear superposition of classical vacua. Such tunneling solutions are most easily obtained by changing to imaginary time with a concomitant change in the time component of the gauge potential

$$t \to -it, \quad A_0 \to -iA_0\,. \tag{146}$$

The metric becomes Euclidean and there is no distinction between covariant and contravariant indices. Tunneling solutions are solutions of the classical field equations derived from the Euclidean action S_E, i.e. the Yang–Mills action (cf. (80)) modified by the substitution (146). We proceed in a by now familiar way and derive the Bogomol'nyi bound for topological excitations in Yang–Mills theories. To this end we rewrite the action (cf. (87))

$$S_E = \frac{1}{4} \int d^4x \, F^a_{\mu\nu} \, F^a_{\mu\nu} = \frac{1}{4} \int d^4x \, \left(\pm F^a_{\mu\nu} \, \tilde{F}^a_{\mu\nu} + \frac{1}{2} (F^a_{\mu\nu} \mp \tilde{F}^a_{\mu\nu})^2 \right) \quad (147)$$

$$\geq \pm \frac{1}{4} \int d^4x \, F^a_{\mu\nu} \, \tilde{F}^a_{\mu\nu} \quad (148)$$

This bound for S_E (Bogomol'nyi bound) is determined by the topological charge ν, i.e. it can be rewritten as a surface term

$$\nu = \frac{g^2}{32\pi^2} \int d^4x \, F^a_{\mu\nu} \, \tilde{F}^a_{\mu\nu} = \int d\sigma_\mu \, K^\mu \quad (149)$$

of the topological current

$$K^\mu = \frac{g^2}{16\pi^2} \, \epsilon^{\mu\alpha\beta\gamma} \left(A^a_\alpha \, \partial_\beta \, A^a_\gamma - \frac{g}{3} \, \epsilon^{abc} \, A^a_\alpha \, A^b_\beta \, A^c_\gamma \right) . \quad (150)$$

Furthermore, if we assume K to vanish at spatial infinity so that

$$\nu = \int_{-\infty}^{+\infty} dt \, \frac{d}{dt} \int K^0 \, d^3x = n_w \, (t = \infty) - n_w(t = -\infty) , \quad (151)$$

the charge ν is seen to be quantized as a difference of two winding numbers.

I first discuss the formal implications of this result. The topological charge has been obtained as a difference of winding numbers of pure (time-independent) gauges (141) satisfying the condition (142). With the winding numbers, also ν is a topological invariant. It characterizes the space-time dependent gauge fields $A_\mu(x)$. Another and more direct approach to the topological charge (149) is provided by the study of cohomology groups. Cohomology groups characterize connectedness properties of topological spaces by properties of differential forms and their integration via Stokes' theorem (cf. Chap. 12 of [58] for an introduction).

Continuous deformations of gauge fields cannot change the topological charge. This implies that ν remains unchanged under continuous gauge transformations. In particular, the $\nu = 0$ equivalence class of gauge fields containing $A_\mu = 0$ as an element cannot be connected to gauge fields with non-vanishing topological charge. Therefore, the gauge orbits can be labeled by ν. Field configurations with $\nu \neq 0$ connect vacua (zero-energy field configurations) with different winding number ((151) and (144)). Therefore, the solutions to the classical Euclidean field equations with non-vanishing topological charge are the tunneling solutions needed for the construction of the semi-classical Yang–Mills ground state.

Like in the examples discussed in the previous sections, the field equations simplify if the Bogomol'nyi bound is saturated. In the case of Yang–Mills theory, the equations of motion can then be solved in closed form. Solutions with topological charge $\nu = 1$ ($\nu = -1$) are called instantons (anti-instantons). Their action is given by

$$S_E = \frac{8\pi^2}{g^2} .$$

By construction, the action of any other field configuration with $|\nu| = 1$ is larger. Solutions with action $S_E = 8\pi^2|\nu|/g^2$ for $|\nu| > 1$ are called multi-instantons. According to (147), instantons satisfy

$$F_{\mu\nu} = \pm\tilde{F}_{\mu\nu}. \tag{152}$$

The interchange $F_{\mu\nu} \leftrightarrow \tilde{F}_{\mu\nu}$ corresponding in Minkowski space to the interchange $\mathbf{E} \to \mathbf{B}$, $\mathbf{B} \to -\mathbf{E}$ is a duality transformation and fields satisfying (152) are said to be selfdual ($+$) or anti-selfdual ($-$) respectively. A spherical Ansatz yields the solutions

$$A_\mu^a = -\frac{2}{g}\frac{\eta_{a\mu\nu}x_\nu}{x^2 + \rho^2} \qquad F_{\mu\nu}^2 = \frac{1}{g^2}\frac{192\rho^4}{(x^2 + \rho^2)^4}, \tag{153}$$

with the 't Hooft symbol [59]

$$\eta_{a\mu\nu} = \begin{cases} \epsilon_{a\mu\nu} & \mu, \nu = 1, 2, 3 \\ \delta_{a\mu} & \nu = 0 \\ -\delta_{a\nu} & \mu = 0 \end{cases}.$$

The size of the instanton ρ can be chosen freely. Asymptotically, gauge potential and field strength behave as

$$A \underset{|x|\to\infty}{\longrightarrow} \frac{1}{x} \qquad F \underset{|x|\to\infty}{\longrightarrow} \frac{1}{x^4}.$$

The unexpectedly strong decrease in the field strength is the result of a partial cancellation of abelian and non-abelian contributions to $F_{\mu\nu}$ (78). For instantons, the asymptotics of the gauge potential is actually gauge dependent. By a gauge transformation, the asymptotics can be changed to x^{-3}. Thereby the gauge fields develop a singularity at $x = 0$, i.e. in the center of the instanton. In this "singular" gauge, the gauge potential is given by

$$A_\mu^a = -\frac{2\rho^2}{gx^2}\frac{\bar{\eta}_{a\mu\nu}x_\nu}{x^2 + \rho^2}, \qquad \bar{\eta}_{a\mu\nu} = \eta_{a\mu\nu}(1 - 2\delta_{\mu,0})(1 - 2\delta_{\nu,0}). \tag{154}$$

7.3   Fermions in Topologically Non-trivial Gauge Fields

Fermions are severely affected by the presence of gauge fields with non-trivial topological properties. A dynamically very important phenomenon is the appearance of fermionic zero modes in certain gauge field configurations. For a variety of low energy hadronic properties, the existence of such zero modes appears to be fundamental. Here I will not enter a detailed discussion of non-trivial fermionic properties induced by topologically non-trivial gauge fields. Rather I will try to indicate the origin of the induced topological fermionic properties in the context of a simple system. I will consider massless fermions in 1+1 dimensions moving in an external (abelian) gauge field. The Lagrangian of this system is (cf. (81))

$$\mathcal{L}_{YM} = -\frac{1}{4}F^{\mu\nu}F_{\mu\nu} + \bar{\psi}i\gamma^\mu D_\mu\psi, \tag{155}$$

with the covariant derivative D_μ given in (5) and ψ denoting a 2-component spinor. The Dirac algebra of the γ matrices

$$\{\gamma^\mu, \gamma^\nu\} = g^{\mu\nu}$$

can be satisfied by the following choice in terms of Pauli-matrices (cf. (50))

$$\gamma^0 = \tau^1, \ \gamma^1 = i\tau^2, \ \gamma^5 = -\gamma^0\gamma^1 = \tau^3.$$

In Weyl gauge, $A_0 = 0$, the Hamiltonian density (cf. (101)) is given by

$$\mathcal{H} = \frac{1}{2}E^2 + \psi^\dagger \mathcal{H}_f \psi, \tag{156}$$

with

$$\mathcal{H}_f = (i\partial_1 - eA_1)\gamma^5. \tag{157}$$

The application of topological arguments is greatly simplified if the spectrum of the fermionic states is discrete. We assume the fields to be defined on a circle and impose antiperiodic boundary conditions for the fermions

$$\psi(x + L) = -\psi(x).$$

The (residual) time-independent gauge transformations are given by (6) and (7) with the Higgs field ϕ replaced by the fermion field ψ. On a circle, the gauge functions $\alpha(x)$ have to satisfy (cf. (6))

$$\alpha(x + L) = \alpha(x) + \frac{2n\pi}{e}.$$

The winding number n_w of the mapping

$$U : S^1 \to S^1$$

partitions gauge transformations into equivalence classes with representatives given by the gauge functions

$$\alpha_n(x) = d_n x, \quad d_n = \frac{2\pi n}{eL}. \tag{158}$$

Large gauge transformations define pure gauges

$$A_1 = U(x)\frac{1}{ie}\partial_1 U^\dagger(x), \tag{159}$$

which inherit the winding number (cf. (144)). For 1+1 dimensional electrodynamics the winding number of a pure gauge is given by

$$n_w = -\frac{e}{2\pi}\int_0^L dx\, A_1(x). \tag{160}$$

As is easily verified, eigenfunctions and eigenvalues of $\mathcal{H}_f$ are given by

$$\psi_n(x) = e^{-ie\int_0^x A_1 dx - iE_n(a)x} u_\pm, \quad E_n(a) = \pm\frac{2\pi}{L}\left(n + \frac{1}{2} - a\right), \qquad (161)$$

with the positive and negative chirality eigenspinors $u_\pm$ of τ^3 and the zero mode of the gauge field

$$a = \frac{e}{2\pi}\int_0^L dx A_1(x).$$

We now consider a change of the external gauge field $A_1(x)$ from $A_1(x) = 0$ to a pure gauge of winding number n_w. The change is supposed to be adiabatic, such that the fermions can adjust at each instance to the changed value of the external field. In the course of this change, a changes continuously from 0 to n_w. Note that adiabatic changes of A_1 generate finite field strengths and therefore do not correspond to gauge transformations. As a consequence we have

$$E_n(n_w) = E_{n-n_w}(0). \qquad (162)$$

As expected, no net change of the spectrum results from this adiabatic changes between two gauge equivalent fields A_1. However, in the course of these changes the labeling of the eigenstates has changed. n_w negative eigenenergies of a certain chirality have become positive and n_w positive eigenenergies of the opposite chirality have become negative. This is called the *spectral flow* associated with this family of Dirac operators. The spectral flow is determined by the winding number of pure gauges and therefore a topological invariant. The presence of pure gauges with non-trivial winding number implies the occurrence of zero modes in the process of adiabatically changing the gauge field. In mathematics, the existence of zero modes of Dirac operators has become an important tool in topological investigations of manifolds ([60]). In physics, the spectral flow of the Dirac operator and the appearance of zero modes induced by topologically non-trivial gauge fields is at the origin of important phenomena like the formation of condensates or the existence of chiral anomalies.

7.4   Instanton Gas

In the semi-classical approximation, as sketched above, the non-perturbative QCD ground state is assumed to be given by topologically distinguished pure gauges and the instantons connecting the different classical vacuum configurations. In the instanton model for the description of low-energy strong interaction physics, one replaces the QCD partition function (134), i.e. the weighted sum over all gauge fields by a sum over (singular gauge) instanton fields (154)

$$A_\mu = \sum_{i=1}^N U(i)\, A_\mu(i)\, U^+(i), \qquad (163)$$

with

$$A_\mu(i) = -\bar{\eta}_{a\mu\nu}\frac{\rho^2}{g[x - z(i)]^2}\frac{x_\nu - z_\nu(i)}{[x - z(i)]^2 + \rho^2}\tau^a.$$

The gauge field is composed of N instantons with their centers located at the positions $z(i)$ and color orientations specified by the $SU(2)$ matrices $U(i)$. The instanton ensemble for calculation of $n-$point functions is obtained by summing over these positions and color orientations

$$Z[J] = \int \prod_{i=1}^{N} [dU(i)dz(i)] \, e^{-S_E[A]+i\int d^4x \, J \cdot A} .$$

Starting point of hadronic phenomenology in terms of instantons are the fermionic zero modes induced by the non-trivial topology of instantons. The zero modes are concentrated around each individual instanton and can be constructed in closed form

$$\not{D}\psi_0 = 0,$$

$$\psi_0 = \frac{\rho}{\pi\sqrt{x^2}} \frac{\gamma x}{(x^2+\rho^2)^{\frac{3}{2}}} \frac{1+\gamma_5}{2} \varphi_0,$$

where φ_0 is an appropriately chosen constant spinor. In the instanton model, the functional integration over quarks is truncated as well and replaced by a sum over the zero modes in a given configuration of non-overlapping instantons. A successful description of low-energy hadronic properties has been achieved [61] although a dilute gas of instantons does not confine quarks and gluons. It appears that the low energy-spectrum of QCD is dominated by the chiral properties of QCD which in turn seem to be properly accounted for by the instanton induced fermionic zero modes. The failure of the instanton model in generating confinement will be analyzed later and related to a deficit of the model in properly accounting for the 'center symmetry' in the confining phase.

To describe confinement, merons have been proposed [62] as the relevant field configurations. Merons are singular solutions of the classical equations of motion [63]. They are literally half-instantons, i.e. up to a factor of $1/2$ the meron gauge fields are identical to the instanton fields in the "regular gauge" (153)

$$A_\mu^{a \, M}(x) = \frac{1}{2} A_\mu^{a \, I}(x) = -\frac{1}{g} \frac{\eta_{a\mu\nu} x_\nu}{x^2},$$

and carry half a unit of topological charge. By this change of normalization, the cancellation between abelian and non-abelian contributions to the field strength is upset and therefore, asymptotically

$$A \sim \frac{1}{x}, \quad F \sim \frac{1}{x^2}.$$

The action

$$S \sim \int d^4x \frac{1}{x^4},$$

exhibits a logarithmic infrared divergence in addition to the ultraviolet divergence. Unlike instantons in singular gauge ($A \sim x^{-3}$), merons always overlap. A dilute gas limit of an ensemble of merons does not exist, i.e. merons are strongly interacting. The absence of a dilute gas limit has prevented development of a quantitative meron model of QCD. Recent investigations [64] in which this strongly interacting system of merons is treated numerically indeed suggest that merons are appropriate effective degrees of freedom for describing the confining phase.

7.5   Topological Charge and Link Invariants

Because of its wide use in the topological analysis of physical systems, I will discuss the topological charge and related topological invariants in the concluding paragraph on instantons.

The quantization of the topological charge ν is a characteristic property of the Yang–Mills theory in 4 dimensions and has its origin in the non-triviality of the mapping (143). Quantities closely related to ν are of topological relevance in other fields of physics. In electrodynamics topologically non-trivial gauge transformations in 3 space dimensions do not exist $\left(\pi_3(S^1) = 0\right)$ and therefore the topological charge is not quantized. Nevertheless, with

$$\tilde{K}^0 = \epsilon^{0ijk} A_i \partial_j A_k \,,$$

the charge

$$h_{\mathbf{B}} = \int d^3x \, \tilde{K}^0 = \int d^3x \, \mathbf{A} \cdot \mathbf{B} \tag{164}$$

describes topological properties of fields. For illustration we consider two linked magnetic flux tubes (Fig. 10) with the axes of the flux tubes forming closed curves $\mathcal{C}_{1,2}$. Since $h_{\mathbf{B}}$ is gauge invariant (the integrand is not, but the integral over the scalar product of the transverse magnetic field and the (longitudinal) change in the gauge field vanishes), we may assume the vector potential to satisfy the Coulomb gauge condition

$$\mathrm{div}\mathbf{A} = 0 \,,$$

which allows us to invert the curl operator

$$(\boldsymbol{\nabla}\times)^{-1} = -\boldsymbol{\nabla} \times \frac{1}{\Delta} \tag{165}$$

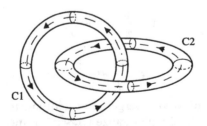

Fig. 10. Linked magnetic flux tubes

and to express $\tilde{K}^0$ uniquely in terms of the magnetic field

$$\tilde{K}^0 = -\left(\boldsymbol{\nabla} \times \frac{1}{\Delta}\mathbf{B}\right) \cdot \mathbf{B} = \frac{1}{4\pi} \int d^3x \int d^3x' \left(\mathbf{B}(\mathbf{x}) \times \mathbf{B}(\mathbf{x}')\right) \cdot \frac{\mathbf{x}' - \mathbf{x}}{|\mathbf{x}' - \mathbf{x}|^3}.$$

For single field lines,

$$\mathbf{B}(\mathbf{x}) = b_1 \frac{d\mathbf{s}_1}{dt}\delta(\mathbf{x} - \mathbf{s}_1(t)) + b_2 \frac{d\mathbf{s}_2}{dt}\delta(\mathbf{x} - \mathbf{s}_2(t))$$

the above integral is given by the linking number of the curves $\mathcal{C}_{1,2}$ (cf. (1)). Integrating finally over the field lines, the result becomes proportional to the magnetic fluxes $\phi_{1,2}$

$$h_{\mathbf{B}} = 2\,\phi_1\phi_2\,lk\{\mathcal{C}_1, \mathcal{C}_2\}. \tag{166}$$

This result indicates that the charge $h_{\mathbf{B}}$, the "magnetic helicity", is a topological invariant. For an arbitrary magnetic field, the helicity $h_{\mathbf{B}}$ can be interpreted as an average linking number of the magnetic field lines [22]. The helicity h_{ω} of vector fields has actually been introduced in hydrodynamics [5] with the vector potential replaced by the velocity field $\mathbf{u}$ of a fluid and the magnetic field by the vorticity $\omega = \boldsymbol{\nabla} \times \mathbf{u}$. The helicity measures the alignment of velocity and vorticity. The prototype of a "helical" flow [65] is

$$\mathbf{u} = \mathbf{u}_0 + \frac{1}{2}\omega_0 \times \mathbf{x}.$$

The helicity density is constant for constant velocity $\mathbf{u}_0$ and vorticity ω_0. For parallel velocity and vorticity, the streamlines of the fluid are right-handed helices. In magnetohydrodynamics, besides $h_{\mathbf{B}}$ and h_{ω}, a further topological invariant the "crossed" helicity can be defined. It characterizes the linkage of ω and $\mathbf{B}$ [66].

Finally, I would like to mention the role of the topological charge in the connection between gauge theories and topological invariants [67,68]. The starting point is the expression (164) for the helicity, which we use as action of the 3-dimensional abelian gauge theory [69], the abelian "Chern–Simons" action

$$S_{CS} = \frac{k}{8\pi} \int_M d^3x\, \mathbf{A} \cdot \mathbf{B},$$

where M is a 3-dimensional manifold and k an integer. One calculates the expectation value of a product of circular Wilson loops

$$W_N = \prod_{i=1}^{N} \exp\left\{i \int_{\mathcal{C}_i} \mathbf{A}\, d\mathbf{s}\right\}.$$

The Gaussian path integral

$$\langle W_N \rangle = \int D[A]e^{iS_{CS}}\, W_N$$

can be performed after inversion of the curl operator (165) in the space of transverse gauge fields. The calculation proceeds along the line of the calculation of $h_\mathbf{B}$ (164) and one finds

$$\langle W_N \rangle \propto \exp\left\{ \frac{2i\pi}{k} \sum_{i \neq j = 1}^{N} lk\{\mathcal{C}_i, \mathcal{C}_j\} \right\}.$$

The path integral for the Chern–Simons theory leads to a representation of a topological invariant. The key property of the Chern–Simons action is its invariance under general coordinate transformations. S_{CS} is itself a topological invariant. As in other evaluations of expectation values of Wilson loops, determination of the proportionality constant in the expression for $\langle W_N \rangle$ requires regularization of the path integral due to the linking of each curve with itself (self linking number). In the extension to non-abelian (3-dimensional) Chern–Simons theory, the very involved analysis starts with K^0 (150) as the non-abelian Chern–Simons Lagrangian. The final result is the Jones–Witten invariant associated with the product of circular Wilson loops [67].

8 Center Symmetry and Confinement

Gauge theories exhibit, as we have seen, a variety of non-perturbative phenomena which are naturally analyzed by topological methods. The common origin of all the topological excitations which I have discussed is vacuum degeneracy, i.e. the existence of a continuum or a discrete set of classical fields of minimal energy. The phenomenon of confinement, the trademark of non-abelian gauge theories, on the other hand, still remains mysterious in spite of large efforts undertaken to confirm or disprove the many proposals for its explanation. In particular, it remains unclear whether confinement is related to the vacuum degeneracy associated with the existence of large gauge transformations or more generally whether classical or semiclassical arguments are at all appropriate for its explanation. In the absence of quarks, i.e. of matter in the fundamental representation, $SU(N)$ gauge theories exhibit a residual gauge symmetry, the center symmetry, which is supposed to distinguish between confined and deconfined phases [70]. Irrespective of the details of the dynamics which give rise to confinement, this symmetry must be realized in the confining phase and spontaneously broken in the "plasma" phase. Existence of a residual gauge symmetry implies certain non-trivial topological properties akin to the non-trivial topological properties emerging in the incomplete spontaneous breakdown of gauge symmetries discussed above. In this and the following chapter I will describe formal considerations and discuss physical consequences related to the center symmetry properties of $SU(2)$ gauge theory. To properly formulate the center symmetry and to construct explicitly the corresponding symmetry transformations and the order parameter associated with the symmetry, the gauge theory has to be formulated on space-time with (at least) one of the space-time directions being compact, i.e. one has to study gauge theories at finite temperature or finite extension.

8.1   Gauge Fields at Finite Temperature and Finite Extension

When heating a system described by a field theory or enclosing it by making a spatial direction compact new phenomena occur which to some extent can be analyzed by topological methods. In relativistic field theories systems at finite temperature and systems at finite extensions with an appropriate choice of boundary conditions are copies of each other. In order to display the physical consequences of this equivalence we consider the Stefan–Boltzmann law for the energy density and pressure for a non-interacting scalar field with the corresponding quantities appearing in the Casimir effect, i.e. the energy density of the system if it is enclosed in one spatial direction by walls. I assume the scalar field to satisfy periodic boundary conditions on the enclosing walls. The comparison

<table>
<tr><td>Stefan–Boltzmann</td><td>Casimir</td></tr>
</table>

$$\epsilon = \frac{\pi^2}{15} T^4 \qquad\qquad p = -\frac{\pi^2}{15} L^{-4}$$

$$p = \frac{\pi^2}{45} T^4 \qquad\qquad \epsilon = -\frac{\pi^2}{45} L^{-4}. \tag{167}$$

expresses a quite general relation between thermal and quantum fluctuations in relativistic field theories [71,72]. This connection is easily established by considering the partition function given in terms of the Euclidean form (cf. (146)) of the Lagrangian

$$Z = \int_{\text{period.}} D[...] e^{-\int_0^\beta dx_0 \int dx_1 dx_2 dx_3 \mathcal{L}_E[...]}$$

which describes a system of infinite extension at temperature $T = \beta^{-1}$. The partition function

$$Z = \int_{\text{period.}} D[...] e^{-\int_0^L dx_3 \int dx_0 dx_1 dx_2 \mathcal{L}_E[...]}$$

describes the same dynamical system in its ground state ($T = 0$) at finite extension L in 3-direction. As a consequence, by interchanging the coordinate labels in the Euclidean, one easily derives allowing for both finite temperature and finite extension

$$Z(\beta, L) = Z(L, \beta)$$

$$\epsilon(\beta, L) = -p(L, \beta). \tag{168}$$

These relations hold irrespective of the dynamics of the system. They apply to non-interacting systems (167) and, more interestingly, they imply that any phase transition taking place when heating up an interacting system has as counterpart a phase transition occurring when compressing the system (*Quantum phase transition* [41] by variation of the size parameter L). Critical temperature and critical length are related by

$$T_c = \frac{1}{L_c}.$$

For QCD with its supposed phase transition at about 150 MeV, this relation predicts the existence of a phase transition when compressing the system beyond 1.3 fm.

Thermodynamic quantities can be calculated as ground state properties of the same system at the corresponding finite extension. This enables us to apply the canonical formalism and with it the standard tools of analyzing the system by symmetry considerations and topological methods. Therefore, in the following a spatial direction, the 3-direction, is chosen to be compact and of extension L

$$0 \leq x_3 \leq L \quad x = (x_\perp, x_3),$$

with

$$x_\perp = (x_0, x_1, x_2).$$

Periodic boundary conditions for gauge and bosonic matter fields

$$A_\mu(x_\perp, x_3 + L) = A_\mu(x_\perp, x_3), \quad \phi(x_\perp, x_3 + L) = \phi(x_\perp, x_3) \tag{169}$$

are imposed, while fermion fields are subject to antiperiodic boundary conditions

$$\psi(x_\perp, x_3 + L) = -\psi(x_\perp, x_3). \tag{170}$$

In finite temperature field theory, i.e. for $T = 1/L$, only this choice of boundary conditions defines the correct partition functions [73]. The difference in sign of fermionic and bosonic boundary conditions reflect the difference in the quantization of the two fields by commutators and anticommutators respectively. The negative sign, appearing when going around the compact direction is akin to the change of sign in a 2π rotation of a spin $1/2$ particle.

At finite extension or finite temperature, the fields are defined on $S^1 \otimes \mathbb{R}^3$ rather than on $\mathbb{R}^4$ if no other compactification is assumed. Non-trivial topological properties therefore emerge in connection with the S^1 component. $\mathbb{R}^3$ can be contracted to a point (cf. (32)) and therefore the cylinder is homotopically equivalent to a circle

$$S^1 \otimes \mathbb{R}^n \sim S^1. \tag{171}$$

Homotopy properties of fields defined on a cylinder (mappings from S^1 to some target space) are therefore given by the fundamental group of the target space. This is illustrated in Fig. 11 which shows two topologically distinct loops. The loop on the surface of the cylinder can be shrunk to a point, while the loop winding around the cylinder cannot.

8.2   Residual Gauge Symmetries in QED

I start with a brief discussion of electrodynamics with the gauge fields coupled to a charged scalar field as described by the Higgs model Lagrangian (2) (cf. [39,74]). Due to the homotopic equivalence (171), we can proceed as in our discussion of 1+1 dimensional electrodynamics and classify gauge transformations according to their winding number and separate the gauge transformations into small and

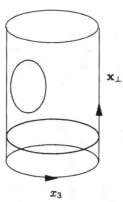

Fig. 11. Polyakov loop (along the compact x_3 direction) and Wilson loop (on the surface of the cylinder) in $S^1 \otimes \mathbb{R}^3$

large ones with representative gauge functions given by (158) (with x replaced by x_3). If we strictly follow the Faddeev–Popov procedure, gauge fixing has to be carried out by allowing for both type of gauge transformations. Most of the gauge conditions employed do not lead to such a complete gauge fixing. Consider for instance within the canonical formalism with $A_0 = 0$ the Coulomb-gauge condition

$$\mathrm{div}\mathbf{A} = 0, \tag{172}$$

and perform a large gauge transformation associated with the representative gauge function (158)

$$\mathbf{A}(x) \to \mathbf{A}(x) + \mathbf{e}_3 d_n \quad \phi(x) \to e^{iex_3 d_n}\phi(x). \tag{173}$$

The transformed gauge field (cf. (7)) is shifted by a constant and therefore satisfies the Coulomb-gauge condition as well. Thus, each gauge orbit $\mathcal{O}$ (cf. (72)) is represented by infinitely many configurations each one representing a suborbit $\mathcal{O}_n$. The suborbits are connected to each other by large gauge transformations, while elements within a suborbit are connected by small gauge transformations. The multiple representation of a gauge orbit implies that the Hamiltonian in Coulomb gauge contains a residual symmetry due to the presence of a residual redundancy. Indeed, the Hamiltonian in Coulomb gauge containing only the transverse gauge fields $\mathbf{A}_{tr}$ and their conjugate momenta $\mathbf{E}_{tr}$ (cf. (10))

$$\mathcal{H} = \frac{1}{2}(\mathbf{E}_{tr}^2 + \mathbf{B}^2) + \pi^*\pi + (D_{tr}\phi)^*(D_{tr}\phi) + V(\phi), \quad H = \int d^3x\,\mathcal{H}(\mathbf{x}) \tag{174}$$

is easily seen to be invariant under the discrete shifts of the gauge fields joined by discrete rotations of the Higgs field

$$[H, e^{iD_3 d_n}] = 0. \tag{175}$$

These transformations are generated by the 3-component of Maxwell's displacement vector

$$\mathbf{D} = \int d^3x(\mathbf{E} + \mathbf{x}j^0),$$

with the discrete set of parameters d_n given in (158). At this point, the analysis of the system via symmetry properties is more or less standard and one can characterize the different phases of the abelian Higgsmodel by their realization of the displacement symmetry. It turns out that the presence of the residual gauge symmetry is necessary to account for the different phases. It thus appears that complete gauge fixing involving also large gauge transformations is not a physically viable option.

Like in the symmetry breakdown occurring in the non-abelian Higgs model, in this procedure of incomplete gauge fixing, the $U(1)$ gauge symmetry has not completely disappeared but the isotropy group H_{lgt} (69) of the large gauge transformations (173) generated by D_3 remains. Denoting with G_1 the (simply connected) group of gauge transformations in (the covering space) $\mathbb{R}^1$ we deduce from (60) the topological relation

$$\pi_1(G_1/H_{lgt}) \sim \mathbb{Z},$$ (176)

which expresses the topological stability of the large gauge transformations. Equation (176) does not translate directly into a topological stability of gauge and matter field configurations. An appropriate Higgs potential is necessary to force the scalar field to assume a non-vanishing value. In this case the topologically non-trivial configurations are strings of constant gauge fields winding around the cylinder with the winding number specifying both the winding of the phase of the matter field and the strength of the gauge field. If, on the other hand, $V(\varphi)$ has just one minimum at $\varphi = 0$ nothing prevents a continuous deformation of a configuration to $A = \varphi = 0$. In such a case, only quantum fluctuations could possibly induce stability.

Consequences of the symmetry can be investigated without such additional assumptions. In the Coulomb phase for instance with the Higgs potential given by the mass term $V(\phi) = m^2\phi\phi^*$, the periodic potential for the gauge field zero-mode

$$a_3^0 = \frac{1}{V} \int d^3x A_3(x)$$ (177)

can be evaluated [75]

$$V_{\text{eff}}(a_3^0) = -\frac{m^2}{\pi^2 L^2} \sum_{n=1}^{\infty} \frac{1}{n^2} \cos(neLa_3^0) K_2(nmL).$$ (178)

The effective potential accounts for the effect of the thermal fluctuations on the gauge field zero-mode. It vanishes at zero temperature ($L \to \infty$). The periodicity of V_{eff} reflects the residual gauge symmetry. For small amplitude oscillations $eLa_3^0 \ll 2\pi$, V_{eff} can be approximated by the quadratic term, which in the small extension or high temperature limit, $mL = m/T \ll 1$, defines the Debye screening mass [73,76]

$$m_D^2 = \frac{1}{3}e^2T^2.$$ (179)

This result can be obtained by standard perturbation theory. We note that this perturbative evaluation of V_{eff} violates the periodicity, i.e. it does not respect the residual gauge symmetry.

8.3   Center Symmetry in SU(2) Yang–Mills Theory

To analyze topological and symmetry properties of gauge fixed $SU(2)$ Yang–Mills theory, we proceed as above, although abelian and non-abelian gauge theories differ in an essential property. Since $\pi_1(SU(2)) = 0$, gauge transformations defined on $S^1 \otimes \mathbb{R}^3$ are topologically trivial. Nevertheless, non-trivial topological properties emerge in the course of an incomplete gauge fixing enforced by the presence of a non-trivial center (62) of $SU(2)$. We will see later that this is actually the correct physical choice for accounting of both the confined and deconfined phases. Before implementing a gauge condition, it is useful to decompose the gauge transformations according to their periodicity properties. Although the gauge fields have been required to be periodic, gauge transformations may not. Gauge transformations preserve periodicity of gauge fields and of matter fields in the adjoint representation (cf. (89) and (90)) if they are periodic up to an element of the center of the gauge group

$$U(x_\perp, L) = c_U \cdot U(x_\perp, 0) . \tag{180}$$

If fields in the fundamental representation are present with their linear dependence on U (89), their boundary conditions require the gauge transformations U to be strictly periodic $c_U = 1$. In the absence of such fields, gauge transformations can be classified according to the value of c_U (± 1 in SU(2)). An important example of an SU(2) (cf. (66)) gauge transformation u_- with $c = -1$ is

$$u_- = e^{i\pi\hat{\psi}\tau x_3/L} = \cos \pi x_3/L + i\hat{\psi}\tau \sin \pi x_3/L. \tag{181}$$

Here $\hat{\psi}(x_\perp)$ is a unit vector in color space. For constant $\hat{\psi}$, it is easy to verify that the transformed gauge fields

$$A_\mu^{[u_-]} = e^{i\pi\hat{\psi}\tau x_3/L} A_\mu e^{-i\pi\hat{\psi}\tau x_3/L} - \frac{\pi}{gL}\hat{\psi}\tau\delta_{\mu 3}$$

indeed remain periodic and continuous. Locally, $c_U = \pm 1$ gauge transformations U cannot be distinguished. Global changes induced by gauge transformations like (181) are detected by loop variables winding around the compact x_3 direction. The Polyakov loop,

$$P(x_\perp) = P \exp\left\{ ig \int_0^L dx_3 \, A_3(x) \right\}, \tag{182}$$

is the simplest of such variables and of importance in finite temperature field theory. The coordinate $x_\perp$ denotes the position of the Polyakov loop in the space transverse to x_3. Under gauge transformations (cf. (94) and (96))

$$P(x_\perp) \rightarrow U(x_\perp, L) P(x_\perp) U^\dagger(x_\perp, 0) .$$

With $x = (x_\perp, 0)$ and $x = (x_\perp, L)$ labeling identical points, the Polyakov loop is seen to distinguish $c_U = \pm 1$ gauge transformations. In particular, we have

$$\text{tr}\{P(x_\perp)\} \rightarrow \text{tr}\{c_U P(x_\perp)\} \stackrel{\text{SU(2)}}{=} \pm\text{tr}\{P(x_\perp)\}.$$

With this result, we now can transfer the classification of gauge transformations to a classification of gauge fields. In $SU(2)$, the gauge orbits $\mathcal{O}$ (cf. (72)) are decomposed according to $c = \pm 1$ into suborbits $\mathcal{O}_\pm$. Thus these suborbits are characterized by the sign of the Polyakov loop at some fixed reference point $x_\perp^0$

$$A(x) \in \mathcal{O}_\pm \ , \quad \text{if} \quad \pm \operatorname{tr}\{P(x_\perp^0)\} \geq 0. \tag{183}$$

Strictly speaking, it is not the trace of the Polyakov loop rather only its modulus $|\operatorname{tr}\{P(x_\perp)\}|$ which is invariant under all gauge transformations. Complete gauge fixing, i.e. a representation of gauge orbits $\mathcal{O}$ by exactly one representative, is only possible if the gauge fixing transformations are not strictly periodic. In turn, if gauge fixing is carried out with strictly periodic gauge fixing transformations $(U, c_U = 1)$ the resulting ensemble of gauge fields contains one representative $A_\pm^f$ for each of the suborbits (183). The label f marks the dependence of the representative on the gauge condition (132). The (large) $c_U = -1$ gauge transformation mapping the representatives of two gauge equivalent suborbits onto each other are called *center reflections*

$$Z : A_+^f \leftrightarrow A_-^f. \tag{184}$$

Under center reflections

$$Z : \quad \operatorname{tr} P(x_\perp) \to -\operatorname{tr} P(x_\perp). \tag{185}$$

The center symmetry is a standard symmetry within the canonical formalism. Center reflections commute with the Hamiltonian

$$[H, Z] = 0. \tag{186}$$

Stationary states in SU(2) Yang–Mills theory can therefore be classified according to their Z-Parity

$$H|n_\pm\rangle = E_{n_\pm}|n_\pm\rangle , \quad Z|n_\pm\rangle = \pm|n_\pm\rangle . \tag{187}$$

The dynamics of the Polyakov loop is intimately connected to confinement. The Polyakov loop is associated with the free energy of a single heavy charge. In electrodynamics, the coupling of a heavy pointlike charge to an electromagnetic field is given by

$$\delta\mathcal{L} = \int d^4x j^\mu A_\mu = e \int d^4x \delta(\mathbf{x} - \mathbf{y}) A_0(x) = e \int_0^L dx_0 A_0(x_0, \mathbf{y}),$$

which, in the Euclidean and after interchange of coordinate labels 0 and 3, reduces to the logarithm of the Polyakov loop. The property of the system to confine can be formulated as a symmetry property. The expected infinite free energy of a static color charge results in a vanishing ground state expectation value of the Polyakov loop

$$\langle 0|\operatorname{tr} P(x_\perp)|0\rangle = 0 \tag{188}$$

in the confined phase. This property is guaranteed if the vacuum is center symmetric. The interaction energy $V(x_\perp)$ of two static charges separated in a transverse direction is, up to an additive constant, given by the Polyakov-loop correlator

$$\langle 0|\mathrm{tr}P(x_\perp)\mathrm{tr}P(0)|0\rangle = e^{-LV(x_\perp)}. \tag{189}$$

Thus, vanishing of the Polyakov-loop expectation values in the center symmetric phase indicates an infinite free energy of static color charges, i.e. confinement. For non-zero Polyakov-loop expectation values, the free energy of a static color charge is finite and the system is deconfined. A non-vanishing expectation value is possible only if the center symmetry is broken. Thus, in the transition from the confined to the plasma phase, the center symmetry, i.e. a discrete part of the underlying gauge symmetry, must be spontaneously broken. As in the abelian case, a complete gauge fixing, i.e. a definition of gauge orbits including large gauge transformations may not be desirable or even possible. It will prevent a characterization of different phases by their symmetry properties.

As in QED, non-trivial residual gauge symmetry transformations do not necessarily give rise to topologically non-trivial gauge fields. For instance, the pure gauge obtained from the non-trivial gauge transformation (181), with constant $\hat{\psi}$, $A_\mu = -\frac{\pi}{gL}\hat{\psi}\tau\delta_{\mu 3}$ is deformed trivially, along a path of vanishing action, into $A_\mu = 0$. In this deformation, the value of the Polyakov loop (182) changes continuously from -1 to 1. Thus a vacuum degeneracy exists with the value of the Polyakov loop labeling the gauge fields of vanishing action. A mechanism, like the Higgs mechanism, which gives rise to the topological stability of excitations built upon the degenerate classical vacuum has not been identified.

8.4 Center Vortices

Here, we again view the (incomplete) gauge fixing process as a symmetry breakdown which is induced by the elimination of redundant variables. If we require the center symmetry to be present after gauge fixing, the isotropy group formed by the center reflections must survive the "symmetry breakdown". In this way, we effectively change the gauge group

$$SU(2) \to SU(2)/Z(2). \tag{190}$$

Since $\pi_1\big(SU(2)/Z_2\big) = Z_2$, as we have seen (63), this space of gauge transformations contains topologically stable defects, line singularities in $\mathbb{R}^3$ or singular sheets in $\mathbb{R}^4$. Associated with such a singular gauge transformation $U_{Z_2}(x)$ are pure gauges (with the singular line or sheet removed)

$$A_{Z_2}^\mu(x) = \frac{1}{ig}\,U_{Z_2}(x)\,\partial^\mu U_{Z_2}^\dagger(x).$$

The following gauge transformation written in cylindrical coordinates ρ, φ, z, t

$$U_{Z_2}(\varphi) = \exp i\frac{\varphi}{2}\tau^3$$

exhibits the essential properties of singular gauge transformations, the *center vortices*, and their associated singular gauge fields . $U_{\mathbb{Z}_2}$ is singular on the sheet $\rho = 0$ (for all z, t). It has the property

$$U_{\mathbb{Z}_2}(2\pi) = -U_{\mathbb{Z}_2}(0),$$

i.e. the gauge transformation is continuous in $SU(2)/\mathbb{Z}_2$ but discontinuous as an element of $SU(2)$. The Wilson loop detects the defect. According to (97) and (98), the Wilson loop, for an arbitrary path C enclosing the vortex, is given by

$$W_{C, \mathbb{Z}_2} = \frac{1}{2} \operatorname{tr} \left\{ U_{\mathbb{Z}_2}(2\pi) U_{\mathbb{Z}_2}^\dagger(0) \right\} = -1. \tag{191}$$

The corresponding pure gauge field has only one non-vanishing space-time component

$$A_{\mathbb{Z}_2}^\varphi(x) = -\frac{1}{2g\rho}\tau^3, \tag{192}$$

which displays the singularity. For calculation of the field strength, we can, with only one color component non-vanishing, apply Stokes theorem. We obtain for the flux through an area of arbitrary size Σ located in the $x - y$ plane

$$\int_\Sigma F_{12}\rho d\rho d\varphi = -\frac{\pi}{g}\tau^3,$$

and conclude

$$F_{12} = -\frac{\pi}{g}\tau^3\delta^{(2)}(x).$$

This divergence in the field strength makes these fields irrelevant in the summation over all configurations. However, minor changes, like replacing the $1/\rho$ in $A_{\mathbb{Z}_2}^\varphi$ by a function interpolating between a constant at $\rho = 0$ and $1/\rho$ at large ρ eliminate this singularity. The modified gauge field is no longer a pure gauge. Furthermore, a divergence in the action from the infinite extension can be avoided by forming closed finite sheets. All these modifications can be carried out without destroying the property (191) that the Wilson loop is -1 if enclosing the vortex. This crucial property together with the assumption of a random distribution of center vortices yields an area law for the Wilson loop. This can be seen (cf. [77]) by considering a large area $\mathcal{A}$ in a certain plane containing a loop of much smaller area $\mathcal{A}_W$. Given a fixed number N of intersection points of vortices with $\mathcal{A}$, the number of intersection points with $\mathcal{A}_W$ will fluctuate and therefore the value W of the Wilson loop. For a random distribution of intersection points, the probability to find n intersection points in $\mathcal{A}_W$ is given by

$$p_n = \binom{N}{n} \left(\frac{\mathcal{A}_W}{\mathcal{A}}\right)^n \left(1 - \frac{\mathcal{A}_W}{\mathcal{A}}\right)^{N-n}.$$

Since, as we have seen, each intersection point contributes a factor -1, one obtains in the limit of infinite $\mathcal{A}$ with the density ν of intersection points, i.e.

vortices per area kept fixed,

$$\langle W \rangle = \sum_{n=1}^{N} (-1)^n p_n \to \exp\left(-2\nu \mathcal{A}_W\right).$$

As exemplified by this simple model, center vortices, if sufficiently abundant and sufficiently disordered, could be responsible for confinement (cf. [78]).

It should be noticed that, unlike the gauge transformation $U_{\mathbb{Z}_2}$, the associated pure gauge $A^\mu_{\mathbb{Z}_2}$ is not topologically stable. It can be deformed into $A^\mu = 0$ by a continuous change of its strength. This deformation, changing the magnetic flux, is not a gauge transformation and therefore the stability of $U_{\mathbb{Z}_2}$ is compatible with the instability of $A_{\mathbb{Z}_2}$. In comparison to nematic substances with their stable $\mathbb{Z}_2$ defects (cf. Fig. 7), the degrees of freedom of Yang–Mills theories are elements of the Lie algebra and not group-elements and it is not unlikely that the stability of $\mathbb{Z}_2$ vortices pertains only to formulations of Yang–Mills theories like lattice gauge theories where the elementary degrees of freedom are group elements.

It is instructive to compare this unstable defect in the gauge field with a topologically stable vortex. In a simple generalization [8] of the non-abelian Higgs model (82) such vortices appear. One considers a system containing two instead of one Higgs field with self-interactions of the type (104)

$$\mathcal{L}_m = \sum_{k=1,2} \left\{ \frac{1}{2} D_\mu \phi_k D^\mu \phi_k - \frac{\lambda_k}{4} (\phi_k^2 - a_k^2)^2 \right\} - V_{12}(\phi_1 \phi_2), \quad \lambda_k > 0. \quad (193)$$

By a choice of the interaction between the two scalar fields which favors the Higgs fields to be orthogonal to each other in color space, a complete spontaneous symmetry breakdown up to multiplication of the Higgs fields with elements of the center of $SU(2)$ can be achieved. The static, cylindrically symmetric Ansatz for such a "$\mathbb{Z}_2$-vortex" solution [79]

$$\phi_1 = \frac{a_1}{2} \tau^3, \quad \phi_2 = \frac{a_2}{2} f(\rho) \left(\cos \varphi \, \tau^1 + \sin \varphi \, \tau^2 \right), \quad A^\varphi = -\frac{1}{2g} \alpha(\rho) \tau^3 \quad (194)$$

leads with $V_{12} \propto (\phi_1 \phi_2)^2$ to a system of equations for the functions $f(\rho)$ and $\alpha(\rho)$ which is almost identical to the coupled system of equations (26) and (27) for the abelian vortex. As for the Nielsen–Olesen vortex or the 't Hooft–Polyakov monopole, the topological stability of this vortex is ultimately guaranteed by the non-vanishing values of the Higgs fields, enforced by the self-interactions and the asymptotic alignment of gauge and Higgs fields. This stability manifests itself in the quantization of the magnetic flux (cf.(125))

$$m = \int_{S^2} \mathbf{B} \cdot d\boldsymbol{\sigma} = -\frac{2\pi}{g}. \quad (195)$$

In this generalized Higgs model, fields can be classified according to their magnetic flux, which either vanishes as for the zero energy configurations or takes on the value (195). With this classification, one can associate a $\mathbb{Z}_2$ symmetry similar to the center symmetry with singular gauge transformations connecting the two classes. Unlike center reflections (181), singular gauge transformations change the value of the action. It has been argued [80] that, within the 2+1 dimensional Higgs model, this "topological symmetry" is spontaneously broken with the vacuum developing a domain structure giving rise to confinement. Whether this happens is a dynamical issue as complicated as the formation of flux tubes in Type II superconductors discussed on p.18. This spontaneous symmetry breakdown requires the center vortices to condense as a result of an attractive vortex–vortex interaction which makes the square of the vortex mass zero or negative. Extensions of such a scenario to pure gauge theories in 3+1 dimensions have been suggested [81,82].

8.5   The Spectrum of the SU(2) Yang–Mills Theory

Based on the results of Sect. 8.3 concerning the symmetry and topology of Yang–Mills theories at finite extension, I will deduce properties of the spectrum of the SU(2) Yang–Mills theory in the confined, center-symmetric phase.

- In the center-symmetric phase,

$$Z|0\rangle = |0\rangle\,,$$

  the vacuum expectation value of the Polyakov loop vanishes (188).
- The correlation function of Polyakov loops yields the interaction energy V of static color charges (in the fundamental representation)

$$\exp\left\{-LV\left(r\right)\right\} = \langle 0|T\left[\operatorname{tr}P\left(x_{\perp}^{E}\right)\operatorname{tr}P\left(0\right)\right]|0\rangle\,,\qquad r^{2} = \left(x_{\perp}^{E}\right)^{2}.\qquad(196)$$

- Due to the rotational invariance in Euclidean space, $x_{\perp}^{E}$ can be chosen to point in the time direction. After insertion of a complete set of excited states

$$\exp\left\{-LV\left(r\right)\right\} = \sum_{n_{-}}|\langle n_{-}|\operatorname{tr}P\left(0\right)|0\rangle|^{2}\,e^{-E_{n_{-}}r}\,.\qquad(197)$$

In the confined phase, the ground state does not contribute (188). Since $P\left(x_{\perp}^{E}\right)$ is odd under reflections only odd excited states,

$$Z|n_{-}\rangle = -|n_{-}\rangle\,,$$

contribute to the above sum. If the spectrum exhibits a gap,

$$E_{n_{-}} \geq E_{1_{-}} > 0,$$

the potential energy V increases linearly with r for large separations,

$$V\left(r\right) \approx \frac{E_{1}}{L}r\quad\text{for}\quad r\to\infty\quad\text{and}\quad L > L_{c}\,.\qquad(198)$$

- The linear rise with the separation, r, of two static charges (cf. (189)) is a consequence of covariance and the existence of a gap in the states excited by the Polyakov-loop operator. The slope of the confining potential is the string tension σ. Thus, in Yang–Mills theory at finite extension, the phenomenon of confinement is connected to the presence of a gap in the spectrum of Z–odd states,

$$E_- \geq \sigma L, \qquad (199)$$

which increases linearly with the extension of the compact direction. When applied to the vacuum, the Polyakov-loop operator generates states which contain a gauge string winding around the compact direction. The lower limit (199) is nothing else than the minimal energy necessary to create such a gauge string in the confining phase. Two such gauge strings, unlike one, are not protected topologically from decaying into the ground state or $Z = 1$ excited states. We conclude that the states in the $Z = -1$ sector contain $\mathbb{Z}_2$- stringlike excitations with excitation energies given by σL. As we have seen, at the classical level, gauge fields with vanishing action exist which wind around the compact direction. Quantum mechanics lifts the vacuum degeneracy and assigns to the corresponding states the energy (199).
- Z–even operators in general will have non-vanishing vacuum expectation values and such operators are expected to generate the hadronic states with the gap determined by the lowest glueball mass $E_+ = m_{gb}$ for sufficiently large extension $m_{gb}L \gg 1$.
- SU(2) Yang–Mills theory contains two sectors of excitations which, in the confined phase, are not connected by any physical process.
    - The hadronic sector, the sector of Z–even states with a mass gap (obtained from lattice calculations) $E_+ = m_{gb} \approx 1.5\,\mathrm{GeV}$
    - The gluonic sector, the sector of Z–odd states with mass gap $E_- = \sigma L$.
- When compressing the system, the gap in the $Z = -1$ sector decreases to about $650\,\mathrm{MeV}$ at $L_c \approx 0.75\mathrm{fm}$, $(T_c \approx 270\,\mathrm{MeV})$. According to SU(3) lattice gauge calculations, when approaching the critical temperature $T_c \approx 220\,\mathrm{MeV}$, the lowest glueball mass decreases. The extent of this decrease is controversial. The value $m_{gb}(T_c) = 770\,\mathrm{MeV}$ has been determined in [83,84] while in a more recent calculation [85] the significantly higher value of $1250\,\mathrm{MeV}$ is obtained for the glueball mass at T_c.
- In the deconfined or plasma phase, the center symmetry is broken. The expectation value of the Polyakov loop does not vanish. Debye screening of the fundamental charges takes place and formation of flux tubes is suppressed. Although the deconfined phase has been subject of numerous numerical investigations, some conceptual issues remain to be clarified. In particular, the origin of the exceptional realization of the center symmetry is not understood. Unlike symmetries of nearly all other systems in physics, the center symmetry is realized in the low temperature phase and broken in the high temperature phase. The confinement–deconfinement transition shares this exceptional behavior with the "inverse melting" process which has been observed in a polymeric system [86] and in a vortex lattice in high-T_c superconductors [87]. In the vortex lattice, the (inverse) melting into a crystalline

state happens as a consequence of the increase in free energy with increasing disorder which, in turn, under special conditions, may favor formation of a vortex lattice. Since nature does not seem to offer a variety of possibilities for inverse melting, one might guess that a similar mechanism is at work in the confinement–deconfinement transition. A solution of this type would be provided if the model of broken topological Z_2 symmetry discussed in Sect. 8.4 could be substantiated. In this model the confinement–deconfinement transition is driven by the dynamics of the "disorder parameter" [80] which exhibits the standard pattern of spontaneous symmetry breakdown.

The mechanism driving the confinement–deconfinement transition must also be responsible for the disparity in the energies involved. As we have seen, glueball masses are of the order of 1.5 GeV. On the other hand, the maximum in the spectrum of the black-body radiation increases with temperature and reaches according to Planck's law at $T = 220$ MeV a value of 620 MeV. A priori one would not expect a dissociation of the glueballs at such low temperatures. According to the above results concerning the $Z = \pm 1$ sectors, the phase transition may be initiated by the gain in entropy through coupling of the two sectors which results in a breakdown of the center symmetry. In this case the relevant energy scale is not the glueball mass but the mass gap of the $Z = -1$ states which, at the extension corresponding to 220 MeV, coincides with the peak in the energy density of the blackbody-radiation.

9   QCD in Axial Gauge

In close analogy to the discussion of the various field theoretical models which exhibit topologically non-trivial excitations, I have described so far $SU(2)$ Yang–Mills theory from a rather general point of view. The combination of symmetry and topological considerations and the assumption of a confining phase has led to intriguing conclusions about the spectrum of this theory. To prepare for more detailed investigations, the process of elimination of redundant variables has to be carried out. In order to make the residual gauge symmetry (the center symmetry) manifest, the gauge condition has to be chosen appropriately. In most of the standard gauges, the center symmetry is hidden and will become apparent in the spectrum only after a complete solution. It is very unlikely that approximations will preserve the center symmetry as we have noticed in the context of the perturbative evaluation of the effective potential in QED (cf. (178) and (179)). Here I will describe $SU(2)$ Yang–Mills theory in the framework of axial gauge, in which the center reflections can be explicitly constructed and approximation schemes can be developed which preserve the center symmetry.

9.1   Gauge Fixing

We now carry out the elimination of redundant variables and attempt to eliminate the 3-component of the gauge field $A_3(x)$. Formally this can be achieved by applying the gauge transformation

$$\Omega(x) = P \exp ig \int_0^{x_3} dz \, A_3 \left(x_\perp, z \right).$$

It is straightforward to verify that the gauge transformed 3-component of the gauge field indeed vanishes (cf. (90))

$$A_3 \left(x \right) \rightarrow \Omega \left(x \right) \left(A_3 \left(x \right) + \frac{1}{ig} \partial_3 \right) \Omega^\dagger \left(x \right) = 0.$$

However, this gauge transformation to axial gauge is not quite legitimate. The gauge transformation is not periodic

$$\Omega(x_\perp, x_3 + L) \neq \Omega(x_\perp, x_3).$$

In general, gauge fields then do not remain periodic either under transformation with Ω. Furthermore, with A_3 also the gauge invariant trace of the Polyakov loop (182) is incorrectly eliminated by Ω. These shortcomings can be cured, i.e. periodicity can be preserved and the loop variables $\operatorname{tr} P(x_\perp)$ can be restored with the following modified gauge transformation

$$\Omega_{ag}(x) = \Omega_D \left(x_\perp \right) \left[P^\dagger(x_\perp) \right]^{x_3/L} \Omega(x) . \tag{200}$$

The gauge fixing to axial gauge thus proceeds in three steps

- Elimination of the 3-component of the gauge field $A_3(x)$
- Restoration of the Polyakov loops $P(x_\perp)$
- Elimination of the gauge variant components of the Polyakov loops $P(x_\perp)$ by diagonalization

$$\Omega_D \left(x_\perp \right) P(x_\perp) \Omega_D^\dagger \left(x_\perp \right) = e^{igLa_3(x_\perp) \tau^3/2} . \tag{201}$$

Generating Functional. With the above explicit construction of the appropriate gauge transformations, we have established that the 3-component of the gauge field indeed can be eliminated in favor of a diagonal x^3-independent field $a_3(x_\perp)$. In the language of the Faddeev–Popov procedure, the axial gauge condition (cf. (132)) therefore reads

$$f[A] = A_3 - \left(a_3 + \frac{\pi}{gL} \right) \frac{\tau^3}{2}. \tag{202}$$

The field $a_3(x_\perp)$ is compact,

$$a_3 = a_3(x_\perp), \quad -\frac{\pi}{gL} \leq a_3(x_\perp) \leq \frac{\pi}{gL}.$$

It is interesting to compare QED and QCD in axial gauge in order to identify already at this level properties which are related to the non-abelian character of

QCD. In QED the same procedure can be carried out with omission of the third step. Once more, a lower dimensional field has to be kept for periodicity and gauge invariance. However, in QED the integer part of $a_3(x_\perp)$ cannot be gauged away; as winding number of the mapping $S^1 \to S^1$ it is protected topologically. In QCD, the appearance of the compact variable is ultimately due to the elimination of the gauge field A_3, an element of the Lie algebra, in favor of $P(x_\perp)$, an element of the compact Lie group.

With the help of the auxiliary field $a_3(x_\perp)$, the generating functional for QCD in axial gauge is written as

$$Z[J] = \int d[a_3]d[A]\,\Delta_f[A]\,\delta\left\{A_3 - \left(a_3 + \frac{\pi}{gL}\right)\frac{\tau^3}{2}\right\}\,e^{iS[A]+i\int d^4x J^\mu A_\mu}. \quad (203)$$

This generating functional contains as dynamical variables the fields $a_3(x_\perp)$, $A_\perp(x)$ with

$$A_\perp(x) = \{A_0(x), A_1(x), A_2(x)\}.$$

It is one of the unique features of axial gauge QCD that the Faddeev–Popov determinant (cf. (136) and 135))

$$\Delta_f[A] = |\det D_3|$$

can be evaluated in closed form

$$\frac{\det D_3}{(\det \partial_3)^3} = \prod_{x_\perp} \frac{1}{L^2}\cos^2 gLa_3(x_\perp)/2\,,$$

and absorbed into the measure

$$Z[J] = \int D[a_3]d[A_\perp]\,e^{iS[A_\perp,a_3-\frac{\pi}{gL}]+i\int d^4x JA}.$$

The measure

$$D[a_3] = \prod_{x_\perp}\cos^2(gLa_3(x_\perp)/2)\,\Theta\left[a_3(x_\perp)^2 - (\pi/gL)^2\right]da_3(x_\perp) \quad (204)$$

is nothing else than the Haar measure of the gauge group. It reflects the presence of variables (a_3) which are built from elements of the Lie group and not of the Lie algebra. Because of the topological equivalence of $SU(2)$ and S^3 (cf. (52)) the Haar measure is the volume element of S^3

$$d\Omega_3 = \cos^2\theta_1 \cos\theta_2\,d\theta^1 d\theta^2 d\varphi\,,$$

with the polar angles defined in the interval $[-\pi/2, \pi/2]$. In the diagonalization of the Polyakov loop (201) gauge equivalent fields corresponding to different values of θ_2 and φ for fixed θ_1 are eliminated as in the example discussed above (cf. (70)). The presence of the Haar measure has far reaching consequences.

Center Reflections. Center reflections Z have been formally defined in (184). They are residual gauge transformations which change the sign of the Polyakov loop (185). These residual gauge transformations are loops in $SU(2)/\mathbb{Z}_2$ (cf. (66)) and, in axial gauge, are given by

$$Z = ie^{i\pi\tau^1/2}e^{i\pi\tau^3 x^3/L}.$$

They transform the gauge fields, and leave the action invariant

$$Z: \quad a_3 \to -a_3, \quad A_\mu^3 \to -A_\mu^3, \quad \Phi_\mu \to \Phi_\mu^\dagger, \quad S[A_\perp a_3] \to S[A_\perp a_3]. \quad (205)$$

The off-diagonal gluon fields have been represented in a spherical basis by the antiperiodic fields

$$\Phi_\mu(x) = \frac{1}{\sqrt{2}}[A_\mu^1(x) + iA_\mu^2(x)]e^{-i\pi x^3/L}. \quad (206)$$

We emphasize that, according to the rules of finite temperature field theory, the bosonic gauge fields $A_\mu^a(x)$ are periodic in the compact variable x_3. For convenience, we have introduced in the definition of Φ an x_3-dependent phase factor which makes these field antiperiodic. With this definition, the action of center reflections simplify, Z becomes a (abelian) charge conjugation with the charged fields $\Phi_\mu(x)$ and the "photons" described by the neutral fields $A_\mu^3(x), a_3(x_\perp)$. Under center reflections, the trace of the Polyakov loop changes sign,

$$\frac{1}{2}\operatorname{tr}P(x_\perp) = -\sin\frac{1}{2}gLa_3(x_\perp). \quad (207)$$

Explicit representations of center reflections are not known in other gauges.

9.2   Perturbation Theory in the Center-Symmetric Phase

The center symmetry protects the Z−odd states with their large excitation energies (199) from mixing with the Z−even ground or excited states. Any approximation compatible with confinement has therefore to respect the center symmetry. I will describe some first attempts towards the development of a perturbative but center-symmetry preserving scheme. In order to display the peculiarities of the dynamics of the Polyakov-loop variables $a_3(x_\perp)$ we disregard in a first step their couplings to the charged gluons Φ_μ (206). The system of decoupled Polyakov-loop variables is described by the Hamiltonian

$$h = \int d^2x_\perp \left[-\frac{1}{2L}\frac{\delta^2}{\delta a_3(\mathbf{x}_\perp)^2} + \frac{L}{2}[\boldsymbol{\nabla}a_3(\mathbf{x}_\perp)]^2 \right] \quad (208)$$

and by the boundary conditions at $a_3 = \pm\frac{\pi}{gL}$ for the "radial" wave function

$$\hat{\psi}[a_3]\big|_{\text{boundary}} = 0. \quad (209)$$

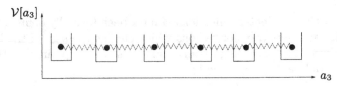

Fig. 12. System of harmonically coupled Polyakov-loop variables (208) trapped by the boundary condition (209) in infinite square wells

This system has a simple mechanical analogy. The Hamiltonian describes a 2 dimensional array of degrees of freedom interacting harmonically with their nearest neighbors (magnetic field energy of the Polyakov-loop variables). If we disregard for a moment the boundary condition, the elementary excitations are "sound waves" which run through the lattice. This is actually the model we would obtain in electrodynamics, with the sound waves representing the massless photons. Mechanically we can interpret the boundary condition as a result of an infinite square well in which each mechanical degree of freedom is trapped, as is illustrated in Fig. 12. This infinite potential is of the same origin as the one introduced in (140) to suppress contributions of fields beyond the Gribov horizon. Considered classically, waves with sufficiently small amplitude and thus with sufficiently small energy can propagate through the system without being affected by the presence of the walls of the potential. Quantum mechanically this may not be the case. Already the zero point oscillations may be changed substantially by the infinite square well. With discretized space (lattice spacing ℓ) and rescaled dynamical variables

$$\tilde{a}_3(x_\perp) = gLa_3(x_\perp)/2\,,$$

it is seen that for $\ell \ll L$ the electric field (kinetic) energy dominates. Dropping the nearest neighbor interaction, the ground state wavefunctional is given by

$$\hat{\Psi}_0\left[\tilde{a}_3\right] = \prod_{\mathbf{x}^\perp}\left[\left(\frac{2}{\pi}\right)^{1/2}\cos\left[\tilde{a}_3(x_\perp)\right]\right]\,.$$

In the absence of the nearest neighbor interaction, the system does not support waves and the excitations remain localized. The states of lowest excitation energy are obtained by exciting a single degree of freedom at one site $\tilde{\mathbf{x}}_\perp$ into its first excited state

$$\cos\left[\tilde{a}_3(\tilde{x}_\perp)\right] \rightarrow \sin\left[2\tilde{a}_3(\tilde{x}_\perp)\right]$$

with excitation energy

$$\Delta E = \frac{3}{8}\frac{g^2 L}{\ell^2}\,. \tag{210}$$

Thus, this perturbative calculation is in agreement with our general considerations and yields excitation energies rising with the extension L. From comparison with (199), the string tension

$$\sigma = \frac{3}{8}\frac{g^2}{\ell^2}$$

is obtained. This value coincides with the strong coupling limit of lattice gauge theory. However, unlike lattice gauge theory in the strong coupling limit, here no confinement-like behavior is obtained in QED. Only in QCD the Polyakov-loop variables a_3 are compact and thereby give rise to localized excitations rather than waves. It is important to realize that in this description of the Polyakov loops and their confinement-like properties we have left completely the familiar framework of classical fields with their well-understood topological properties. Classically the fields $a_3 = $ const. have zero energy. The quantum mechanical zero point motion raises this energy insignificantly in electrodynamics and dramatically for chromodynamics. The confinement-like properties are purely quantum mechanical in origin. Within quantum mechanics, they are derived from the "geometry" (the Haar measure) of the kinetic energy of the momenta conjugate to the Polyakov loop variables, the chromo-electric fluxes around the compact direction.

Perturbative Coupling of Gluonic Variables. In the next step, one may include coupling of the Polyakov-loop variables to each other via the nearest neighbor interactions. As a result of this coupling, the spectrum contains bands of excited states centered around the excited states in absence of the magnetic coupling [88]. The width of these bands is suppressed by a factor ℓ^2/L^2 as compared to the excitation energies (210) and can therefore be neglected in the continuum limit. Significant changes occur by the coupling of the Polyakov-loop variables to the charged gluons Φ_μ. We continue to neglect the magnetic coupling $(\partial_\mu a_3)^2$. The Polyakov-loop variables a_3 appearing at most quadratically in the action can be integrated out in this limit and the following effective action is obtained

$$S_{\text{eff}}\left[A_\perp\right] = S\left[A_\perp\right] + \frac{1}{2}M^2 \sum_{a=1,2} \int d^4x A_\mu^a(x) A^{a,\mu}(x) . \qquad (211)$$

The antiperiodic boundary conditions of the charged gluons, which have arisen in the change of field variables (206) reflect the mean value of A_3 in the center-symmetric phase

$$A_3 = a_3 + \frac{\pi}{gL},$$

while the geometrical (g−independent) mass

$$M^2 = \left(\frac{\pi^2}{3} - 2\right)\frac{1}{L^2} \qquad (212)$$

arises from their fluctuations. Antiperiodic boundary conditions describe the appearance of Aharonov–Bohm fluxes in the elimination of the Polyakov-loop variables. The original periodic charged gluon fields may be continued to be used if the partial derivative ∂_3 is replaced by the covariant one

$$\partial_3 \to \partial_3 + \frac{i\pi}{2L}[\tau^3, \cdot] .$$

Such a change of boundary conditions is a phenomenon well known in quantum mechanics. It occurs for a point particle moving on a circle (with circumference L) in the presence of a magnetic flux generated by a constant vector potential along the compact direction. With the transformation of the wave function

$$\psi(x) \rightarrow e^{ieAx}\psi(x),$$

the covariant derivative

$$(\frac{d}{dx} - ieA)\psi(x) \rightarrow \frac{d}{dx}\psi(x)$$

becomes an ordinary derivative at the expense of a change in boundary conditions at $x = L$. Similarly, the charged massive gluons move in a constant color neutral gauge field of strength $\frac{\pi}{gL}$ pointing in the spatial 3 direction. With x_3 compact, a color-magnetic flux is associated with this gauge field,

$$\Phi_{\text{mag}} = \frac{\pi}{g} \,, \tag{213}$$

corresponding to a magnetic field of strength

$$B = \frac{1}{gL^2} \,. $$

Also quark boundary conditions are changed under the influence of the color-magnetic fluxes

$$\psi(x) \rightarrow \exp\left[-ix_3\frac{\pi}{2L}\tau^3\right]\psi(x) \,. \tag{214}$$

Depending on their color they acquire a phase of $\pm i$ when transported around the compact direction. Within the effective theory, the Polyakov-loop correlator can be calculated perturbatively. As is indicated in the diagram of Fig. 13, Polyakov loops propagate only through their coupling to the charged gluons. Confinement-like properties are preserved when coupling to the Polyakov loops to the charged gluons. The linear rise of the interaction energy of fundamental charges obtained

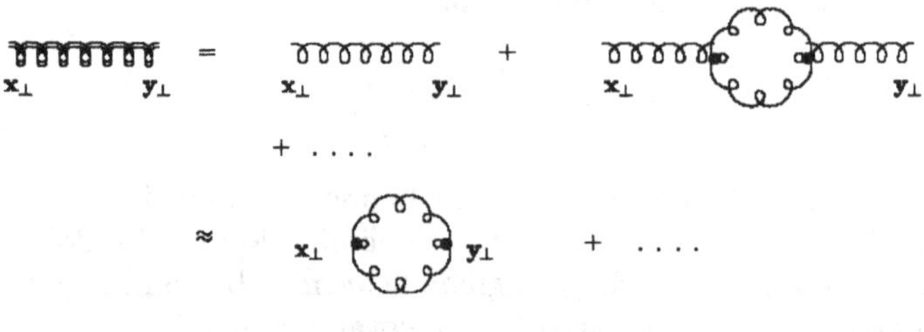

Fig. 13. One loop contribution from charged gluons to the propagator of Polyakov loops (external lines)

in leading order persist. As a consequence of the coupling of the Polyakov loops to the charged gluons, the value of the string constant is now determined by the threshold for charged gluon pair production

$$\sigma_{pt} = \frac{2}{L^2} \sqrt{\frac{4\pi^2}{3} - 2} \,, \tag{215}$$

i.e. the perturbative string tension vanishes in limit $L \to \infty$. This deficiency results from the perturbative treatment of the charged gluons. A realistic string constant will arise only if the threshold of a Z−odd pair of charged gluons increases linearly with the extension L (199).

Within this approximation, also the effect of dynamical quarks on the Polyakov-loop variables can be calculated by including quark loops besides the charged gluon loop in the calculation of the Polyakov-loop propagator (cf. Fig. 13). As a result of this coupling, the interaction energy of static charges ceases to rise linearly; it saturates for asymptotic distances at a value of

$$V(r) \approx 2m \,.$$

Thus, string breaking by dynamical quarks is obtained. This is a remarkable and rather unexpected result. Even though perturbation theory has been employed, the asymptotic value of the interaction energy is independent of the coupling constant g in contradistinction to the e^4 dependence of the Uehling potential in QED which accounts e.g. for the screening of the proton charge in the hydrogen atom by vacuum polarization [89]. Furthermore, the quark loop contribution vanishes if calculated with anti-periodic or periodic boundary conditions. A finite result only arises with the boundary conditions (214) modified by the Aharonov–Bohm fluxes. The $1/g$ dependence of the strength of these fluxes (213) is responsible for the coupling constant independence of the asymptotic value of $V(r)$.

9.3 Polyakov Loops in the Plasma Phase

If the center-symmetric phase would persist at high temperatures or small extensions, charged gluons with their increasing geometrical mass (212) and the increasing strength of the interaction (206) with the Aharonov–Bohm fluxes, would decouple

$$\Delta E \approx \frac{\pi}{L} \to \infty.$$

Only neutral gluon fields are periodic in the compact x_3 direction and therefore possess zero modes. Thus, at small extension or high temperature $L \to 0$, only neutral gluons would contribute to thermodynamic quantities. This is in conflict with results of lattice gauge calculations [90] and we therefore will assume that the center symmetry is spontaneously broken for $L \leq L_c = 1/T_c$. In the high-temperature phase, Aharonov–Bohm fluxes must be screened and the geometrical mass must be reduced. Furthermore, with the string tension vanishing in the plasma phase, the effects of the Haar measure must be effectively suppressed and the Polyakov-loop variables may be treated as classical fields. On the basis of this assumption, I now describe the development of a phenomenological treatment of the plasma phase [91]. For technical simplicity, I will neglect the space time dependence of a_3 and describe the results for vanishing geometrical mass M. For the description of the high-temperature phase it is more appropriate to use the variables

$$\chi = gLa_3 + \pi$$

with vanishing average Aharonov–Bohm flux. Charged gluons satisfy quasi-periodic boundary conditions

$$A_\mu^{1,2}(x_\perp, x_3 + L) = e^{i\chi} A_\mu^{1,2}(x_\perp, x_3). \tag{216}$$

Furthermore, we will calculate the thermodynamic properties by evaluation of the energy density in the Casimir effect (cf. (167) and (168)). In the Casimir effect, the central quantity to be calculated is the ground state energy of gluons between plates on which the fields have to satisfy appropriate boundary conditions. In accordance with our choice of boundary conditions (169), we assume the enclosing plates to extend in the x_1 and x_2 directions and to be separated in the x_3 direction. The essential observation for the following phenomenological description is the dependence of the Casimir energy on the boundary conditions and therefore on the presence of Aharonov–Bohm fluxes. The Casimir energy of the charged gluons is obtained by summing, after regularization, the zero point energies

$$\varepsilon(L, \chi) = \frac{1}{2} \sum_{n=-\infty}^{\infty} \int \frac{d^2 k_\perp}{(2\pi)^2} \left[\frac{k_\perp^2 + (2\pi n + \chi)^2}{L^2} \right]^{1/2} = \frac{4\pi^2}{3L^4} B_4\left(\frac{\chi}{2\pi}\right) \tag{217}$$

with

$$B_4(x) = -\frac{1}{30} + x^2(1 - x)^2.$$

Thermodynamic stability requires positive pressure at finite temperature and thus, according to (168), a negative value for the Casimir energy density. This requirement is satisfied if

$$\chi \leq 1.51.$$

For complete screening ($\chi = 0$) of the Aharonov–Bohm fluxes, the expression for the pressure in black-body radiation is obtained (the factor of two difference between (167) and (217) accounts for the two charged gluonic states). Unlike QED, QCD is not stable for vanishing Aharonov–Bohm fluxes. In QCD the perturbative ground state energy can be lowered by spontaneous formation of magnetic fields. Magnetic stability can be reached if the strength of the Aharonov–Bohm fluxes does not decrease beyond a certain minimal value. By calculating the Casimir effect in the presence of an external, homogeneous color-magnetic field

$$B_i^a = \delta^{a3} \delta_{i3} B,$$

this minimal value can be determined. The energy of a single quantum state is given in terms of the oscillator quantum number m for the Landau orbits,

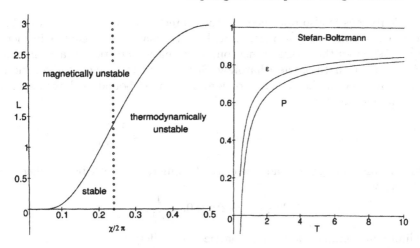

Fig. 14. Left: Regions of stability and instability in the (L, χ) plane. To the right of the circles, thermodynamic instability; above the solid line, magnetic instability. Right: Energy density and pressure normalized to Stefan–Boltzmann values vs. temperature in units of Λ_{MS}

in terms of the momentum quantum number n for the motion in the (compact) direction of the magnetic field, and by a magnetic moment contribution ($s = \pm 1$)

$$E_{mns} = \left[2gH(m + 1/2) + \frac{(2\pi n + \chi)^2}{L^2} + 2sgH \right]^{1/2} .$$

This expression shows that the destabilizing magnetic moment contribution $2sgH$ in the state with

$$s = -1, m = 0, n = 0$$

can be compensated by a non-vanishing Aharonov–Bohm flux χ of sufficient strength. For determination of the actual value of χ, the sum over these energies has to be performed. After regularizing the expression, the Casimir energy density can be computed numerically. The requirement of magnetic stability yields a lower limit on χ. As Fig. 14 shows, the Stefan–Boltzmann limit $\chi = 0$ is not compatible with magnetic stability for any value of the temperature. Identification of the Aharonov–Bohm flux with the minimal allowed values sets upper limits to energy density and pressure which are shown in Fig. 14. These results are reminiscent of lattice data [92] in the slow logarithmic approach of energy density and pressure

$$\chi(T) \geq \frac{11}{12} g^2(T), \; T \to \infty$$

to the Stefan–Boltzmann limit.

It appears that the finite value of the Aharonov–Bohm flux accounts for interactions present in the deconfined phase fairly well; qualitative agreement with lattice calculations is also obtained for the "interaction measure" $\epsilon - 3P$. Furthermore, these limits on χ also yield a realistic estimate for the change in energy density $-\Delta\epsilon$ across the phase transition. The phase transition is accompanied by a change in strength of the Aharonov–Bohm flux from the center symmetric value π to a value in the stability region. The lower bound is determined by thermodynamic stability

$$-\Delta\epsilon \geq \epsilon(L_c, \chi = \pi) - \epsilon(L_c, \chi = 1.51) = \frac{7\pi^2}{180}\frac{1}{L_c^4}.$$

For establishing an upper bound, the critical temperature must be specified. For $T_c \approx 270\,\mathrm{MeV}$,

$$0.38\,\frac{1}{L_c^4} \leq -\Delta\epsilon \leq 0.53\,\frac{1}{L_c^4}.$$

These limits are compatible with the lattice result [93]

$$\Delta\epsilon = -0.45\frac{1}{L_c^4} .$$

The picture of increasing Debye screening of the Aharonov–Bohm fluxes with increasing temperature seems to catch the essential physics of the thermodynamic quantities. It is remarkable that the requirement of magnetic stability, which prohibits complete screening, seems to determine the temperature dependence of the Aharonov–Bohm fluxes and thereby to simulate the non-perturbative dynamics in a semiquantitative way.

9.4　Monopoles

The discussion of the dynamics of the Polyakov loops has demonstrated that significant changes occur if compact variables are present. The results discussed strongly suggest that confinement arises naturally in a setting where the dynamics is dominated by such compact variables. The Polyakov-loop variables $a_3(x_\perp)$ constitute only a small set of degrees of freedom in gauge theories. In axial gauge, the remaining degrees of freedom $A_\perp(x)$ are standard fields which, with interactions neglected, describe freely propagating particles. As a consequence, the coupling of the compact variables to the other degrees of freedom almost destroys the confinement present in the system of uncoupled Polyakov-loop variables. This can be prevented to happen only if mechanisms are operative by which all the gluon fields acquire infrared properties similar to those of the Polyakov-loop variables. In the axial gauge representation it is tempting to connect such mechanisms to the presence of monopoles whose existence is intimately linked to the compactness of the Polyakov-loop variables. In analogy to the abelian Higgs model, condensation of magnetic monopoles could be be a first and crucial element of a mechanism for confinement. It would correspond to the formation of the charged Higgs condensate $|\phi| = a$ (13) enforced by the Higgs self-interaction (3). Furthermore, this magnetically charged medium should display excitations which behave as chromo-electric vortices. Concentration of the electric field lines to these vortices finally would give rise to a linear increase in the interaction energy of two chromo-electric charges with their separation as in (31). These phenomena actually happen in the Seiberg–Witten theory [94]. The Seiberg–Witten theory is a supersymmetric generalization of the non-abelian Higgs model. Besides gauge and Higgs fields it contains fermions in the adjoint representation. It exhibits vacuum degeneracy enlarged by supersymmetry and contains topologically non-trivial excitations, both monopoles and instantons. The monopoles can become massless and when partially breaking the supersymmetry, condensation of monopoles occurs that induces confinement of the gauge degrees of freedom.

In this section I will sketch the emergence of monopoles in axial gauge and discuss some elements of their dynamics. Singular field arise in the last step of the gauge fixing procedure (200), where the variables characterizing the orientation of the Polyakov loops in color space are eliminated as redundant variables by diagonalization of the Polyakov loops. The diagonalization of group elements is achieved by the unitary matrix

$$\Omega_D = e^{i\boldsymbol{\omega}\boldsymbol{\tau}} = \cos\omega + i\tau\hat{\omega}\sin\omega\,,$$

with $\omega(x_\perp)$ depending on the Polyakov loop $P(x_\perp)$ to be diagonalized. This diagonalization is ill defined if

$$P(x_\perp) = \pm 1\,, \tag{218}$$

i.e. if the Polyakov loop is an element of the center of the group (cf. (62)). Diagonalization of an element in the neighborhood of the center of the group is akin to the definition of the azimuthal angle on the sphere close to the north or south pole. With Ω_D ill defined, the transformed fields

$$A'_\mu(x) = \Omega_D(x_\perp)\left[A_\mu(x) + \frac{1}{ig}\partial_\mu\right]\Omega_D^\dagger(x_\perp)$$

develop singularities. The most singular piece arises from the inhomogeneous term in the gauge transformation

$$s_\mu(x_\perp) = \Omega_D(x_\perp)\frac{1}{ig}\partial_\mu\Omega_D^\dagger(x_\perp)\,.$$

For a given $a_3(x_\perp)$ with orientation described by polar $\theta(x_\perp)$ and azimuthal angles $\varphi(x_\perp)$ in color space, the matrix diagonalizing $a_3(x_\perp)$ can be represented as

$$\Omega_D = \begin{pmatrix} e^{i\varphi}\cos(\theta/2) & \sin(\theta/2) \\ -\sin(\theta/2) & e^{-i\varphi}\cos(\theta/2) \end{pmatrix}$$

and therefore the nature of the singularities can be investigated in detail. The condition for the Polyakov loop to be in the center of the group, i.e. at a definite point on S^3 (218), determines in general uniquely the corresponding position in $\mathbb{R}^3$ and therefore the singularities form world-lines in 4-dimensional space-time.

The singularities are "monopoles" with topologically quantized charges. Ω_D is determined only up to a gauge transformation

$$\Omega_D(x_\perp) \to e^{i\tau^3 \psi(x_\perp)} \Omega_D(x_\perp)$$

and is therefore an element of $SU(2)/U(1)$. The mapping of a sphere S^2 around the monopole in $x_\perp$ space to $SU(2)/U(1)$ is topologically non-trivial $\pi_2[SU(2)/U(1)] = \mathbb{Z}$ (67). This argument is familiar to us from the discussion of the 't Hooft–Polyakov monopole (cf. (125) and (126)). Also here we identify the winding number associated with this mapping as the magnetic charge of the monopole.

Properties of Singular Fields

- Dirac monopoles, extended to include color, constitute the simplest examples of singular fields (Euclidean $x_\perp = \mathbf{x}$)

$$\mathbf{A} \sim \frac{m}{2gr} \left\{ \frac{1 + \cos\theta}{\sin\theta} \, \hat{\varphi}\tau^3 + [(\hat{\varphi} + i\hat{\theta})e^{-i\varphi}(\tau^1 - i\tau^2) + \text{h.c.}] \right\}. \quad (219)$$

  In addition to the pole at $r = 0$, the fields contain a Dirac string in 3-space (here chosen along $\theta = 0$) and therefore a sheet-like singularity in 4-space which emanates from the monopole word-line.
- Monopoles are characterized by two charges, the "north-south" charge for the two center elements of SU(2) (218),

$$z = \pm 1, \quad (220)$$

  and the quantized strength of the singularity

$$m = \pm 1, \pm 2, \dots . \quad (221)$$

- The topological charge (149) is determined by the two charges of the monopoles present in a given configuration [95–97]

$$\nu = \frac{1}{2} \sum_i m_i z_i. \quad (222)$$

  Thus, after elimination of the redundant variables, the topological charge resides exclusively in singular field configurations.
- The action of singular fields is in general finite and can be arbitrarily small for $\nu = 0$. The singularities in the abelian and non-abelian contributions to the field strength cancel since by gauge transformations singularities in gauge covariant quantities cannot be generated.

9.5 Monopoles and Instantons

By the gauge choice, i.e. by the diagonalization of the Polyakov loop by Ω_D in (200), monopoles appear; instantons, which in (singular) Lorentz gauge have a point singularity (154) at the center of the instanton, must possess according to the relation (222) at least two monopoles with associated strings (cf. (219)). Thus, in axial gauge, an instanton field becomes singular on world lines and world sheets. To illustrate the connection between topological charges and monopole charges (222), we consider the singularity content of instantons in axial gauge [64] and calculate the Polyakov loop of instantons. To this end, the generalization of the instantons (154) to finite temperature (or extension) is needed. The so-called "calorons" are known explicitly [98]

$$A_\mu = \frac{1}{g}\bar{\eta}_{\mu\nu}\nabla_\nu \ln\left\{1 + \gamma\frac{(\sinh u)/u}{\cosh u - \cos v}\right\} \tag{223}$$

where

$$u = 2\pi|\mathbf{x}_\perp - \mathbf{x}_\perp^0|/L, \quad v = 2\pi x_3/L, \quad \gamma = 2(\pi\rho/L)^2.$$

The topological charge and the action are independent of the extension,

$$\nu = 1, \quad S = \frac{8\pi^2}{g^2}.$$

The Polyakov loops can be evaluated in closed form

$$P(\mathbf{x}) = \exp\left\{i\pi\frac{(\mathbf{x}_\perp - \mathbf{x}_\perp^0)\,\tau}{|\mathbf{x}_\perp - \mathbf{x}_\perp^0|}\chi(u)\right\}, \tag{224}$$

with

$$\chi(u) = 1 - \frac{(1 - \gamma/u^2)\sinh u + \gamma/u\cosh(u)}{\sqrt{(\cosh u + \gamma/u\sinh u)^2 - 1}}.$$

As Fig. 15 illustrates, instantons contain a $z = -1$ monopole at the center and a $z = 1$ monopole at infinity; these monopoles carry the topological charge of the instanton. Furthermore, tunneling processes represented by instantons connect field configurations of different winding number (cf. (151)) but with the same value for the Polyakov loop. In the course of the tunneling, the Polyakov loop of the instanton may pass through or get close to the center element $z = -1$, it however always returns to its original value $z = +1$. Thus, instanton ensembles in the dilute gas limit are not center symmetric and therefore cannot give rise to confinement. One cannot rule out that the $z = -1$ values of the Polyakov loop are encountered more and more frequently with increasing instanton density. In this way, a center-symmetric ensemble may finally be reached in the high-density limit. This however seems to require a fine tuning of instanton size and the average distance between instantons.

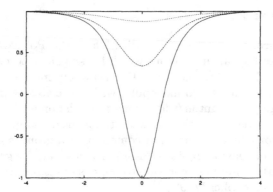

Fig. 15. Polyakov loop (224) of an instanton (223) of "size" $\gamma = 1$ as a function of time $t = 2\pi x_0/L$ for minimal distance to the center $2\pi x_1/L = 0$ (solid line), $L = 1$ (dashed line), $L = 2$ (dotted line), $x_2 = 0$

9.6   Elements of Monopole Dynamics

In axial gauge, instantons are composed of two monopoles. An instanton gas (163) of finite density n_I therefore contains field configurations with infinitely many monopoles. The instanton density in 4-space can be converted approximately to a monopole density in 3-space [97]

$$n_M \sim (L n_I \rho)^{3/2} \quad , \quad \rho \ll L,$$

$$n_M \sim L n_I \quad , \quad \rho \geq L.$$

With increasing extension or equivalently decreasing temperature, the monopole density diverges for constant instanton density. Nevertheless, the action density of an instanton gas remains finite. This is in accordance with our expectation that production of monopoles is not necessarily suppressed by large values of the action. Furthermore, a finite or possibly even divergent density of monopoles as in the case of the dilute instanton gas does not imply confinement.

Beyond the generation of monopoles via instantons, the system has the additional option of producing one type ($z = +1$ or $z = -1$) of poles and corresponding antipoles only. No topological charge is associated with such singular fields and their occurrence is not limited by the instanton bound ((147) and (152)) on the action as is the case for a pair of monopoles of opposite z-charge. Thus, entropy favors the production of such configurations. The entropy argument also applies in the plasma phase. For purely kinematical reasons, a decrease in the monopole density must be expected as the above estimates within the instanton model show. This decrease is counteracted by the enhanced probability to produce monopoles when, with decreasing L, the Polyakov loop approaches more and more the center of the group, as has been discussed above (cf. left part of Fig. 14). A finite density of singular fields is likely to be present also in the deconfined phase. In order for this to be compatible with the partially perturbative nature of the plasma phase and with dimensional reduction to QCD_{2+1}, poles and antipoles may have to be strongly correlated with each other and to form effectively a gas of dipoles.

Since entropy favors proliferate production of monopoles and monopoles may be produced with only a small increase in the total action, the coupling of the monopoles to the quantum fluctuations must ultimately prevent unlimited increase in the number of monopoles. A systematic study of the relevant dynamics has not been carried out. Monopoles are not solutions to classical field equations. Therefore, singular fields are mixed with quantum fluctuations even on the level of bilinear terms in the action. Nevertheless, two mechanisms can be identified which might limit the production of monopoles.

- The 4-gluon vertex couples pairs of monopoles to charged and neutral gluons and can generate masses for all the color components of the gauge fields. A simple estimate yields

$$\delta m^2 = -\frac{\pi}{V} \sum_{\substack{i,j=1 \\ i<j}}^{N} m_i m_j |\mathbf{x}_{\perp i} - \mathbf{x}_{\perp j}|$$

  with the monopole charges m_i and positions $\mathbf{x}_{\perp i}$. If operative also in the deconfined phase, this mechanism would give rise to a magnetic gluon mass.
- In general, fluctuations around singular fields generate an infinite action. Finite values of the action result only if the fluctuations $\delta\phi, \delta A^3$ satisfy the boundary conditions,

$$\delta\phi(x)\, e^{2i\varphi(x_\perp)} \quad \text{continuous along the strings},$$

$$\delta\phi(x)\Big|_{\text{at pole}} = \delta A^3\Big|_{\text{at pole}} = 0.$$

For a finite monopole density, long wave-length fluctuations cannot simultaneously satisfy boundary conditions related to monopoles or strings which are close to each other. One therefore might suspect quantum fluctuations with wavelengths

$$\lambda \geq \lambda_{\max} = n_{\mathrm{M}}^{-1/3}$$

to be suppressed.

We note that both mechanisms would also suppress the propagators of the quantum fluctuations in the infrared. Thereby, the decrease in the string constant by coupling Polyakov loops to charged gluons could be alleviated if not cured.

9.7 Monopoles in Diagonalization Gauges

In axial gauge, monopoles appear in the gauge fixing procedure (200) as defects in the diagonalization of the Polyakov loops. Although the choice was motivated by the distinguished role of the Polyakov-loop variables as order parameters, formally one may choose any quantity ϕ which, if local, transforms under gauge transformations U as

$$\phi \to U\phi U^\dagger,$$

where ϕ could be either an element of the algebra or of the group. In analogy to (202), the gauge condition can be written as

$$f[\phi] = \phi - \varphi \frac{\tau^3}{2}, \tag{225}$$

with arbitrary φ to be integrated in the generating functional. A simple illustrative example is [99]

$$\phi = F_{12}. \tag{226}$$

The analysis of the defects and the resulting properties of the monopoles can be taken over with minor modifications from the procedure described above. Defects occur if

$$\phi = 0$$

(or $\phi = \pm 1$ for group elements). The condition for the defect is gauge invariant. Generically, the three defect conditions determine for a given gauge field the world-lines of the monopoles generated by the gauge condition (225). The quantization of the monopole charge is once more derived from the topological identity (67) which characterizes the mapping of a (small) sphere in the space transverse to the monopole world-line and enclosing the defect. The coset space, appears as above since the gauge condition leaves a $U(1)$ gauge symmetry related to the rotations around the direction of ϕ unspecified. With ϕ being an element of the Lie algebra, only one sort of monopoles appears. The characterization as $z = \pm 1$ monopoles requires ϕ to be an element of the group. As a consequence, the generalization of the connection between monopoles and topological charges is not straightforward. It has been established [20] with the help of the Hopf-invariant (cf. (45)) and its generalization.

It will not have escaped the attention of the reader that the description of Yang–Mills theories in diagonalization gauges is almost in one to one correspondence to the description of the non-abelian Higgs model in the unitary gauge. In particular, the gauge condition (225) is essentially identical to the unitary gauge condition (115). However, the physics content of these gauge choices is very different. The unitary gauge is appropriate if the Higgs potential forces the Higgs field to assume (classically) a value different from zero. In the classical limit, no monopoles related to the vanishing of the Higgs field appear in unitary gauge and one might expect that quantum fluctuations will not change this qualitatively. Associated with the unspecified $U(1)$ are the photons in the Georgi–Glashow model. In pure Yang–Mills theory, gauge conditions like (226) are totally inappropriate in the classical limit, where vanishing action produces defects filling the whole space. Therefore, in such gauges a physically meaningful condensate of magnetic monopoles signaling confinement can arise only if quantum fluctuations change the situation radically. Furthermore, the unspecified $U(1)$ does not indicate the presence of massless vector particles, it rather reflects an incomplete gauge fixing. Other diagonalization gauges may be less singular in the classical limit, like the axial gauge. However, independent of the gauge choice, defects in the gauge condition have not been related convincingly to physical properties of the system. They exist as as coordinate singularities and their physical significance remains enigmatic.

10   Conclusions

In these lecture notes I have described the instanton, the 't Hooft–Polyakov monopole, and the Nielsen–Olesen vortex which are the three paradigms of topological objects appearing in gauge theories. They differ from each other in the dimensionality of the core of these objects, i.e. in the dimension of the submanifold of space-time on which gauge and/or matter fields are singular. This dimension is determined by the topological properties of the spaces in which these fields take their values and dictates to a large extent the dynamical role these objects can play. 't Hooft–Polyakov monopoles are singular along a world-line and therefore describe particles. I have presented the strong theoretical evidence based on topological arguments that these particles have been produced most likely in phase transitions of the early universe. These relics of the big bang have not been and most likely cannot be observed. Their abundance has been diluted in the inflationary phase. Nielsen–Olesen vortices are singular on lines in space or equivalently on world-sheets in space-time. Under suitable conditions such objects occur in Type II superconductors. They give rise to various phases and a wealth of phenomena in superconducting materials. Instantons become singular on a point in Euclidean 4-space and they therefore represent tunneling processes. In comparison to monopoles and vortices, the manifestation of these objects is only indirect. They cannot be observed but are supposed to give rise to non-perturbative properties of the corresponding quantum mechanical ground state.

Despite their difference in dimensionality, these topological objects have many properties in common. They are all solutions of the non-linear field equations of gauge theories. They owe their existence and topological stability to vacuum degeneracy, i.e. the presence of a continuous or discrete set of distinct solutions with minimal energy. They can be classified according to a charge, which is quantized as a consequence of the non-trivial topology. Their non-trivial properties leave a topological imprint on fermionic or bosonic degrees of freedom when coupled to these objects. Among the topological excitations of a given type, a certain class is singled out by their energy determined by the quantized charge.

In these lecture notes I also have described efforts in the topological analysis of QCD. A complete picture about the role of topologically non-trivial field configurations has not yet emerged from such studies. With regard to the breakdown of chiral symmetry, the formation of quark condensates and other chiral properties, these efforts have met with success. The relation between the topological charge and fermionic properties appears to be at the origin of these phenomena. The instanton model incorporates this connection explicitly by reducing the quark and gluon degrees of freedom to instantons and quark zero modes generated by the topological charge of the instantons. However, a generally accepted topological explanation of confinement has not been achieved nor have field configurations been identified which are relevant for confinement. The negative outcome of such investigations may imply that, unlike mass generation by the Higgs mechanism, confinement does not have an explanation within the context of classical field theory. Such a conclusion is supported by the simple explanation of confinement in the strong coupling limit of lattice gauge theory. In this limit, confinement results from the kinetic energy [100] of the compact link variables. The potential energy generated by the magnetic field, which has been the crucial ingredient in the construction of the Nielsen–Olesen Vortex and the 't Hooft–Polyakov monopole, is negligible in this limit. It is no accident that, as we have seen, Polyakov-loop variables, which as group elements are compact, also exhibit confinement-like behavior.

Apart from instantons as the genuine topological objects, Yang–Mills theories exhibit non-trivial topological properties related to the center of the gauge group. The center symmetry as a residual gauge symmetry offers the possibility to formulate confinement as a symmetry property and to characterize confined and deconfined phases. The role of the center vortices (gauge transformations which are singular on a two dimensional space-time sheet) remains to be clarified. The existence of obstructions in imposing gauge conditions is another non-trivial property of non-abelian gauge theories which might be related to confinement. I have described the appearance of monopoles as the results of such obstructions in so-called diagonalization or abelian gauges. These singular fields can be characterized by topological methods and, on a formal level, are akin to the 't Hooft–Polyakov monopole. I have described the difficulties in developing a viable framework for formulating their dynamics which is supposed to yield confinement via a dual Meissner effect.

Acknowledgment

I thank M. Thies, L.v. Smekal, and J. Pawlowski for discussions on the various subjects of these notes. I'm indebted to J. Jäckel and F. Steffen for their meticulous reading of the manuscript and for their many valuable suggestions for improvement.

References

1. C. F. Gauß , Werke, Vol. 5, Göttingen, Königliche Gesellschaft der Wissenschaften 1867, p. 605
2. B. A. Dubrovin, A. T. Fomenko, and S. P. Novikov, Modern Geometry, Part II. Springer Verlag 1985
3. T. Frankel, The Geometry of Physics, Cambridge University Press, 1997
4. P.G. Tait, Collected Scientific Papers, 2 Vols., Cambridge University Press, 1898/1900

5. H. K. Moffat, The Degree of Knottedness of Tangled Vortex Lines, *J. Fluid Mech.* **35**, 117 (1969)

6. P. A. M. Dirac, Quantised Singularities in the Electromagnetic Field, *Proc. Roy. Soc. A* **133**, 60 (1931)

7. C. N. Yang and R. L. Mills, Conservation of Isotopic Spin and Isotopic Gauge Invariance Phys. Rev. **96**, 191 (1954)

8. N. K. Nielsen and P. Olesen, Vortex-Line Models for Dual Strings, *Nucl. Phys. B* **61**, 45 (1973)

9. P. G. de Gennes, Superconductivity of Metals and Alloys, W. A. Benjamin 1966

10. M. Tinkham, Introduction to Superconductivity, McGraw-Hill 1975

11. G. Blatter, M. V. Feigel'man, V. B. Geshkenbein, A. I. Larkin, and V. M. Minokur, *Rev. Mod. Phys.* **66**, 1125 (1994)

12. D. Nelson, Defects and Geometry in Condensed Matter Physics, Cambridge University Press, 2002

13. C. P. Poole, Jr., H. A. Farach and R. J. Creswick, Superconductivity, Academic Press, 1995

14. E. B. Bogomol'nyi, The Stability of Classical Solutions, *Sov. J. Nucl. Phys.* **24**, 449 (1976)

15. R. Jackiw and P. Rossi, Zero Modes of the Vortex-Fermion System, *Nucl. Phys. B* **252**, 343 (1991)

16. E. Weinberg, Index Calculations for the Fermion-Vortex System, Phys. Rev. D **24**, 2669 (1981)

17. C. Nash and S. Sen, Topology and Geometry for Physicists, Academic Press 1983

18. M. Nakahara, Geometry, Topology and Physics, Adam Hilger 1990

19. J. R. Munkres, Topology, Prentice Hall 2000

20. O. Jahn, Instantons and Monopoles in General Abelian Gauges, *J. Phys.* **A33**, 2997 (2000)

21. T. W. Gamelin and R. E. Greene, Introduction to Topology, Dover 1999

22. V. I. Arnold, B. A. Khesin, Topological Methods in Hydrodynamics, Springer 1998

23. D. J. Thouless, Topological Quantum Numbers in Nonrelativistic Physics, World Scientific 1998

24. N. Steenrod, The Topology of Fiber Bundels, Princeton University Press 1951

25. G. Morandi, The Role of Topology in Classical and Quantum Physics, Springer 1992

26. W. Miller, Jr., Symmetry Groups and Their Applications, Academic Press 1972

27. N. D. Mermin, The Topological Theory of Defects in Ordered Media, *Rev. Mod. Phys.* **51**, 591 (1979)

28. V. P. Mineev, Topological Objects in Nematic Liquid Crystals, Appendix A, in: V. G. Boltyanskii and V. A. Efremovich, Intuitive Combinatorial Topology, Springer 2001

29. S. Chandrarsekhar, Liquid Crystals, Cambridge University Press 1992

30. P. G. de Gennes and J. Prost, The Physics of Liquid Crystals, Clarendon Press 1993

31. P. Poulin, H. Stark, T. C. Lubensky and D.A. Weisz, Novel Colloidal Interactions in Anisotropic Fluids, Science **275** 1770 (1997)

32. H. Georgi and S. Glashow, Unified Weak and Electromagnetic Interactions without Neutral Currents, *Phys. Rev. Lett.* **28**, 1494 (1972)

33. H. Weyl, Gruppentheorie und Quantenmechanik, Hirzel Verlag 1928.

34. R. Jackiw, Introduction to the Yang–Mills Quantum Theory, *Rev. Mod. Phys.* **52**, 661 (1980)

35. F. Lenz, H. W. L. Naus and M. Thies, QCD in the Axial Gauge Representation, *Ann. Phys.* **233**, 317 (1994)

36. F. Lenz and S. Wörlen, Compact variables and Singular Fields in QCD, in: at the frontier of Particle Physics, handbook of QCD edited by M. Shifman, Vol. 2, p. 762, World Scientific 2001

37. G.'t Hooft, Magnetic Monopoles in Unified Gauge Models, *Nucl. Phys.* B **79**, 276 (1974)

38. A.M. Polyakov, Particle Spectrum in Quantum Field Theory, *JETP Lett.* **20**, 194 (1974); Isometric States in Quantum Fields, *JETP Lett.* **41**, 988 (1975)

39. F. Lenz, H. W. L. Naus, K. Ohta, and M. Thies, Quantum Mechanics of Gauge Fixing, *Ann. Phys.* **233**, 17 (1994)

40. A. Vilenkin and E. P. S. Shellard, Cosmic Strings and Other Topological Defects, Cambridge University Press 1994

41. S.L.Sondhi, S. M. Girvin, J. P. Carini, and D. Shahar, Continuous Quantum Phase Transitions, *Rev.Mod.Phys.* **69** 315, (1997)

42. R. Rajaraman, Solitons and Instantons, North Holland 1982

43. B. Julia and A. Zee, Poles with Both Electric and Magnetic Charges in Nonabelian Gauge Theory, *Phys. Rev.* D **11**, 2227 (1975)

44. E. Tomboulis and G. Woo, Soliton Quantization in Gauge Theories, *Nucl. Phys.* B **107**, 221 (1976); J. L. Gervais, B. Sakita and S. Wadia, The Surface Term in Gauge Theories, *Phys. Lett.* B **63 B**, 55 (1999)

45. C. Callias, Index Theorems on Open Spaces, *Commun. Mat. Phys.* **62**, 213 (1978)

46. R. Jackiw and C. Rebbi, Solitons with Fermion Number 1/2, *Phys. Rev.* D **13**, 3398 (1976)

47. R. Jackiw and C. Rebbi, Spin from Isospin in Gauge Theory, *Phys. Rev. Lett.* **36**, 1116 (1976)

48. P. Hasenfratz and G. 't Hooft, Fermion-Boson Puzzle in a Gauge Theory, *Phys. Rev. Lett.* **36**, 1119 (1976)

49. E. W. Kolb and M. S. Turner, The Early Universe, Addison-Wesley 1990

50. J. A. Peacock, Cosmological Physics, Cambridge University Press 1999

51. V. N. Gribov, Quantization of Non-Abelian Gauge Theories, *Nucl. Phys.* B **139**, 1 (1978)

52. I. M. Singer, Some Remarks on the Gribov Ambiguity, *Comm. Math. Phys.* **60**, 7 (1978)

53. T. T. Wu and C. N. Yang, Concept of Non-Integrable Phase Factors and Global Formulations of Gauge Fields, *Phys. Rev.* D **12**, 3845 (1975)

54. A. A. Belavin, A. M. Polyakov, A. S. Schwartz and Yu. S. Tyupkin, Pseudoparticle solutions of the Yang–Mills equations, *Phys. Lett.* B **59**, 85 (1975)

55. J. D. Bjorken, in: Lectures on Lepton Nucleon Scattering and Quantum Chromodynamics, W. Atwood *et al.* , Birkhäuser 1982

56. R. Jackiw, Topological Investigations of Quantized Gauge theories, in: Current Algebra and Anomalies, edt. by S. Treiman et al., Princeton University Press, 1985

57. A. S. Schwartz, Quantum Field Theory and Topology, Springer 1993

58. M. B. Green, J. H. Schwarz and E. Witten, Superstring Theory, Vol. 2, Cambridge University Press 1987

59. G. 't Hooft, Computation of the Quantum Effects Due to a Four Dimensional Quasiparticle, *Phys. Rev.* D **14**, 3432 (1976)

60. G. Esposito, Dirac Operators and Spectral Geometry, Cambridge University Press 1998

61. T. Schäfer and E. V. Shuryak, Instantons in QCD, *Rev. Mod. Phys.* **70**, 323 (1998)
62. C. G. Callan, R. F. Dashen and D. J. Gross, Toward a Theory of Strong Interactions, *Phys. Rev.* D **17**, 2717 (1978)
63. V. de Alfaro, S. Fubini and G. Furlan, A New Classical Solution Of The Yang–Mills Field Equations, *Phys. Lett.* B **65**, 163 (1976).
64. F. Lenz, J. W. Negele and M. Thies, Confinement from Merons, **hep-th/0306105** to appear in *Phys. Rev. D*
65. H. K. Moffat and A. Tsinober, Helicity in Laminar and Turbulent Flow, *Ann. Rev. Fluid Mech.* **24** 281, (1992)
66. P. A. Davidson, An Introduction to Magnetohydrodynamics, Cambridge University Press, 2001
67. E. Witten, Some Geometrical Applications of Quantum Field Theory, in *Swansea 1988, Proceedings of the IX th International Congr. on Mathematical Physics, p.77
68. L. H. Kauffman, Knots and Physics, World Scientific 1991
69. A. M. Polyakov, Fermi-Bose Transmutation Induced by Gauge Fields, *Mod. Phys. Lett.* **A 3**, 325 (1988)
70. B. Svetitsky, Symmetry Aspects of Finite Temperature Confinement Transitions, *Phys. Rep.* **132**, 1 (1986)
71. D. J. Toms, Casimir Effect and Topological Mass, *Phys. Rev.* D **21**, 928 (1980)
72. F. Lenz and M. Thies, Polyakov Loop Dynamics in the Center Symmetric Phase, *Ann. Phys.* **268**, 308 (1998)
73. J. I. Kapusta, Finite-temperature field theory, Cambridge University Press 1989
74. F. Lenz, H. W. L. Naus, K. Ohta, and M. Thies, Zero Modes and Displacement Symmetry in Electrodynamics, *Ann. Phys.* **233**, 51 (1994)
75. F. Lenz, J. W. Negele, L. O'Raifeartaigh and M. Thies, Phases and Residual Gauge Symmetries of Higgs Models, *Ann. Phys.* **285**, 25 (2000)
76. M. Le Bellac, Thermal field theory, Cambridge University Press 1996
77. H. Reinhardt, M. Engelhardt, K. Langfeld, M. Quandt, and A. Schäfke, Magnetic Monopoles, Center Vortices, Confinement and Topology of Gauge Fields, **hep-th/ 9911145**
78. J. Greensite, The Confinement Problem in Lattice Gauge Theory, **hep-lat/ 0301023**
79. H. J. de Vega and F. A. Schaposnik, Electrically Charged Vortices in Non-Abelian Gauge Theories, *Phys. Rev. Lett.* **56**, 2564 (1986)
80. G. 't Hooft, On the Phase Transition Towards Permanent Quark Confinement, *Nucl. Phys.* B **138**, 1 (1978)
81. S. Samuel, Topological Symmetry Breakdown and Quark Confinement, *Nucl. Phys.* B **154**, 62 (1979)
82. A. Kovner, Confinement, Z_N Symmetry and Low-Energy Effective Theory of Gluodynamicsagnetic, in: at the frontier of Particle Physics, handbook of QCD edited by M. Shifman, Vol. 3, p. 1778, World Scientific 2001
83. J. Fingberg, U. Heller, and F. Karsch, Scaling and Asymptotic Scaling in the SU(2) Gauge Theory, *Nucl. Phys.* B **392**, 493 (1993)
84. B. Grossman, S. Gupta, U. M. Heller, and F. Karsch, Glueball-Like Screening Masses in Pure SU(3) at Finite Temperatures, *Nucl. Phys.* B **417**, 289 (1994)
85. M. Ishii, H. Suganuma and H. Matsufuru, Scalar Glueball Mass Reduction at Finite Temperature in $SU(3)$ Anisotropic Lattice QCD, *Phys. Rev.* D **66**, 014507 (2002); Glueball Properties at Finite Temperature in $SU(3)$ Anisotropic Lattice QCD, *Phys. Rev. D* **66**, 094506 (2002)

86. S. Rastogi, G. W. Höhne and A. Keller, Unusual Pressure-Induced Phase Behavior in Crystalline Poly(4-methylpenthene-1): Calorimetric and Spectroscopic Results and Further Implications, *Macromolecules* **32** 8897 (1999)

87. N. Avraham, B. Kayhkovich, Y. Myasoedov, M. Rappaport, H. Shtrikman, D. E. Feldman, T. Tamegai, P. H. Kes, Ming Li, M. Konczykowski, Kees van der Beek, and Eli Zeldov, ' Inverse' Melting of a Vortex Lattice, *Nature* **411**, 451, (2001)

88. F. Lenz, E. J. Moniz and M. Thies, Signatures of Confinement in Axial Gauge QCD, *Ann. Phys.* **242**, 429 (1995)

89. M. E. Peskin and D. V. Schroeder, An Introduction to Quantum Field Theory, Addison-Wesley Publishing Company, 1995

90. T. Reisz, Realization of Dimensional Reduction at High Temperature, *Z. Phys.* C **53**, 169 (1992)

91. V. L. Eletsky, A. C. Kalloniatis, F. Lenz, and M. Thies, Magnetic and Thermodynamic Stability of $SU(2)$ Yang–Mills Theory, *Phys. Rev.* D **57**, 5010 (1998)

92. F. Karsch, E. Laermann, and A. Peikert, The Pressure in 2, 2 + 1 and 3 Flavor QCD, *Phys. Lett.* B **478**, 447 (2000)

93. J. Engels, F. Karsch and K. Redlich, Scaling Properties of the Energy Density in $SU(2)$ Lattice Gauge Theory, *Nucl. Phys.* B**435**, 295 (1995)

94. N. Seiberg, E. Witten, Monopole Condensation, and Confinement in N = 2 Supersymmetric QCD, *Nucl. Phys.* B **426**, 19 (1994); Monopoles, Duality and Chiral Symmetry Breaking in N = 2 supersymmetric QCD *Nucl. Phys.* B **431**, 484 (1995)

95. M. Quandt, H. Reinhardt and A. Schäfke, Magnetic Monopoles and Topology of Yang–Mills Theory in Polyakov Gauge, *Phys. Lett.* B **446**, 290 (1999)

96. C. Ford, T. Tok and A. Wipf, $SU(N)$ Gauge Theories in Polyakov Gauge on the Torus, *Phys. Lett.* B **456**, 155 (1999)

97. O. Jahn and F. Lenz, Structure and Dynamics of Monopoles in Axial Gauge QCD, *Phys. Rev.* D **58**, 85006 (1998)

98. B. J. Harrington and H. K. Shepard, Periodic Euclidean Solutions and the Finite-Temperature Yang–Mills Gas, *Phys. Rev.* D **17**, 2122 (1978)

99. G. 't Hooft, Topology of the Gauge Condition and New Confinement Phases in Non-Abelian Gauge Theories, *Nucl. Phys.* B **190**, 455 (1981)

100. J. Kogut and L. Susskind, Hamiltonian Formulation of Wilson's Lattice Gauge Theories, *Phys. Rev.* D **11**, 395 (1975)

Aspects of BRST Quantization

J.W. van Holten

National Institute for Nuclear and High-Energy Physics (NIKHEF) P.O. Box 41882, 1009 DB Amsterdam, The Netherlands, and Department of Physics and Astronomy, Faculty of Science, Vrije Universiteit Amsterdam

Abstract. BRST-methods provide elegant and powerful tools for the construction and analysis of constrained systems, including models of particles, strings and fields. These lectures provide an elementary introduction to the ideas, illustrated with some important physical applications.

1 Symmetries and Constraints

The time evolution of physical systems is described mathematically by differential equations of various degree of complexity, such as Newton's equation in classical mechanics, Maxwell's equations for the electro-magnetic field, or Schrödinger's equation in quantum theory. In most cases these equations have to be supplemented with additional constraints, like initial conditions and/or boundary conditions, which select only one – or sometimes a restricted subset – of the solutions as relevant to the physical system of interest.

Quite often the preferred dynamical equations of a physical system are not formulated directly in terms of observable degrees of freedom, but in terms of more primitive quantities, such as potentials, from which the physical observables are to be constructed in a second separate step of the analysis. As a result, the interpretation of the solutions of the evolution equation is not always straightforward. In some cases certain solutions have to be excluded, as they do not describe physically realizable situations; or it may happen that certain classes of apparently different solutions are physically indistinguishable and describe the same actual history of the system.

The BRST-formalism [1,2] has been developed specifically to deal with such situations. The roots of this approach to constrained dynamical systems are found in attempts to quantize General Relativity [3,4] and Yang–Mills theories [5]. Out of these roots has grown an elegant and powerful framework for dealing with quite general classes of constrained systems using ideas borrowed from algebraic geometry.[1]

In these lectures we are going to study some important examples of constrained dynamical systems, and learn how to deal with them so as to be able to extract relevant information about their observable behaviour. In view of the applications to fundamental physics at microscopic scales, the emphasis is on quantum theory. Indeed, this is the domain where the full power and elegance of our methods become most apparent. Nevertheless, many of the ideas and

---

[1] Some reviews can be found in [6–14].

J.W. van Holten, Aspects of BRST Quantization, Lect. Notes Phys. **659**, 99–166 (2005)
http://www.springerlink.com/

results are applicable in classical dynamics as well, and wherever possible we treat classical and quantum theory in parallel. Our conventions and notations are summarized at the end of these notes.

1.1 Dynamical Systems with Constraints

Before delving into the general theory of constrained systems, it is instructive to consider some examples; they provide a background for both the general theory and the applications to follow later.

The Relativistic Particle. The motion of a relativistic point particle is specified completely by its world line $x^\mu(\tau)$, where x^μ are the position co-ordinates of the particle in some fixed inertial frame, and τ is the proper time, labeling successive points on the world line. All these concepts must and can be properly defined; in these lectures I trust you to be familiar with them, and my presentation only serves to recall the relevant notions and relations between them.

In the absence of external forces, the motion of a particle with respect to an inertial frame satisfies the equation

$$\frac{d^2 x^\mu}{d\tau^2} = 0. \tag{1}$$

It follows that the four-velocity $u^\mu = dx^\mu/d\tau$ is constant. The complete solution of the equations of motion is

$$x^\mu(\tau) = x^\mu(0) + u^\mu \tau. \tag{2}$$

A most important observation is, that the four-velocity u^μ is not completely arbitrary, but must satisfy the *physical* requirement

$$u_\mu u^\mu = -c^2, \tag{3}$$

where c is a *universal* constant, equal to the velocity of light, for all particles irrespective of their mass, spin, charge or other physical properties. Equivalently, (3) states that the proper time is related to the space-time interval travelled by

$$c^2 d\tau^2 = -dx_\mu dx^\mu = c^2 dt^2 - d\boldsymbol{x}^2, \tag{4}$$

independent of the physical characteristics of the particle.

The universal condition (3) is required not only for free particles, but also in the presence of interactions. When subject to a four-force f^μ the equation of motion (1) for a relativistic particle becomes

$$\frac{dp^\mu}{d\tau} = f^\mu, \tag{5}$$

where $p^\mu = mu^\mu$ is the four-momentum. Physical forces – e.g., the Lorentz force in the case of the interaction of a charged particle with an electromagnetic field – satisfy the condition

$$p \cdot f = 0. \tag{6}$$

This property together with the equation of motion (5) are seen to imply that $p^2 = p_\mu p^\mu$ is a constant along the world line. The constraint (3) is then expressed by the statement that

$$p^2 + m^2 c^2 = 0, \tag{7}$$

with c the same universal constant. Equation (7) defines an invariant hypersurface in momentum space for any particle of given rest mass m, which the particle can never leave in the course of its time-evolution.

Returning for simplicity to the case of the free particle, we now show how the equation of motion (1) and the constraint (3) can both be derived from a single action principle. In addition to the co-ordinates x^μ, the action depends on an auxiliary variable e; it reads

$$S[x^\mu; e] = \frac{m}{2} \int_1^2 \left(\frac{1}{e} \frac{dx_\mu}{d\lambda} \frac{dx^\mu}{d\lambda} - ec^2 \right) d\lambda. \tag{8}$$

Here λ is a real parameter taking values in the interval $[\lambda_1, \lambda_2]$, which is mapped by the functions $x^\mu(\lambda)$ into a curve in Minkowski space with fixed end points (x_1^μ, x_2^μ), and $e(\lambda)$ is a nowhere vanishing real function of λ on the same interval.

Before discussing the equations that determine the stationary points of the action, we first observe that by writing it in the equivalent form

$$S[x^\mu; e] = \frac{m}{2} \int_1^2 \left(\frac{dx_\mu}{ed\lambda} \frac{dx^\mu}{ed\lambda} - c^2 \right) ed\lambda, \tag{9}$$

it becomes manifest that the action is invariant under a change of parametrization of the real interval $\lambda \to \lambda'(\lambda)$, if the variables (x^μ, e) are transformed simultaneously to (x'^μ, e') according to the rule

$$x'^\mu(\lambda') = x^\mu(\lambda), \qquad e'(\lambda') \, d\lambda' = e(\lambda) \, d\lambda. \tag{10}$$

Thus, the co-ordinates $x^\mu(\lambda)$ transform as scalar functions on the real line $\mathbf{R}^1$, whilst $e(\lambda)$ transforms as the (single) component of a covariant vector (1-form) in one dimension. For this reason, it is often called the *einbein*. For obvious reasons, the invariance of the action (8) under the transformations (10) is called reparametrization invariance.

The condition of stationarity of the action S implies the functional differential equations

$$\frac{\delta S}{\delta x^\mu} = 0, \qquad \frac{\delta S}{\delta e} = 0. \tag{11}$$

These equations are equivalent to the ordinary differential equations

$$\frac{1}{e} \frac{d}{d\lambda} \left(\frac{1}{e} \frac{dx^\mu}{d\lambda} \right) = 0, \qquad \left(\frac{1}{e} \frac{dx^\mu}{d\lambda} \right)^2 = -c^2. \tag{12}$$

The equations coincide with the equation of motion (1) and the constraint (3) upon the identification

$$d\tau = ed\lambda, \tag{13}$$

a manifestly reparametrization invariant definition of proper time. Recall, that after this identification the constraint (3) automatically implies (4), hence this definition of proper time coincides with the standard geometrical one.

Remark. One can use the constraint (12) to eliminate e from the action; with the choice $e > 0$ (which implies that τ increases with increasing λ) the action reduces to the Einstein form

$$S_{\rm E} = -mc \int_1^2 \sqrt{-\frac{dx_\mu}{d\lambda} \frac{dx^\mu}{d\lambda}}\, d\lambda = -mc^2 \int_1^2 d\tau,$$

where $d\tau$ given by (4). As a result one can deduce that the solutions of the equations of motion are time-like geodesics in Minkowski space. The solution with $e < 0$ describes particles for which proper time runs counter to physical laboratory time; this action can therefore be interpreted as describing anti-particles of the same mass.

The Electro-magnetic Field. In the absence of charges and currents the evolution of electric and magnetic fields $(\boldsymbol{E}, \boldsymbol{B})$ is described by the equations

$$\frac{\partial \boldsymbol{E}}{\partial t} = \boldsymbol{\nabla} \times \boldsymbol{B}, \quad \frac{\partial \boldsymbol{B}}{\partial t} = -\boldsymbol{\nabla} \times \boldsymbol{E}. \tag{14}$$

Each of the electric and magnetic fields has three components, but only two of them are independent: physical electro-magnetic fields in vacuo are transverse polarized, as expressed by the conditions

$$\boldsymbol{\nabla} \cdot \boldsymbol{E} = 0, \qquad \boldsymbol{\nabla} \cdot \boldsymbol{B} = 0. \tag{15}$$

The set of the four equations (14) and (15) represents the standard form of Maxwell's equations in empty space.

Repeated use of (14) yields

$$\frac{\partial^2 \boldsymbol{E}}{\partial t^2} = -\boldsymbol{\nabla} \times (\boldsymbol{\nabla} \times \boldsymbol{E}) = \Delta \boldsymbol{E} - \boldsymbol{\nabla}\boldsymbol{\nabla} \cdot \boldsymbol{E}, \tag{16}$$

and an identical equation for $\boldsymbol{B}$. However, the transversality conditions (15) simplify these equations to the linear wave equations

$$\Box \boldsymbol{E} = 0, \qquad \Box \boldsymbol{B} = 0, \tag{17}$$

with $\Box = \Delta - \partial_t^2$. It follows immediately that free electromagnetic fields satisfy the superposition principle and consist of transverse waves propagating at the speed of light ($c = 1$, in natural units).

Again both the time evolution of the fields and the transversality constraints can be derived from a single action principle, but it is a little bit more subtle than in the case of the particle. For electrodynamics we only introduce auxiliary fields $\boldsymbol{A}$ and ϕ to impose the equation of motion and constraint for the electric field; those for the magnetic field then follow automatically. The action is

$$S_{\text{EM}}[\boldsymbol{E}, \boldsymbol{B}; \boldsymbol{A}, \phi] = \int_1^2 dt\, L_{\text{EM}}(\boldsymbol{E}, \boldsymbol{B}; \boldsymbol{A}, \phi),$$

$$L_{\text{EM}} = \int d^3x \left(-\frac{1}{2} \left(\boldsymbol{E}^2 - \boldsymbol{B}^2\right) + \boldsymbol{A} \cdot \left(\frac{\partial \boldsymbol{E}}{\partial t} - \boldsymbol{\nabla} \times \boldsymbol{B}\right) - \phi \boldsymbol{\nabla} \cdot \boldsymbol{E}\right). \tag{18}$$

Obviously, stationarity of the action implies

$$\frac{\delta S}{\delta \boldsymbol{A}} = \frac{\partial \boldsymbol{E}}{\partial t} - \boldsymbol{\nabla} \times \boldsymbol{B} = 0, \qquad \frac{\delta S}{\delta \phi} = -\boldsymbol{\nabla} \cdot \boldsymbol{E} = 0, \tag{19}$$

reproducing the equation of motion and constraint for the electric field. The other two stationarity conditions are

$$\frac{\delta S}{\delta \boldsymbol{E}} = -\boldsymbol{E} - \frac{\partial \boldsymbol{A}}{\partial t} + \boldsymbol{\nabla}\phi = 0, \qquad \frac{\delta S}{\delta \boldsymbol{B}} = \boldsymbol{B} - \boldsymbol{\nabla} \times \boldsymbol{A} = 0, \tag{20}$$

or equivalently

$$\boldsymbol{E} = -\frac{\partial \boldsymbol{A}}{\partial t} + \boldsymbol{\nabla}\phi, \qquad \boldsymbol{B} = \boldsymbol{\nabla} \times \boldsymbol{A}. \tag{21}$$

The second equation (21) directly implies the transversality of the magnetic field: $\boldsymbol{\nabla} \cdot \boldsymbol{B} = 0$. Taking its time derivative one obtains

$$\frac{\partial \boldsymbol{B}}{\partial t} = \boldsymbol{\nabla} \times \left(\frac{\partial \boldsymbol{A}}{\partial t} - \boldsymbol{\nabla}\phi\right) = -\boldsymbol{\nabla} \times \boldsymbol{E}, \tag{22}$$

where in the middle expression we are free to add the gradient $\boldsymbol{\nabla}\phi$, as $\boldsymbol{\nabla} \times \boldsymbol{\nabla}\phi = 0$ identically.

An important observation is, that the expressions (21) for the electric and magnetic fields are invariant under a redefinition of the potentials $\boldsymbol{A}$ and ϕ of the form

$$\boldsymbol{A}' = \boldsymbol{A} + \boldsymbol{\nabla}\Lambda, \qquad \phi' = \phi + \frac{\partial \Lambda}{\partial t}, \tag{23}$$

where $\Lambda(x)$ is an arbitrary scalar function. The transformations (23) are the well-known gauge transformations of electrodynamics.

It is easy to verify, that the Lagrangean L_{EM} changes only by a total time derivative under gauge transformations, modulo boundary terms which vanish if the fields vanish sufficiently fast at spatial infinity:

$$L'_{\text{EM}} = L_{\text{EM}} - \frac{d}{dt} \int d^3x\, \Lambda \boldsymbol{\nabla} \cdot \boldsymbol{E}. \tag{24}$$

As a result the action S_{EM} itself is strictly invariant under gauge transformations, provided $\int d^3x \Lambda \boldsymbol{\nabla} \cdot \boldsymbol{E}|_{t_1} = \int d^3x \Lambda \boldsymbol{\nabla} \cdot \boldsymbol{E}|_{t_2}$; however, no physical principle requires such strict invariance of the action. This point we will discuss later in more detail.

We finish this discussion of electro-dynamics by recalling how to write the equations completely in relativistic notation. This is achieved by first collecting the electric and magnetic fields in the anti-symmetric field-strength tensor

$$F_{\mu\nu} = \begin{pmatrix} 0 & -E_1 & -E_2 & -E_3 \\ E_1 & 0 & B_3 & -B_2 \\ E_2 & -B_3 & 0 & B_1 \\ E_3 & B_2 & -B_1 & 0 \end{pmatrix}, \tag{25}$$

and the potentials in a four-vector:

$$A_\mu = (\phi, \mathbf{A}). \tag{26}$$

Equations (21) then can be written in covariant form as

$$F_{\mu\nu} = \partial_\mu A_\nu - \partial_\nu A_\mu, \tag{27}$$

with the electric field equations (19) reading

$$\partial_\mu F^{\mu\nu} = 0. \tag{28}$$

The magnetic field equations now follow trivially from (27) as

$$\varepsilon^{\mu\nu\kappa\lambda} \partial_\nu F_{\kappa\lambda} = 0. \tag{29}$$

Finally, the gauge transformations can be written covariantly as

$$A'_\mu = A_\mu + \partial_\mu \Lambda. \tag{30}$$

The invariance of the field strength tensor $F_{\mu\nu}$ under these transformations follows directly from the commutativity of the partial derivatives.

Remark. Equations (27)–(29) can also be derived from the action

$$S_{\text{cov}} = \int d^4x \left(\frac{1}{4} F^{\mu\nu} F_{\mu\nu} - F^{\mu\nu} \partial_\mu A_\nu \right).$$

This action is equivalent to S_{EM} modulo a total divergence. Eliminating $F_{\mu\nu}$ as an independent variable gives back the usual standard action

$$S[A_\mu] = -\frac{1}{4} \int d^4x \, F^{\mu\nu}(A) F_{\mu\nu}(A),$$

with $F_{\mu\nu}(A)$ given by the right-hand side of (27).

1.2   Symmetries and Noether's Theorems

In the preceding section we have presented two elementary examples of systems whose complete physical behaviour was described conveniently in terms of one or more evolution equations plus one or more constraints. These constraints are needed to select a subset of solutions of the evolution equation as the physically relevant solutions. In both examples we found, that the full set of equations could be derived from an action principle. Also, in both examples the additional (auxiliary) degrees of freedom, necessary to impose the constraints, allowed non-trivial local (space-time dependent) redefinitions of variables leaving the lagrangean invariant, at least up to a total time-derivative.

The examples given can easily be extended to include more complicated but important physical models: the relativistic string, Yang–Mills fields, and general relativity are all in this class. However, instead of continuing to produce more examples, at this stage we turn to the general case to derive the relation between local symmetries and constraints, as an extension of Noether's well-known theorem relating (rigid) symmetries and conservation laws.

Before presenting the more general analysis, it must be pointed out that our approach distinguishes in an important way between time- and space-like dimensions; indeed, we have emphasized from the start the distinction between equations of motion (determining the behaviour of a system as a function of time) and constraints, which impose additional requirements. e.g. restricting the spatial behaviour of electro-magnetic fields. This distinction is very natural in the context of hamiltonian dynamics, but potentially at odds with a covariant lagrangean formalism. However, in the examples we have already observed that the non-manifestly covariant treatment of electro-dynamics could be translated without too much effort into a covariant one, and that the dynamics of the relativistic particle, including its constraints, was manifestly covariant throughout.

In quantum theory we encounter similar choices in the approach to dynamics, with the operator formalism based on equal-time commutation relations distinguishing space- and time-like behaviour of states and observables, whereas the covariant path-integral formalism allows treatment of space- and time-like dimensions on an equal footing; indeed, upon the analytic continuation of the path-integral to euclidean time the distinction vanishes altogether. In spite of these differences, the two approaches are equivalent in their physical content.

In the analysis presented here we continue to distinguish between time and space, and between equations of motion and constraints. This is convenient as it allows us to freely employ hamiltonian methods, in particular Poisson brackets in classical dynamics and equal-time commutators in quantum mechanics. Nevertheless, as we hope to make clear, all applications to relativistic models allow a manifestly covariant formulation.

Consider a system described by generalized coordinates $q^i(t)$, where i labels the complete set of physical plus auxiliary degrees of freedom, which may be infinite in number. For the relativistic particle in n-dimensional Minkowski space the $q^i(t)$ represent the n coordinates $x^\mu(\lambda)$ plus the auxiliary variable $e(\lambda)$ (sometimes called the 'einbein'), with λ playing the role of time; for the case of a field theory with N fields $\varphi^a(\boldsymbol{x}; t)$, $a = 1, ..., N$, the $q^i(t)$ represent the infinite set of field amplitudes $\varphi^a_{\boldsymbol{x}}(t)$ at fixed location $\boldsymbol{x}$ as function of time t, i.e. the dependence on the spatial co-ordinates $\boldsymbol{x}$ is included in the labels i. In such a case summation over i is understood to include integration over space.

Assuming the classical dynamical equations to involve at most second-order time derivatives, the action for our system can now be represented quite generally by an integral

$$S[q^i] = \int_1^2 L(q^i, \dot{q}^i)\, dt, \tag{31}$$

where in the case of a field theory L itself is to be represented as an integral of some density over space. An arbitrary variation of the co-ordinates leads to a variation of the action of the form

$$\delta S = \int_1^2 dt\, \delta q^i \left(\frac{\partial L}{\partial q^i} - \frac{d}{dt} \frac{\partial L}{\partial \dot{q}^i} \right) + \left[\delta q^i \frac{\partial L}{\partial \dot{q}^i} \right]_1^2, \tag{32}$$

with the boundary terms due to an integration by parts. As usual we define generalized canonical momenta as

$$p_i = \frac{\partial L}{\partial \dot{q}^i}. \tag{33}$$

From (32) two well-known important consequences follow:
- the action is stationary under variations vanishing at initial and final times: $\delta q^i(t_1) = \delta q^i(t_2) = 0$, if the Euler–Lagrange equations are satisfied:

$$\frac{dp_i}{dt} = \frac{d}{dt} \frac{\partial L}{\partial \dot{q}^i} = \frac{\partial L}{\partial q^i}. \tag{34}$$

- let $q^i_c(t)$ and its associated momentum $p_{c\,i}(t)$ represent a solution of the Euler–Lagrange equations; then for arbitrary variations around the classical paths $q^i_c(t)$ in configuration space: $q^i(t) = q^i_c(t) + \delta q^i(t)$, the total variation of the action is

$$\delta S_c = \left[\delta q^i(t) p_{c\,i}(t) \right]_1^2. \tag{35}$$

We now define an infinitesimal *symmetry* of the action as a set of continuous transformations $\delta q^i(t)$ (smoothly connected to zero) such that the lagrangean L transforms to first order into a total time derivative:

$$\delta L = \delta q^i \frac{\partial L}{\partial q^i} + \delta \dot{q}^i \frac{\partial L}{\partial \dot{q}^i} = \frac{dB}{dt}, \tag{36}$$

where B obviously depends in general on the co-ordinates and the velocities, but also on the variation δq^i. It follows immediately from the definition that

$$\delta S = [B]_1^2. \tag{37}$$

Observe, that according to our definition a symmetry does *not* require the action to be invariant in a strict sense. Now comparing (35) and (37) we establish the result that, whenever there exists a set of symmetry transformations δq^i, the physical motions of the system satisfy

$$\left[\delta q^i p_{ci} - B_c\right]_1^2 = 0. \tag{38}$$

Since the initial and final times (t_1, t_2) on the particular orbit are arbitrary, the result can be stated equivalently in the form of a conservation law for the quantity inside the brackets.

To formulate it more precisely, let the symmetry variations be parametrized by k linearly independent parameters ϵ^α, $\alpha = 1, ..., k$, possibly depending on time:

$$\delta q^i = R^i[\alpha] = \epsilon^\alpha R_\alpha^{(0)i} + \dot{\epsilon}^\alpha R_\alpha^{(1)i} + ... + \overset{(n)}{\epsilon}{}^\alpha R_\alpha^{(n)\,i} + ..., \tag{39}$$

where $\overset{(n)}{\epsilon}{}^\alpha$ denotes the nth time derivative of the parameter. Correspondingly, the lagrangean transforms into the derivative of a function $B[\epsilon]$, with

$$B[\epsilon] = \epsilon^\alpha B_\alpha^{(0)} + \dot{\epsilon}^\alpha B_\alpha^{(1)} + ... + \overset{(n)}{\epsilon}{}^\alpha B_\alpha^{(n)} + ... \tag{40}$$

With the help of these expressions we define the 'on shell' quantity[2]

$$G[\epsilon] = p_{ci} R_c^i[\epsilon] - B_c[\epsilon]$$

$$= \epsilon^\alpha G_\alpha^{(0)} + \dot{\epsilon}^\alpha G_\alpha^{(1)} + ... + \overset{(n)}{\epsilon}{}^\alpha G_\alpha^{(n)} + ..., \tag{41}$$

with component by component $G_\alpha^{(n)} = p_{ci} R_{c\alpha}^{(n)\,i} - B_{c\alpha}^{(n)}$. The conservation law (38) can now be stated equivalently as

$$\frac{dG[\epsilon]}{dt} = \epsilon^\alpha \dot{G}_\alpha^{(0)} + \dot{\epsilon}^\alpha \left(G_\alpha^{(0)} + \dot{G}_\alpha^{(1)}\right) + ... + \overset{(n)}{\epsilon}{}^\alpha \left(G_\alpha^{(n-1)} + \dot{G}_\alpha^{(n)}\right) + ... = 0. \tag{42}$$

We can now distinguish various situations, of which we consider only the two extreme cases here. First, if the symmetry exists only for ϵ = constant (a *rigid* symmetry), then all time derivatives of ϵ vanish and $G_\alpha^{(n)} \equiv 0$ for $n \geq 1$, whilst for the lowest component

$$G_\alpha^{(0)} = g_\alpha = \text{constant}, \qquad G[\epsilon] = \epsilon^\alpha g_\alpha, \tag{43}$$

as defined on a particular classical trajectory (the value of g_α may be different on different trajectories). Thus, rigid symmetries imply constants of motion; this is Noether's theorem.

Second, if the symmetry exists for *arbitrary* time-dependent $\epsilon(t)$ (a *local* symmetry), then $\epsilon(t)$ and all its time derivatives at the same instant are independent.

---

[2] An 'on shell' quantity is a quantity defined on a classical trajectory.

As a result

$$\dot{G}_\alpha^{(0)} = 0,$$

$$\dot{G}_\alpha^{(1)} = -G_\alpha^{(0)},$$

$$\dots \tag{44}$$

$$\dot{G}_\alpha^{(n)} = -G_\alpha^{(n-1)},$$

$$\dots$$

Now in general the transformations (39) do not depend on arbitrarily high-order derivatives of ϵ, but only on a *finite* number of them: there is some finite N such that $R_\alpha^{(n)} = 0$ for $n \geq N$. Typically, transformations depend at most on the first derivative of ϵ, and $R_\alpha^{(n)} = 0$ for $n \geq 2$. In general, for any finite N all quantities $R^{(n)\,i}$, $B^{(n)}$, $G^{(n)}$ then vanish identically for $n \geq N$. But then $G_\alpha^{(n)} = 0$ for $n = 0, ..., N-1$ as well, as a result of (44). Therefore $G[\epsilon] = 0$ at all times. This is a set of *constraints* relating the coordinates and velocities on a classical trajectory. Moreover, as $dG/dt = 0$, these constraints have the nice property that they are preserved during the time-evolution of the system.

The upshot of this analysis therefore is that local symmetries imply time-independent constraints. This result is sometimes referred to as Noether's second theorem.

Remark. If there is *no* upper limit on the order of derivatives in the transformation rule (no finite N), one reobtains a conservation law

$$G[\epsilon] = g_\alpha \, \epsilon^\alpha(0) = \text{constant}.$$

To show this, observe that $G_\alpha^{(n)} = ((-t)^n/n!)\, g_\alpha$, with g_α a constant; then comparison with the Taylor expansion for $\epsilon(0) = \epsilon(t - t)$ around $\epsilon(t)$ leads to the above result.

Group Structure of Symmetries. To round off our discussion of symmetries, conservation laws, and constraints in the lagrangean formalism, we show that symmetry transformations as defined by (36) possess an infinitesimal group structure, i.e. they have a closed commutator algebra (a Lie algebra or some generalization thereof). The proof is simple. First observe, that performing a second variation of δL gives

$$\delta_2 \delta_1 L = \delta_2 q^j \delta_1 q^i \frac{\partial^2 L}{\partial q^j \partial q^i} + \delta_2 \dot{q}^j \delta_1 q^i \frac{\partial^2 L}{\partial \dot{q}^j \partial q^i} + (\delta_2 \delta_1 q^i)\frac{\partial L}{\partial q^i}$$

$$+ \delta_2 \dot{q}^j \delta_1 \dot{q}^i \frac{\partial^2 L}{\partial \dot{q}^j \partial \dot{q}^i} + \delta_2 q^j \delta_1 \dot{q}^i \frac{\partial^2 L}{\partial q^j \partial \dot{q}^i} + (\delta_2 \delta_1 \dot{q}^i)\frac{\partial L}{\partial \dot{q}^i} = \frac{d(\delta_2 B_1)}{dt}. \tag{45}$$

By antisymmetrization this immediately gives

$$[\delta_2, \delta_1] L = \left([\delta_1, \delta_2] q^i\right) \frac{\partial L}{\partial q^i} + \left([\delta_2, \delta_1] \dot{q}^i\right) \frac{\partial L}{\partial \dot{q}^i} = \frac{d}{dt} (\delta_2 B_1 - \delta_1 B_2). \qquad (46)$$

By assumption of the completeness of the set of symmetry transformations it follows, that there must exist a symmetry transformation

$$\delta_3 q^i = [\delta_2, \delta_1] q^i, \qquad \delta_3 \dot{q}^i = [\delta_2, \delta_1] \dot{q}^i, \qquad (47)$$

with the property that the associated $B_3 = \delta_2 B_1 - \delta_1 B_2$. Implementing these conditions gives

$$[\delta_2, \delta_1] q^i = R_2^j \frac{\partial R_1^i}{\partial q^j} + \dot{q}^k \frac{\partial R_2^j}{\partial q^k} \frac{\partial R_1^i}{\partial \dot{q}^j} + \ddot{q}^k \frac{\partial R_2^j}{\partial \dot{q}^k} \frac{\partial R_1^i}{\partial \dot{q}^j} - [1 \leftrightarrow 2] = R_3^i, \qquad (48)$$

where we use a condensed notation $R_a^i \equiv R^i[\epsilon_a]$, $a = 1, 2, 3$. In all standard cases, the symmetry transformations $\delta q^i = R^i$ involve only the coordinates and velocities: $R^i = R^i(q, \dot{q})$. Then R_3 cannot contain terms proportional to $\ddot{q}$, and the conditions (48) reduce to two separate conditions

$$R_2^j \frac{\partial R_1^i}{\partial q^j} - R_1^j \frac{\partial R_2^i}{\partial q^j} + \dot{q}^k \left(\frac{\partial R_2^j}{\partial q^k} \frac{\partial R_1^i}{\partial \dot{q}^j} - \frac{\partial R_1^j}{\partial q^k} \frac{\partial R_2^i}{\partial \dot{q}^j} \right) = R_3^i,$$

$$(49)$$

$$\frac{\partial R_2^j}{\partial \dot{q}^k} \frac{\partial R_1^i}{\partial \dot{q}^j} - \frac{\partial R_1^j}{\partial \dot{q}^k} \frac{\partial R_2^i}{\partial \dot{q}^j} = 0.$$

Clearly, the parameter ϵ_3 of the transformation on the right-hand side must be an antisymmetric bilinear combination of the other two parameters:

$$\epsilon_3^\alpha = f^\alpha(\epsilon_1, \epsilon_2) = -f^\alpha(\epsilon_2, \epsilon_1). \qquad (50)$$

1.3   Canonical Formalism

The canonical formalism describes dynamics in terms of phase-space coordinates (q^i, p_i) and a hamiltonian $H(q, p)$, starting from an action

$$S_{\text{can}}[q, p] = \int_1^2 \left(p_i \dot{q}^i - H(q, p)\right) dt. \qquad (51)$$

Variations of the phase-space coordinates change the action to first order by

$$\delta S_{\text{can}} = \int_1^2 dt \left[\delta p_i \left(\dot{q}^i - \frac{\partial H}{\partial p_i}\right) - \delta q^i \left(\dot{p}_i + \frac{\partial H}{\partial q^i}\right) + \frac{d}{dt} \left(p_i \delta q^i\right)\right]. \qquad (52)$$

The action is stationary under variations vanishing at times (t_1, t_2) if Hamilton's equations of motion are satisfied:

$$\dot{p}_i = \frac{\partial H}{\partial q^i}, \qquad \dot{q}^i = -\frac{\partial H}{\partial p_i}. \qquad (53)$$

This motivates the introduction of the Poisson brackets

$$\{F, G\} = \frac{\partial F}{\partial q^i} \frac{\partial G}{\partial p_i} - \frac{\partial F}{\partial p_i} \frac{\partial G}{\partial q^i}, \tag{54}$$

which allow us to write the time derivative of any phase-space function $G(q, p)$ as

$$\dot{G} = \dot{q}^i \frac{\partial G}{\partial q^i} + \dot{p}_i \frac{\partial G}{\partial p_i} = \{G, H\}. \tag{55}$$

It follows immediately that G is a constant of motion if and only if

$$\{G, H\} = 0 \tag{56}$$

everywhere along the trajectory of the physical system in phase space. This is guaranteed to be the case if (56) holds everywhere in phase space, but as we discuss below, more subtle situations can arise.

Suppose (56) is satisfied; then we can construct variations of (q, p) defined by

$$\delta q^i = \{q^i, G\} = \frac{\partial G}{\partial p_i}, \qquad \delta p_i = \{p_i, G\} = -\frac{\partial G}{\partial q^i}, \tag{57}$$

which leave the hamiltonian invariant:

$$\delta H = \delta q^i \frac{\partial H}{\partial q^i} + \delta p_i \frac{\partial H}{\partial p_i} = \frac{\partial G}{\partial p_i} \frac{\partial H}{\partial q^i} - \frac{\partial G}{\partial q^i} \frac{\partial H}{\partial p_i} = \{H, G\} = 0. \tag{58}$$

They represent infinitesimal symmetries of the theory provided (56), and hence (58), is satisfied as an identity, irrespective of whether or not the phase-space coordinates (q, p) satisfy the equations of motion. To see this, consider the variation of the action (52) with $(\delta q, \delta p)$ given by (57) and $\delta H = 0$ by (58):

$$\delta S_{\text{can}} = \int_1^2 dt \left[-\frac{\partial G}{\partial q^i} \dot{q}^i - \frac{\partial G}{\partial p_i} \dot{p}_i + \frac{d}{dt} \left(\frac{\partial G}{\partial p_i} p_i \right) \right] = \int_1^2 dt \frac{d}{dt} \left(\frac{\partial G}{\partial p_i} p_i - G \right). \tag{59}$$

If we call the quantity inside the parentheses $B(q, p)$, then we have rederived (37) and (38); indeed, we then have

$$G = \frac{\partial G}{\partial p_i} p_i - B = \delta q^i p_i - B, \tag{60}$$

where we know from (55), that G is a constant of motion on classical trajectories (on which Hamilton's equations of motion are satisfied). Observe that – whereas in the lagrangean approach we showed that symmetries imply constants of motion – here we have derived the inverse Noether theorem: constants of motion generate symmetries. An advantage of this derivation over the lagrangean one is, that we have also found explicit expressions for the variations $(\delta q, \delta p)$.

A further advantage is, that the infinitesimal group structure of the tranformations (the commutator algebra) can be checked directly. Indeed, if two symmetry generators G_α and G_β both satisfy (56), then the Jacobi identity for Poisson brackets implies

$$\{\{G_\alpha, G_\beta\}, H\} = \{G_\alpha, \{G_\beta, H\}\} - \{G_\beta, \{G_\alpha, H\}\} = 0. \qquad (61)$$

Hence if the set of generators $\{G_\alpha\}$ is complete, we must have an identity of the form

$$\{G_\alpha, G_\beta\} = P_{\alpha\beta}(G) = -P_{\beta\alpha}(G), \qquad (62)$$

where the $P_{\alpha\beta}(G)$ are polynomials in the constants of motion G_α:

$$P_{\alpha\beta}(G) = c_{\alpha\beta} + f_{\alpha\beta}{}^\gamma G_\gamma + \frac{1}{2} g_{\alpha\beta}{}^{\gamma\delta} G_\gamma G_\delta + ... \qquad (63)$$

The coefficients $c_{\alpha\beta}$, $f_{\alpha\beta}{}^\gamma$, $g_{\alpha\beta}{}^{\gamma\delta}$, ... are constants, having zero Poisson brackets with any phase-space function. As such the first term $c_{\alpha\beta}$ may be called a central charge.

It now follows that the transformation of any phase-space function $F(q,p)$, given by

$$\delta_\alpha F = \{F, G_\alpha\}, \qquad (64)$$

satisfies the commutation relation

$$[\delta_\alpha, \delta_\beta] F = \{\{F, G_\beta\}, G_\alpha\} - \{\{F, G_\alpha\}, G_\beta\} = \{F, \{G_\beta, G_\alpha\}\}$$
$$= C_{\beta\alpha}{}^\gamma(G) \, \delta_\gamma F, \qquad (65)$$

where we have introduced the notation

$$C_{\beta\alpha}{}^\gamma(G) = \frac{\partial P_{\beta\alpha}(G)}{\partial G_\gamma} = f_{\alpha\beta}{}^\gamma + g_{\alpha\beta}{}^{\gamma\delta} G_\delta + ... \qquad (66)$$

In particular this holds for the coordinates and momenta (q,p) themselves; taking F to be another constraint G_γ, we find from the Jacobi identity for Poisson brackets the consistency condition

$$C_{[\alpha\beta}{}^\delta P_{\gamma]\delta} = f_{[\alpha\beta}{}^\delta c_{\gamma]\delta} + \left(f_{[\alpha\beta}{}^\delta f_{\gamma]\delta}{}^\varepsilon + g_{[\alpha\beta}{}^{\delta\varepsilon} c_{\gamma]\delta} \right) G_\varepsilon + ... = 0. \qquad (67)$$

By the same arguments as in Sect. 1.2 (cf. (41 and following), it is established, that whenever the theory generated by G_α is a *local* symmetry with time-dependent parameters, the generator G_α turns into a constraint:

$$G_\alpha(q,p) = 0. \qquad (68)$$

However, compared to the case of rigid symmetries, a subtlety now arises: the constraints $G_\alpha = 0$ define a hypersurface in the phase space to which all physical trajectories of the system are confined. This implies that it is sufficient for the constraints to commute with the hamiltonian (in the sense of Poisson brackets) on the physical hypersurface (i.e. *on shell*). Off the hypersurface (i.e. *off shell*), the bracket of the hamiltonian with the constraints can be anything, as the physical trajectories never enter this part of phase space. Thus, the most general allowed algebraic structure defined by the hamiltonian and the constraints is

$$\{G_\alpha, G_\beta\} = P_{\alpha\beta}(G), \qquad \{H, G_\alpha\} = Z_\alpha(G), \tag{69}$$

where both $P_{\alpha\beta}(G)$ and $Z_\alpha(G)$ are polynomials in the constraints with the property that $P_{\alpha\beta}(0) = Z_\alpha(0) = 0$. This is sufficient to guarantee that in the physical sector of the phase space $\{H, G_\alpha\}|_{G=0} = 0$. Note, that in the case of local symmetries with generators G_α defining constraints, the central charge in the bracket of the constraints must vanish: $c_{\alpha\beta} = 0$. This is a genuine restriction on the existence of local symmetries. A dynamical system with constraints and hamiltonian satisfying (69) is said to be *first class*. Actually, it is quite easy to see that the general first-class algebra of Poisson brackets is more appropriate for systems with local symmetries. Namely, even if the brackets of the constraints and the hamiltonian genuinely vanish on and off shell, one can always change the hamiltonian of the system by adding a polynomial in the constraints:

$$H' = H + R(G), \qquad R(G) = \rho_0 + \rho_1^\alpha G_\alpha + \frac{1}{2} \rho_2^{\alpha\beta} G_\alpha G_\beta + \dots \tag{70}$$

This leaves the hamiltonian on the physical shell in phase space invariant (up to a constant ρ_0), and therefore the physical trajectories remain the same. Furthermore, even if $\{H, G_\alpha\} = 0$, the new hamiltonian satisfies

$$\{H', G_\alpha\} = \{R(G), G_\alpha\} = Z_\alpha^{(R)}(G) \equiv \rho_1^\beta P_{\beta\alpha}(G) + \dots, \tag{71}$$

which is of the form (69). In addition the equations of motion for the variables (q, p) are changed by a local symmetry transformation only, as

$$(\dot{q}^i)' = \{q^i, H'\} = \{q^i, H\} + \{q^i, G_\alpha\} \frac{\partial R}{\partial G_\alpha} = \dot{q}^i + \varepsilon^\alpha \delta_\alpha q^i, \tag{72}$$

where ε^α are some – possibly complicated – local functions which may depend on the phase-space coordinates (q, p) themselves. A similar observation holds of course for the momenta p_i. We can actually allow the coefficients $\rho_1^\alpha, \rho_2^{\alpha\beta}, \dots$ to be space-time dependent variables themselves, as this does not change the general form of the equations of motion (72), whilst variation of the action with respect to these new variables will only impose the constraints as equations of motion:

$$\frac{\delta S}{\delta \rho_1^\alpha} = G_\alpha(q, p) = 0, \tag{73}$$

in agreement with the dynamics already established.

The same argument shows however, that the part of the hamiltonian depending on the constraints in not unique, and may be changed by terms like $R(G)$. In many cases this allows one to get rid of all or part of $Z_\alpha(G)$.

1.4 Quantum Dynamics

In quantum dynamics in the canonical operator formalism, one can follow largely the same lines of argument as presented for classical theories in Sect. 1.3. Consider a theory of canonical pairs of operators $(\hat{q}, \hat{p})$ with commutation relations

$$[\hat{q}^i, \hat{p}_j] = i\delta^i_j, \tag{74}$$

and a hamiltonian $\hat{H}(\hat{q}, \hat{p})$ such that

$$i\frac{d\hat{q}^i}{dt} = [\hat{q}^i, \hat{H}], \qquad i\frac{d\hat{p}_i}{dt} = [\hat{p}_i, \hat{H}]. \tag{75}$$

The δ-symbol on the right-hand side of (74) is to be interpreted in a generalized sense: for continuous parameters (i, j) it represents a Dirac delta-function rather than a Kronecker delta.

In the context of quantum theory, constants of motion become operators $\hat{G}$ which commute with the hamiltonian:

$$[\hat{G}, \hat{H}] = i\frac{d\hat{G}}{dt} = 0, \tag{76}$$

and therefore can be diagonalized on stationary eigenstates. We henceforth assume we have at our disposal a complete set $\{\hat{G}_\alpha\}$ of such constants of motion, in the sense that any operator satisfying (76) can be expanded as a polynomial in the operators $\hat{G}_\alpha$.

In analogy to the classical theory, we define infinitesimal symmetry transformations by

$$\delta_\alpha \hat{q}^i = -i[\hat{q}^i, \hat{G}_\alpha], \qquad \delta_\alpha \hat{p}_i = -i[\hat{p}_i, \hat{G}_\alpha]. \tag{77}$$

By construction they have the property of leaving the hamiltonian invariant:

$$\delta_\alpha \hat{H} = -i[\hat{H}, \hat{G}_\alpha] = 0. \tag{78}$$

Therefore, the operators $\hat{G}_\alpha$ are also called symmetry generators. It follows by the Jacobi identity, analogous to (61), that the commutator of two such generators commutes again with the hamiltonian, and therefore

$$-i[\hat{G}_\alpha, \hat{G}_\beta] = P_{\alpha\beta}(\hat{G}) = c_{\alpha\beta} + f_{\alpha\beta}{}^\gamma \hat{G}_\gamma + \dots. \tag{79}$$

A calculation along the lines of (65) then shows, that for any operator $\hat{F}(\hat{q}, \hat{p})$ one has

$$\delta_\alpha \hat{F} = -i[\hat{F}, \hat{G}_\alpha], \qquad [\delta_\alpha, \delta_\beta]\hat{F} = if_{\alpha\beta}{}^\gamma \delta_\gamma \hat{F} + \dots \tag{80}$$

Observe, that compared to the classical theory, in the quantum theory there is an additional potential source for the appearance of central charges in (79), to wit the operator ordering on the right-hand side. As a result, even when no central charge is present in the classical theory, such central charges can arise in the quantum theory. This is a source of anomalous behaviour of symmetries in quantum theory.

As in the classical theory, local symmetries impose additional restrictions; if a symmetry generator $\hat{G}[\epsilon]$ involves time-dependent parameters $\epsilon^a(t)$, then its evolution equation (76) is modified to:

$$i\frac{d\hat{G}[\epsilon]}{dt} = \left[\hat{G}[\epsilon], \hat{H}\right] + i\frac{\partial\hat{G}[\epsilon]}{\partial t}, \tag{81}$$

where

$$\frac{\partial\hat{G}[\epsilon]}{\partial t} = \frac{\partial\epsilon^a}{\partial t}\frac{\delta\hat{G}[\epsilon]}{\delta\epsilon^a}. \tag{82}$$

It follows, that $\hat{G}[\epsilon]$ can generate symmetries of the hamiltonian and be conserved at the same time for arbitrary $\epsilon^a(t)$ only if the functional derivative vanishes:

$$\frac{\delta\hat{G}[\epsilon]}{\delta\epsilon^a(t)} = 0, \tag{83}$$

which defines a set of operator constraints, the quantum equivalent of (44). The important step in this argument is to realize, that the transformation properties of the evolution operator should be consistent with the Schrödinger equation, which can be true only if both conditions (symmetry and conservation law) hold. To see this, recall that the evolution operator

$$\hat{U}(t, t') = e^{-i(t-t')\hat{H}}, \tag{84}$$

is the formal solution of the Schrödinger equation

$$\left(i\frac{\partial}{\partial t} - \hat{H}\right)\hat{U} = 0, \tag{85}$$

satisfying the initial condition $\hat{U}(t, t) = \hat{1}$. Now under a symmetry transformation (77) and (80), this equation transforms into

$$\delta\left[\left(i\frac{\partial}{\partial t} - \hat{H}\right)\hat{U}\right] = -i\left[\left(i\frac{\partial}{\partial t} - \hat{H}\right)\hat{U}, \hat{G}[\epsilon]\right]$$

$$= -i\left(i\frac{\partial}{\partial t} - \hat{H}\right)\left[\hat{U}, \hat{G}[\epsilon]\right] - i\left[\left(i\frac{\partial}{\partial t} - \hat{H}\right), \hat{G}[\epsilon]\right]\hat{U} \tag{86}$$

For the transformations to respect the Schrödinger equation, the left-hand side of this identity must vanish, hence so must the right-hand side. But the right-hand side vanishes for arbitrary $\epsilon(t)$ if and only if both conditions are met:

$$\left[\hat{H}, \hat{G}[\epsilon]\right] = 0, \quad \text{and} \quad \frac{\partial\hat{G}[\epsilon]}{\partial t} = 0.$$

This is what we set out to prove. Of course, like in the classical hamiltonian formulation, we realize that for generators of local symmetries a more general first-class algebra of commutation relations is allowed, along the lines of (69). Also here, the hamiltonian may then be modified by terms involving only the constraints and, possibly, corresponding Lagrange multipliers. The discussion parallels that for the classical case.

1.5   The Relativistic Particle

In this section and the next we revisit the two examples of constrained systems discussed in Sect. 1.1 to illustrate the general principles of symmetries, conservation laws, and constraints. First we consider the relativistic particle.

The starting point of the analysis is the action (8):

$$S[x^\mu; e] = \frac{m}{2} \int_1^2 \left(\frac{1}{e} \frac{dx_\mu}{d\lambda} \frac{dx^\mu}{d\lambda} - ec^2 \right) d\lambda.$$

Here λ plays the role of system time, and the hamiltonian we construct is the one generating time-evolution in this sense. The canonical momenta are given by

$$p_\mu = \frac{\delta S}{\delta(dx^\mu/d\lambda)} = \frac{m}{e} \frac{dx_\mu}{d\lambda}, \qquad p_e = \frac{\delta S}{\delta(de/d\lambda)} = 0. \tag{87}$$

The second equation is a constraint on the extended phase space spanned by the canonical pairs $(x^\mu, p_\mu; e, p_e)$. Next we perform a Legendre transformation to obtain the hamiltonian

$$H = \frac{e}{2m} \left(p^2 + m^2 c^2 \right) + p_e \frac{de}{d\lambda}. \tag{88}$$

The last term obviously vanishes upon application of the constraint $p_e = 0$. The canonical (hamiltonian) action now reads

$$S_{\text{can}} = \int_1^2 d\lambda \left(p_\mu \frac{dx^\mu}{d\lambda} - \frac{e}{2m} \left(p^2 + m^2 c^2 \right) \right). \tag{89}$$

Observe, that the dependence on p_e has dropped out, irrespective of whether we constrain it to vanish or not. The role of the einbein is now clear: it is a Lagrange multiplier imposing the dynamical constraint (7):

$$p^2 + m^2 c^2 = 0.$$

Note, that in combination with $p_e = 0$, this constraint implies $H = 0$, i.e. the hamiltonian consists *only* of a polynomial in the constraints. This is a general feature of systems with reparametrization invariance, including for example the theory of relativistic strings and general relativity.

In the example of the relativistic particle, we immediately encounter a generic phenomenon: any time we have a constraint on the dynamical variables imposed by a Lagrange multiplier (here: e), its associated momentum (here: p_e) is constrained to vanish. It has been shown in a quite general context, that one may always reformulate hamiltonian theories with constraints such that all constraints appear with Lagrange multipliers [16]; therefore this pairing of constraints is a generic feature in hamiltonian dynamics. However, as we have already discussed in Sect. 1.3, Lagrange multiplier terms do not affect the dynamics, and the multipliers as well as their associated momenta can be eliminated from the physical hamiltonian.

The non-vanishing Poisson brackets of the theory, including the Lagrange multipliers, are

$$\{x^\mu, p_\nu\} = \delta^\mu_\nu, \qquad \{e, p_e\} = 1. \tag{90}$$

As follows from the hamiltonian treatment, all equations of motion for any quantity $\Phi(x, p; e, p_e)$ can then be obtained from a Poisson bracket with the hamiltonian:

$$\frac{d\Phi}{d\lambda} = \{\Phi, H\}, \tag{91}$$

although this equation does not imply any non-trivial information on the dynamics of the Lagrange multipliers. Nevertheless, in this formulation of the theory it must be assumed *a priori* that (e, p_e) are allowed to vary; the dynamics can be projected to the hypersurface $p_e = 0$ only after computing the Poisson brackets. The alternative is to work with a restricted phase space spanned only by the physical co-ordinates and momenta (x^μ, p_μ). This is achieved by performing a Legendre transformation only with respect to the physical velocities[3]. We first explore the formulation of the theory in the extended phase space.

All possible symmetries of the theory can be determined by solving (56):

$$\{G, H\} = 0.$$

Among the solutions we find the generators of the Poincaré group: translations p_μ and Lorentz transformations $M_{\mu\nu} = x_\nu p_\mu - x_\mu p_\nu$. Indeed, the combination of generators

$$G[\epsilon] = \epsilon^\mu p_\mu + \frac{1}{2} \epsilon^{\mu\nu} M_{\mu\nu}. \tag{92}$$

with constant $(\epsilon^\mu, \epsilon^{\mu\nu})$ produces the expected infinitesimal transformations

$$\delta x^\mu = \{x^\mu, G[\epsilon]\} = \epsilon^\mu + \epsilon^\mu{}_\nu x^\nu, \qquad \delta p_\mu = \{p_\mu, G[\epsilon]\} = \epsilon_\mu{}^\nu p_\nu. \tag{93}$$

The commutator algebra of these transformations is well-known to be closed: it is the Lie algebra of the Poincaré group.

For the generation of constraints the local reparametrization invariance of the theory is the one of interest. The infinitesimal form of the transformations (10) is obtained by taking $\lambda' = \lambda - \epsilon(\lambda)$, with the result

$$\delta x^\mu = x'^\mu(\lambda) - x^\mu(\lambda) = \epsilon \frac{dx^\mu}{d\lambda}, \qquad \delta p_\mu = \epsilon \frac{dp_\mu}{d\lambda},$$

$$\delta e = e'(\lambda) - e(\lambda) = \frac{d(e\epsilon)}{d\lambda}. \tag{94}$$

---

[3] This is basically a variant of Routh's procedure; see e.g. Goldstein [15], Chap. 7.

Now recall that $ed\lambda = d\tau$ is a reparametrization-invariant form. Furthermore, $\epsilon(\lambda)$ is an arbitrary local function of λ. It follows, that without loss of generality we can consider an equivalent set of *covariant* transformations with parameter $\sigma = e\epsilon$:

$$\delta_{\text{cov}}\, x^\mu = \frac{\sigma}{e}\,\frac{dx^\mu}{d\lambda}, \qquad \delta_{\text{cov}}\, p_\mu = \frac{\sigma}{e}\,\frac{dp_\mu}{d\lambda},$$

$$\delta_{\text{cov}}\, e = \frac{d\sigma}{d\lambda}. \tag{95}$$

It is straightforward to check that under these transformations the canonical lagrangean (the integrand of (89)) transforms into a total derivative, and $\delta_{\text{cov}}\, S_{\text{can}} = [B_{\text{cov}}]_1^2$ with

$$B_{\text{cov}}[\sigma] = \sigma\left(p_\mu\,\frac{dx^\mu}{ed\lambda} - \frac{1}{2m}(p^2 + m^2 c^2)\right). \tag{96}$$

Using (60), we find that the generator of the local transformations (94) is given by

$$G_{\text{cov}}[\sigma] = (\delta_{\text{cov}} x^\mu)p_\mu + (\delta_{\text{cov}} e)p_e - B_{\text{cov}} = \frac{\sigma}{2m}\,(p^2 + m^2 c^2) + p_e\frac{d\sigma}{d\lambda}. \tag{97}$$

It is easily verified that $dG_{\text{cov}}/d\lambda = 0$ on physical trajectories for arbitrary $\sigma(\lambda)$ if and only if the two earlier constraints are satisfied at all times:

$$p^2 + m^2 c^2 = 0, \qquad p_e = 0. \tag{98}$$

It is also clear that the Poisson brackets of these constraints among themselves vanish. On the canonical variables, G_{cov} generates the transformations

$$\delta_G\, x^\mu = \{x^\mu, G_{\text{cov}}[\sigma]\} = \frac{\sigma p^\mu}{m},\ \delta_G\, p_\mu = \{p_\mu, G_{\text{cov}}[\sigma]\} = 0,$$

$$\delta_G\, e = \{e, G_{\text{cov}}[\sigma]\} = \frac{d\sigma}{d\lambda}, \qquad \delta_G\, p_e = \{p_e, G_{\text{cov}}[\sigma]\} = 0. \tag{99}$$

These transformation rules actually differ from the original ones, cf. (95). However, all differences vanish when applying the equations of motion:

$$\delta' x^\mu = (\delta_{\text{cov}} - \delta_G)x^\mu = \frac{\sigma}{m}\left(\frac{m}{e}\,\frac{dx^\mu}{d\lambda} - p^\mu\right) \approx 0,$$

$$\delta' p_\mu = (\delta_{\text{cov}} - \delta_G)p_\mu = \frac{\sigma}{e}\,\frac{dp_\mu}{d\lambda} \approx 0. \tag{100}$$

The transformations δ' are in fact themselves symmetry transformations of the canonical action, but of a trivial kind: as they vanish on shell, they do not imply any conservation laws or constraints [17]. Therefore, the new transformations δ_G are physically equivalent to δ_{cov}.

The upshot of this analysis is, that we can describe the relativistic particle by the hamiltonian (88) and the Poisson brackets (90), provided we impose on all physical quantities in phase space the constraints (98).

A few comments are in order. First, the hamiltonian is by construction the generator of translations in the time coordinate (here: λ); therefore, after the general exposure in Sects. 1.2 and 1.3, it should not come as a surprise, that when promoting such translations to a local symmetry, the hamiltonian is constrained to vanish.

Secondly, we briefly discuss the other canonical procedure, which takes direct advantage of the the local parametrization invariance (10) by using it to fix the einbein; in particular, the choice $e = 1$ leads to the identification of λ with proper time: $d\tau = ed\lambda \to d\tau = d\lambda$. This procedure is called gauge fixing. Now the canonical action becomes simply

$$S_{\text{can}}|_{e=1} = \int_1^2 d\tau \left(p \cdot \dot{x} - \frac{1}{2m} \left(p^2 + m^2 c^2 \right) \right). \tag{101}$$

This is a regular action for a hamiltonian system. It is completely Lorentz covariant, only the local reparametrization invariance is lost. As a result, the constraint $p^2 + m^2 c^2 = 0$ can no longer be derived from the action; it must now be imposed separately as an external condition. Because we have fixed e, we do not need to introduce its conjugate momentum p_e, and we can work in a restricted physical phase space spanned by the canonical pairs (x^μ, p_μ). Thus, a second consistent way to formulate classical hamiltonian dynamics for the relativistic particle is to use the gauge-fixed hamiltonian and Poisson brackets

$$H_f = \frac{1}{2m} \left(p^2 + m^2 c^2 \right), \qquad \{x^\mu, p_\nu\} = \delta^\mu_\nu, \tag{102}$$

whilst adding the constraint $H_f = 0$ to be satisfied at all (proper) times. Observe, that the remaining constraint implies that one of the momenta p_μ is not independent:

$$p_0^2 = \boldsymbol{p}^2 + m^2 c^2. \tag{103}$$

As this defines a hypersurface in the restricted phase space, the dimensionality of the physical phase space is reduced even further. To deal with this situation, we can again follow two different routes; the first one is to solve the constraint and work in a reduced phase space. The standard procedure for this is to introduce light-cone coordinates $x^\pm = (x^0 \pm x^3)/\sqrt{2}$, with canonically conjugate momenta $p_\pm = (p_0 \pm p_3)/\sqrt{2}$, such that

$$\{x^\pm, p_\pm\} = 1, \qquad \{x^\pm, p_\mp\} = 0. \tag{104}$$

The constraint (103) can then be written

$$2p_+ p_- = p_1^2 + p_2^2 + m^2 c^2, \tag{105}$$

which allows us to eliminate the light-cone co-ordinate x_- and its conjugate momentum $p_- = (p_1^2 + p_2^2 + m^2 c^2)/2p_+$. Of course, by this procedure the manifest

Lorentz-covariance of the model is lost. Therefore one often prefers an alternative route: to work in the covariant phase space (102), and impose the constraint on physical phase space functions only after solving the dynamical equations.

1.6   The Electro-magnetic Field

The second example to be considered here is the electro-magnetic field. As our starting point we take the action of (18) modified by a total time-derivative, and with the magnetic field written as usual in terms of the vector potential as $B(A) = \nabla \times A$:

$$
S_{\text{em}}[\phi, A, E] = \int_1^2 dt\, L_{\text{em}}(\phi, A, E),
$$

$$
L_{\text{em}} = \int d^3x \left(-\frac{1}{2} \left(E^2 + [B(A)]^2 \right) - \phi \nabla \cdot E - E \cdot \frac{\partial A}{\partial t} \right)
$$

(106)

It is clear, that $(A, -E)$ are canonically conjugate; by adding the time derivative we have chosen to let A play the role of co-ordinates, whilst the components of $-E$ represent the momenta:

$$
\pi_A = -E = \frac{\delta S_{\text{em}}}{\delta(\partial A/\partial t)} \tag{107}
$$

Also, like the einbein in the case of the relativistic particle, here the scalar potential $\phi = A_0$ plays the role of Lagrange multiplier to impose the constraint $\nabla \cdot E = 0$; therefore its canonical momentum vanishes:

$$
\pi_\phi = \frac{\delta S_{\text{em}}}{\delta(\partial \phi/\partial t)} = 0. \tag{108}
$$

This is the generic type of constraint for Lagrange multipliers, which we encountered also in the case of the relativistic particle. Observe, that the lagrangean (106) is already in the canonical form, with the hamiltonian given by

$$
H_{\text{em}} = \int d^3x \left(\frac{1}{2} \left(E^2 + B^2 \right) + \phi \nabla \cdot E + \pi_\phi \frac{\partial \phi}{\partial t} \right). \tag{109}
$$

Again, as in the case of the relativistic particle, the last term can be taken to vanish upon imposing the constraint (108), but in any case it cancels in the canonical action

$$
S_{\text{em}} = \int_1^2 dt \left(\int d^3x \left[-E \cdot \frac{\partial A}{\partial t} + \pi_\phi \frac{\partial \phi}{\partial t} \right] - H_{\text{em}}(E, A, \pi_\phi, \phi) \right)
$$

$$
= \int_1^2 dt \left(\int d^3x \left[-E \cdot \frac{\partial A}{\partial t} \right] - H_{\text{em}}(E, A, \phi)|_{\pi_\phi = 0} \right)
$$

(110)

To proceed with the canonical analysis, we have the same choice as in the case of the particle: to keep the full hamiltonian, and include the canonical pair (ϕ, π_ϕ) in an extended phase space; or to use the local gauge invariance to remove ϕ by fixing it at some particular value.

In the first case we have to introduce Poisson brackets

$$\{A_i(\boldsymbol{x}, t), E_j(\boldsymbol{y}, t)\} = -\delta_{ij}\, \delta^3(\boldsymbol{x} - \boldsymbol{y}), \qquad \{\phi(\boldsymbol{x}, t), \pi_\phi(\boldsymbol{y}, t)\} = \delta^3(\boldsymbol{x} - \boldsymbol{y}). \quad (111)$$

It is straightforward to check, that the Maxwell equations are reproduced by the brackets with the hamiltonian:

$$\dot{\Phi} = \{\Phi, H\}, \quad (112)$$

where Φ stands for any of the fields $(\boldsymbol{A}, \boldsymbol{E}, \phi, \pi_\phi)$ above, although in the sector of the scalar potential the equations are empty of dynamical content.

Among the quantities commuting with the hamiltonian (in the sense of Poisson brackets), the most interesting for our purpose is the generator of the gauge transformations

$$\delta\boldsymbol{A} = \boldsymbol{\nabla}\Lambda, \qquad \delta\phi = \frac{\partial\Lambda}{\partial t}, \qquad \delta\boldsymbol{E} = \delta\boldsymbol{B} = 0. \quad (113)$$

Its construction proceeds according to (60). Actually, the action (106) is gauge invariant provided the gauge parameter vanishes sufficiently fast at spatial infinity, as $\delta L_{\mathrm{em}} = -\int d^3x\, \boldsymbol{\nabla} \cdot (\boldsymbol{E}\partial\Lambda/\partial t)$. Therefore the generator of the gauge transformations is

$$G[\Lambda] = \int d^3x\, (-\delta\boldsymbol{A} \cdot \boldsymbol{E} + \delta\phi\, \pi_\phi)$$

$$= \int d^3x\, \left(-\boldsymbol{E} \cdot \boldsymbol{\nabla}\Lambda + \pi_\phi\frac{\partial\Lambda}{\partial t}\right) = \int d^3x\, \left(\Lambda\boldsymbol{\nabla} \cdot \boldsymbol{E} + \pi_\phi\frac{\partial\Lambda}{\partial t}\right). \quad (114)$$

The gauge transformations (113) are reproduced by the Poisson brackets

$$\delta\Phi = \{\Phi, G[\Lambda]\}. \quad (115)$$

From the result (114) it follows, that conservation of $G[\Lambda]$ for arbitrary $\Lambda(\boldsymbol{x}, t)$ is due to the constraints

$$\boldsymbol{\nabla} \cdot \boldsymbol{E} = 0, \qquad \pi_\phi = 0, \quad (116)$$

which are necessary and sufficient. These in turn imply that $G[\Lambda] = 0$ itself.

One reason why this treatment might be preferred, is that in a relativistic notation $\phi = A_0$, $\pi_\phi = \pi^0$, the brackets (111) take the quasi-covariant form

$$\{A_\mu(\boldsymbol{x}, t), \pi^\nu(\boldsymbol{y}, t)\} = \delta^\nu_\mu\, \delta^3(\boldsymbol{x} - \boldsymbol{y}), \quad (117)$$

and similarly for the generator of the gauge transformations :

$$G[\Lambda] = -\int d^3x\, \pi^\mu\partial_\mu\Lambda. \quad (118)$$

Of course, the three-dimensional δ-function and integral show, that the covariance of these equations is not complete.

The other procedure one can follow, is to use the gauge invariance to set $\phi = \phi_0$, a constant. Without loss of generality this constant can be chosen equal to zero, which just amounts to fixing the zero of the electric potential. In any case, the term $\phi \, \boldsymbol{\nabla} \cdot \boldsymbol{E}$ vanishes from the action and for the dynamics it suffices to work in the reduced phase space spanned by $(\boldsymbol{A}, \boldsymbol{E})$. In particular, the hamiltonian and Poisson brackets reduce to

$$H_{\text{red}} = \int d^3 x \, \frac{1}{2} \left(\boldsymbol{E}^2 + \boldsymbol{B}^2 \right), \qquad \{A_i(\boldsymbol{x}, t), E_j(\boldsymbol{y}, t)\} = -\delta_{ij} \delta^3(\boldsymbol{x} - \boldsymbol{y}). \quad (119)$$

The constraint $\boldsymbol{\nabla} \cdot \boldsymbol{E} = 0$ is no longer a consequence of the dynamics, but has to be imposed separately. Of course, its bracket with the hamiltonian still vanishes: $\{H_{\text{red}}, \boldsymbol{\nabla} \cdot \boldsymbol{E}\} = 0$. The constraint actually signifies that one of the components of the canonical momenta (in fact an infinite set: the longitudinal electric field at each point in space) is to vanish; therefore the dimensionality of the physical phase space is again reduced by the constraint. As the constraint is preserved in time (its Poisson bracket with H vanishes), this reduction is consistent. Again, there are two options to proceed: solve the constraint and obtain a phase space spanned by the physical degrees of freedom only, or keep the constraint as a separate condition to be imposed on all solutions of the dynamics. The explicit solution in this case consists of splitting the electric field in transverse and longitudinal parts by projection operators:

$$\boldsymbol{E} = \boldsymbol{E}_T + \boldsymbol{E}_L = \left(1 - \boldsymbol{\nabla} \frac{1}{\Delta} \boldsymbol{\nabla} \right) \cdot \boldsymbol{E} + \boldsymbol{\nabla} \frac{1}{\Delta} \boldsymbol{\nabla} \cdot \boldsymbol{E}, \quad (120)$$

and similarly for the vector potential. One can now restrict the phase space to the transverse parts of the fields only; this is equivalent to requiring $\boldsymbol{\nabla} \cdot \boldsymbol{E} = 0$ and $\boldsymbol{\nabla} \cdot \boldsymbol{A} = 0$ simultaneously. In practice it is much more convenient to use these constraints as such in computing physical observables, instead of projecting out the longitudinal components explicitly at all intermediate stages. Of course, one then has to check that the final result does not depend on any arbitrary choice of dynamics attributed to the longitudinal fields.

1.7 Yang–Mills Theory

Yang–Mills theory is an important extension of Maxwell theory, with a very similar canonical structure. The covariant action is a direct extension of the covariant electro-magnetic action used before:

$$S_{YM} = -\frac{1}{4} \int d^4 x \, F_{\mu\nu}^a F_a^{\mu\nu}, \quad (121)$$

where $F_{\mu\nu}^a$ is the field strength of the Yang–Mills vector potential A_μ^a:

$$F_{\mu\nu}^a = \partial_\mu A_\nu^a - \partial_\nu A_\mu^a - g f_{bc}{}^a A_\mu^b A_\nu^c. \quad (122)$$

Here g is the coupling constant, and the coefficients $f_{bc}{}^a$ are the structure constant of a compact Lie algebrag with (anti-hermitean) generators T_a:

$$[T_a, T_b] = f_{ab}{}^c T_c. \tag{123}$$

The Yang–Mills action (121) is invariant under (infinitesimal) local gauge transformations with parameters $\Lambda^a(x)$:

$$\delta A_\mu^a = (D_\mu \Lambda)^a = \partial_\mu \Lambda^a - g f_{bc}{}^a A_\mu^b \Lambda^c, \tag{124}$$

under which the field strength $F_{\mu\nu}^a$ transforms as

$$\delta F_{\mu\nu}^a = g f_{bc}{}^a \Lambda^b F_{\mu\nu}^c. \tag{125}$$

To obtain a canonical description of the theory, we compute the momenta

$$\pi_a^\mu = \frac{\delta S_{YM}}{\delta \partial_0 A_\mu^a} = -F_a^{0\mu} = \begin{cases} -E_a^i, & \mu = i = (1,2,3); \\ 0, & \mu = 0. \end{cases} \tag{126}$$

Clearly, the last equation is a constraint of the type we have encountered before; indeed, the time component of the vector field, A_0^a, plays the same role of Lagrange mutiplier for a Gauss-type constraint as the scalar potential $\phi = A_0$ in electro-dynamics, to which the theory reduces in the limit $g \to 0$. This is brought out most clearly in the hamiltonian formulation of the theory, with action

$$S_{YM} = \int_1^2 dt \left(\int d^3x \left[-\boldsymbol{E}_a \cdot \frac{\partial \boldsymbol{A}^a}{\partial t} \right] - H_{YM} \right), \tag{127}$$

$$H_{YM} = \int d^3x \left(\frac{1}{2} (\boldsymbol{E}_a^2 + \boldsymbol{B}_a^2) + A_0^a (\boldsymbol{D} \cdot \boldsymbol{E})_a \right).$$

Here we have introduced the notation $\boldsymbol{B}^a$ for the magnetic components of the field strength:

$$B_i^a = \frac{1}{2} \varepsilon_{ijk} F_{jk}^a. \tag{128}$$

In (127) we have left out all terms involving the time-component of the momentum, since they vanish as a result of the constraint $\pi_a^0 = 0$, cf. (126). Now A_0^a appearing only linearly, its variation leads to another constraint

$$(\boldsymbol{D} \cdot \boldsymbol{E})^a = \boldsymbol{\nabla} \cdot \boldsymbol{E}^a - g f_{bc}{}^a \boldsymbol{A}^b \cdot \boldsymbol{E}^c = 0. \tag{129}$$

As in the other theories we have encountered so far, the constraints come in pairs: one constraint, imposed by a Lagrange multiplier, restricts the physical degrees of freedom; the other constraint is the vanishing of the momentum associated with the Lagrange multiplier.

To obtain the equations of motion, we need to specify the Poisson brackets:

$$\{A_i^a(\boldsymbol{x}, t), E_{jb}(\boldsymbol{y}, t)\} = -\delta_{ij}\delta_b^a \, \delta^3(\boldsymbol{x}-\boldsymbol{y}), \quad \{A_0^a, (\boldsymbol{x}, t), \pi_b^0(\boldsymbol{y}, t)\} = \delta_{ij}\delta_b^a \, \delta^3(\boldsymbol{x}-\boldsymbol{y}), \tag{130}$$

or in quasi-covariant notation

$$\{A_\mu^a(\boldsymbol{x}, t), \pi_b^\nu(\boldsymbol{y}, t)\} = \delta_\mu^\nu \delta_b^a \, \delta^3(\boldsymbol{x} - \boldsymbol{y}). \tag{131}$$

Provided the gauge parameter vanishes sufficiently fast at spatial infinity, the canonical action is gauge invariant:

$$\delta S_{YM} = -\int_1^2 dt \int d^3x \, \boldsymbol{\nabla} \cdot \left(\boldsymbol{E}_a \frac{\partial \Lambda^a}{\partial t}\right) \simeq 0. \tag{132}$$

Therefore it is again straightforward to construct the generator for the local gauge transformations:

$$G[\Lambda] = \int d^3x \, \left(-\delta \boldsymbol{A}^a \cdot \boldsymbol{E}_a + \delta A_0^a \, \pi_a^0\right)$$

$$= \int d^3x \, \pi_a^\mu (D_\mu \Lambda)^a \simeq \int d^3x \, \left(\Lambda_a (\boldsymbol{D} \cdot \boldsymbol{E})^a + \pi_a^0 \, (D_0 \Lambda)^a\right). \tag{133}$$

The new aspect of the gauge generators in the case of Yang–Mills theory is, that the constraints satisfy a non-trivial Poisson bracket algebra:

$$\{G[\Lambda_1], G[\Lambda_2]\} = G[\Lambda_3], \tag{134}$$

where the parameter on the right-hand side is defined by

$$\Lambda_3 = g f_{bc}{}^a \, \Lambda_1^b \, \Lambda_2^c. \tag{135}$$

We can also write the physical part of the constraint algebra in a local form; indeed, let

$$G_a(x) = (\boldsymbol{D} \cdot \boldsymbol{E})_a(x). \tag{136}$$

Then a short calculation leads to the result

$$\{G_a(\boldsymbol{x}, t), G_b(\boldsymbol{y}, t)\} = g f_{ab}{}^c \, G_c(\boldsymbol{x}, t) \, \delta^3(\boldsymbol{x} - \boldsymbol{y}). \tag{137}$$

We observe, that the condition $G[\Lambda] = 0$ is satisfied for arbitrary $\Lambda(x)$ if and only if the two local constraints hold:

$$(\boldsymbol{D} \cdot \boldsymbol{E})^a = 0, \qquad \pi_a^0 = 0. \tag{138}$$

This is sufficient to guarantee that $\{G[\Lambda], H\} = 0$ holds as well. Together with the closure of the algebra of constraints (134) this guarantees that the constraints $G[\Lambda] = 0$ are consistent both with the dynamics and among themselves.

Equation (138) is the generalization of the transversality condition (116) and removes the same number of momenta (electric field components) from the physical phase space. Unlike the case of electrodynamics however, it is non-linear and cannot be solved explicitly. Moreover, the constraint does not determine in closed form the conjugate co-ordinate (the combination of gauge potentials) to be removed from the physical phase space with it. A convenient possibility to impose in classical Yang–Mills theory is the transversality condition $\boldsymbol{\nabla} \cdot \boldsymbol{A}^a = 0$, which removes the correct number of components of the vector potential and still respects the rigid gauge invariance (with constant parameters Λ^a).

1.8   The Relativistic String

As the last example in this section we consider the massless relativistic (bosonic) string, as described by the Polyakov action

$$S_{str} = \int d^2\xi \left(-\frac{1}{2} \sqrt{-g} \, g^{ab} \, \partial_a X^\mu \partial_b X_\mu \right), \tag{139}$$

where $\xi^a = (\xi^0, \xi^1) = (\tau, \sigma)$ are co-ordinates parametrizing the two-dimensional world sheet swept out by the string, g_{ab} is a metric on the world sheet, with g its determinant, and $X^\mu(\xi)$ are the co-ordinates of the string in the D-dimensional embedding space-time (the target space), which for simplicity we take to be flat (Minkowskian). As a generally covariant two-dimensional field theory, the action is manifestly invariant under reparametrizations of the world sheet:

$$X'_\mu(\xi') = X_\mu(\xi), \qquad g'_{ab}(\xi') = g_{cd}(\xi) \frac{\partial \xi^c}{\partial \xi'^a} \frac{\partial \xi^d}{\partial \xi'^b}. \tag{140}$$

The canonical momenta are

$$\Pi_\mu = \frac{\delta S_{str}}{\delta \partial_0 X^\mu} = -\sqrt{-g} \, \partial^0 X_\mu, \qquad \pi_{ab} = \frac{\delta S_{str}}{\delta \partial_0 g^{ab}} = 0. \tag{141}$$

The latter equation brings out, that the inverse metric g^{ab}, or rather the combination $h^{ab} = \sqrt{-g} \, g^{ab}$, acts as a set of Lagrange multipliers, imposing the vanishing of the symmetric energy-momentum tensor:

$$T_{ab} = \frac{2}{\sqrt{-g}} \frac{\delta S_{str}}{\delta g^{ab}} = -\partial_a X^\mu \partial_b X_\mu + \frac{1}{2} g_{ab} g^{cd} \partial_c X^\mu \partial_d X_\mu = 0. \tag{142}$$

Such a constraint arises because of the local reparametrization invariance of the action. Note, however, that the energy-momentum tensor is traceless:

$$T_a{}^a = g^{ab} T_{ab} = 0. \tag{143}$$

and as a result it has only two independent components. The origin of this reduction of the number of constraints is the local Weyl invariance of the action (139)

$$g_{ab}(\xi) \to \bar{g}_{ab}(\xi) = e^{\Lambda(\xi)} g_{ab}(\xi), \qquad X^\mu(\xi) \to \bar{X}^\mu(\xi) = X^\mu(\xi), \tag{144}$$

which leaves h^{ab} invariant: $\bar{h}^{ab} = h^{ab}$. Indeed, h^{ab} itself also has only two independent components, as the negative of its determinant is unity: $-h = -\det h^{ab} = 1$.

The hamiltonian is obtained by Legendre transformation, and taking into account $\pi^{ab} = 0$, it reads

$$H = \frac{1}{2} \int d\sigma \left(\sqrt{-g} \left(-g^{00} [\partial_0 X]^2 + g^{11} [\partial_1 X]^2 \right) + \pi^{ab} \partial_0 g_{ab} \right) \tag{145}$$

$$= \int d\sigma \left(T^0{}_0 + \pi^{ab} \partial_0 g_{ab} \right).$$

The Poisson brackets are

$$\{X^\mu(\tau,\sigma), \Pi_\nu(\tau,\sigma')\} = \delta^\mu_\nu \, \delta(\sigma - \sigma'),$$

$$\{g_{ab}(\tau,\sigma), \pi^{cd}(\tau,\sigma')\} = \frac{1}{2} \left(\delta^c_a \delta^d_b + \delta^d_a \delta^c_b \right) \delta(\sigma - \sigma'). \tag{146}$$

The constraints (142) are most conveniently expressed in the hybrid forms (using relations $g = g_{00}g_{11} - g_{01}^2$ and $g_{11} = gg^{00}$):

$$gT^{00} = -T_{11} = \frac{1}{2} \left(\Pi^2 + [\partial_1 X]^2 \right) = 0,$$

$$\sqrt{-g}\, T^0{}_1 = \Pi \cdot \partial_1 X = 0. \tag{147}$$

These results imply, that the hamiltonian (145) actually vanishes, as in the case of the relativistic particle. The reason is also the same: reparametrization invariance, now on a two-dimensional world sheet rather than on a one-dimensional world line.

The infinitesimal form of the transformations (140) with $\xi' = \xi - \Lambda(\xi)$ is

$$\delta X^\mu(\xi) = X'^\mu(\xi) - X^\mu(\xi) = \Lambda^a \partial_a X^\mu = \frac{1}{gg^{00}} \left(\sqrt{-g}\, \Lambda^0 \Pi^\mu + \Lambda_1 \partial_\sigma X^\mu \right),$$

$$\delta g_{ab}(\xi) = (\partial_a \Lambda^c) g_{cb} + (\partial_b \Lambda^c) g_{ac} + \Lambda^c \partial_c g_{ab} = D_a \Lambda_b + D_b \Lambda_a, \tag{148}$$

where we use the covariant derivative $D_a \Lambda_b = \partial_a \Lambda_b - \Gamma_{ab}{}^c \Lambda_c$. The generator of these transformations as constructed by our standard procedure now becomes

$$G[\Lambda] = \int d\sigma \left(\Lambda^a \partial_a X \cdot \Pi + \frac{1}{2} \Lambda^0 \sqrt{-g}\, g^{ab} \partial_a X \cdot \partial_b X + \pi^{ab} (D_a \Lambda_b + D_b \Lambda_a) \right)$$

$$= \int d\sigma \left(-\sqrt{-g}\, \Lambda^a T^0{}_a + 2\pi^{ab} D_a \Lambda_b \right). \tag{149}$$

which has to vanish in order to represent a canonical symmetry: the constraint $G[\Lambda] = 0$ summarizes all constraints introduced above. The brackets of $G[\Lambda]$ now take the form

$$\{X^\mu, G[\Lambda]\} = \Lambda^a \partial_a X^\mu = \delta X^\mu, \qquad \{g_{ab}, G[\Lambda]\} = D_a \Lambda_b + D_b \Lambda_a = \delta g_{ab}, \tag{150}$$

and, in particular,

$$\{G[\Lambda_1], G[\Lambda_2]\} = G[\Lambda_3], \qquad \Lambda_3^a = \Lambda_{[1}^b \partial_b \Lambda_{2]}^a. \tag{151}$$

It takes quite a long and difficult calculation to check this result.

Most practitioners of string theory prefer to work in the restricted phase space, in which the metric g_{ab} is not a dynamical variable, and there is no need to introduce its conjugate momentum π^{ab}. Instead, g_{ab} is chosen to have a convenient value by exploiting the reparametrization invariance (140) or (148):

$$g_{ab} = \rho \eta_{ab} = \rho \begin{pmatrix} -1 & 0 \\ 0 & 1 \end{pmatrix}. \tag{152}$$

Because of the Weyl invariance (144), ρ never appears explicitly in any physical quantity, so it does not have to be fixed itself. In particular, the hamiltonian becomes

$$H_{\text{red}} = \frac{1}{2} \int d\sigma \left([\partial_0 X]^2 + [\partial_1 X]^2 \right) = \frac{1}{2} \int d\sigma \left(\Pi^2 + [\partial_\sigma X]^2 \right), \tag{153}$$

whilst the constrained gauge generators (149) become

$$G_{\text{red}}[\Lambda] = \int d\sigma \left(\frac{1}{2} \Lambda^0 \left(\Pi^2 + [\partial_\sigma X]^2 \right) + \Lambda^1 \Pi \cdot \partial_\sigma X \right). \tag{154}$$

Remarkably, these generators still satisfy a closed bracket algebra:

$$\{ G_{\text{red}}[\Lambda_1], G_{\text{red}}[\Lambda_2] \} = G_{\text{red}}[\Lambda_3], \tag{155}$$

but the structure constants have changed, as becomes evident from the expressions for Λ_3^a:

$$\Lambda_3^0 = \Lambda_{[1}^1 \partial_\sigma \Lambda_{2]}^0 + \Lambda_{[1}^0 \partial_\sigma \Lambda_{2]}^1,$$
$$\Lambda_3^1 = \Lambda_{[1}^0 \partial_\sigma \Lambda_{2]}^0 + \Lambda_{[1}^1 \partial_\sigma \Lambda_{2]}^1 \tag{156}$$

The condition for $G_{\text{red}}[\Lambda]$ to generate a symmetry of the hamiltonian H_{red} (and hence to be conserved), is again $G_{\text{red}}[\Lambda] = 0$. Observe, that these expressions reduce to those of (151) when the Λ^a satisfy

$$\partial_\sigma \Lambda^1 = \partial_\tau \Lambda^0, \qquad \partial_\sigma \Lambda^0 = \partial_\tau \Lambda^1. \tag{157}$$

In terms of the light-cone co-ordinates $u = \tau - \sigma$ or $v = \tau + \sigma$ this can be written:

$$\partial_u (\Lambda^1 + \Lambda^0) = 0, \qquad \partial_v (\Lambda^1 - \Lambda^0) = 0. \tag{158}$$

As a result, the algebras are identical for parameters living on only one branch of the (two-dimensional) light-cone:

$$\Lambda^0(u, v) = \Lambda_+(v) - \Lambda_-(u), \qquad \Lambda^1(u, v) = \Lambda_+(v) + \Lambda_-(u), \tag{159}$$

with $\Lambda_\pm = (\Lambda_1 \pm \Lambda_0)/2$.

2   Canonical BRST Construction

Many interesting physical theories incorporate constraints arising from a local gauge symmetry, which forces certain components of the momenta to vanish in the physical phase space. For reparametrization-invariant systems (like the relativistic particle or the relativistic string) these constraints are quadratic in the momenta, whereas in abelian or non-abelian gauge theories of Maxwell–Yang–Mills type they are linear in the momenta (i.e., in the electric components of the field strength).

There are several ways to deal with such constraints. The most obvious one is to solve them and formulate the theory purely in terms of physical degrees of freedom. However, this is possible only in the simplest cases, like the relativistic particle or an unbroken abelian gauge theory (electrodynamics). And even then, there can arise complications such as non-local interactions. Therefore, an alternative strategy is more fruitful in most cases and for most applications; this preferred strategy is to keep (some) unphysical degrees of freedom in the theory in such a way that desirable properties of the description – like locality, and rotation or Lorentz-invariance – can be preserved at intermediate stages of the calculations. In this section we discuss methods for dealing with such a situation, when unphysical degrees of freedom are taken along in the analysis of the dynamics.

The central idea of the BRST construction is to identify the solutions of the constraints with the cohomology classes of a certain nilpotent operator, the BRST operator Ω. To construct this operator we introduce a new class of variables, the ghost variables. For the theories we have discussed in Sect. 1, which do not involve fermion fields in an essential way (at least from the point of view of constraints), the ghosts are anticommuting variables: odd elements of a Grassmann algebra. However, theories with more general types of gauge symmetries involving fermionic degrees of freedom, like supersymmetry or Siegel's κ-invariance in the theory of superparticles and superstrings, or theories with reducible gauge symmetries, require commuting ghost variables as well. Nevertheless, to bring out the central ideas of the BRST construction as clearly as possible, here we discuss theories with bosonic symmetries only.

2.1 Grassmann Variables

The BRST construction involves anticommuting variables, which are odd elements of a Grassmann algebra. The theory of such variables plays an important role in quantum field theory, most prominently in the description of fermion fields as they naturally describe systems satisfying the Pauli exclusion principle. For these reasons we briefly review the basic elements of the theory of anticommuting variables at this point. For more detailed expositions we refer to the references [18,19].

A Grassmann algebra of rank n is the set of polynomials constructed from elements $\{e, \theta_1, ..., \theta_n\}$ with the properties

$$e^2 = e, \qquad e\theta_i = \theta_i e = \theta_i, \qquad \theta_i\theta_j + \theta_j\theta_i = 0. \qquad (160)$$

Thus, e is the identity element, which will often not be written out explicitly. The elements θ_i are nilpotent, $\theta_i^2 = 0$, whilst for $i \neq j$ the elements θ_i and θ_j

anticommute. As a result, a general element of the algebra consists of 2^n terms and takes the form

$$g = \alpha e + \sum_{i=1}^{n} \alpha^i \, \theta_i + \sum_{(i,j)=1}^{n} \frac{1}{2!} \, \alpha^{ij} \, \theta_i \theta_j + \ldots + \tilde{\alpha} \, \theta_1 \ldots \theta_n, \qquad (161)$$

where the coefficients $\alpha^{i_1 \ldots i_p}$ are completely antisymmetric in the indices. The elements $\{\theta_i\}$ are called the generators of the algebra. An obvious example of a Grassmann algebra is the algebra of differential forms on an n-dimensional manifold.

On the Grassmann algebra we can define a co-algebra of polynomials in elements $\{\bar{\theta}^1, \ldots, \bar{\theta}^n\}$, which together with the unit element e is a Grassmann algebra by itself, but which in addition has the property

$$[\bar{\theta}^i, \theta_j]_+ = \bar{\theta}^i \, \theta_j + \theta_j \, \bar{\theta}^i = \delta_j^i \, e. \qquad (162)$$

This algebra can be interpreted as the algebra of derivations on the Grassmann algebra spanned by (e, θ_i).

By the property (162), the complete set of elements $\{e; \theta_i; \bar{\theta}^i\}$ is actually turned into a Clifford algebra, which has a (basically unique) representation in terms of Dirac matrices in $2n$-dimensional space. The relation can be established by considering the following complex linear combinations of Grassmann generators:

$$\Gamma_i = \gamma_i = \bar{\theta}^i + \theta_i, \qquad \tilde{\Gamma}_i = \gamma_{i+n} = i\left(\bar{\theta}^i - \theta_i\right), \qquad i = 1, \ldots, n. \qquad (163)$$

By construction, these elements satisfy the relation

$$[\gamma_a, \gamma_b]_+ = 2\,\delta_{ab}\, e, \qquad (a,b) = 1, \ldots, 2n, \qquad (164)$$

but actually the subsets $\{\Gamma_i\}$ and $\{\tilde{\Gamma}_i\}$ define two mutually anti-commuting Clifford algebras of rank n:

$$[\Gamma_i, \Gamma_j]_+ = [\tilde{\Gamma}_i, \tilde{\Gamma}_j]_+ = 2\,\delta_{ij}, \qquad [\Gamma_i, \tilde{\Gamma}_j]_+ = 0. \qquad (165)$$

Of course, the construction can be turned around to construct a Grassmann algebra of rank n and its co-algebra of derivations out of a Clifford algebra of rank $2n$.

In field theory applications we are mostly interested in Grassmann algebras of infinite rank, not only $n \to \infty$, but particularly also the continuous case

$$[\bar{\theta}(t), \theta(s)]_+ = \delta(t - s), \qquad (166)$$

where (s, t) are real-valued arguments. Obviously, a Grassmann *variable* ξ is a quantity taking values in a set of linear Grassmann forms $\sum_i \alpha^i \theta_i$ or its continuous generalization $\int_t \alpha(t) \, \theta(t)$. Similarly, one can define derivative operators $\partial/\partial\xi$ as linear operators mapping Grassmann forms of rank p into forms of rank $p-1$, by

$$\frac{\partial}{\partial\xi} \xi = 1 - \xi\frac{\partial}{\partial\xi}, \qquad (167)$$

and its generalization for systems of multi-Grassmann variables. These derivative operators can be constructed as linear forms in $\bar{\theta}^i$ or $\bar{\theta}(t)$.

In addition to differentiation one can also define Grassmann integration. In fact, Grassmann integration is defined as identical with Grassmann differentiation. For a single Grassmann variable, let $f(\xi) = f_0 + \xi f_1$; then one defines

$$\int d\xi \, f(\xi) = f_1. \tag{168}$$

This definition satisfies all standard properties of indefinite integrals:

1. linearity:

$$\int d\xi \, [\alpha f(\xi) + \beta g(\xi)] = \alpha \int d\xi \, f(\xi) + \beta \int d\xi \, g(\xi); \tag{169}$$

2. translation invariance:

$$\int d\xi \, f(\xi + \eta) = \int d\xi \, f(\xi); \tag{170}$$

3. fundamental theorem of calculus (Gauss–Stokes):

$$\int d\xi \frac{\partial f}{\partial \xi} = 0; \tag{171}$$

4. reality: for *real* functions $f(\xi)$ (i.e. $f_{0,1} \in \mathbf{R}$)

$$\int d\xi f(\xi) = f_1 \in \mathbf{R}. \tag{172}$$

A particularly useful result is the evaluation of Gaussian Grassmann integrals. First we observe that

$$\int [d\xi_1...d\xi_n] \, \xi_{\alpha_1}...\xi_{\alpha_n} = \varepsilon_{\alpha_1...\alpha_n}. \tag{173}$$

From this it follows, that a general Gaussian Grassmann integral is

$$\int [d\xi_1...d\xi_n] \exp\left(\frac{1}{2} \xi_\alpha A_{\alpha\beta} \xi_\beta\right) = \pm\sqrt{|\det A|}. \tag{174}$$

This is quite obvious after bringing A into block-diagonal form:

$$A = \begin{pmatrix} \begin{matrix} 0 & \omega_1 \\ -\omega_1 & 0 \end{matrix} & & 0 \\ & \begin{matrix} 0 & \omega_2 \\ -\omega_2 & 0 \end{matrix} & \\ 0 & & \ddots \end{pmatrix}. \tag{175}$$

There are then two possibilities:
(i) If the dimensionality of the matrix A is even $[(\alpha, \beta) = 1, ..., 2r]$ and none of the characteristic values ω_i vanishes, then every 2×2 block gives a contribution $2\omega_i$ to the exponential:

$$\exp\left(\frac{1}{2} \xi_\alpha A_{\alpha\beta} \xi_\beta\right) = \exp\left(\sum_{i=1}^{r} \omega_i \xi_{2i-1} \xi_{2i}\right) = 1 + ... + \prod_{i=1}^{r} (\omega_i \xi_{2i-1} \xi_{2i}). \quad (176)$$

The final result is then established by performing the Grassmann integrations, which leaves a non-zero contribution only from the last term, reading

$$\prod_{i=1}^{r} \omega_i = \pm\sqrt{|\det A|} \quad (177)$$

with the sign depending on the number of negative characteristic values ω_i.
(ii) If the dimensionality of A is odd, the last block is one-dimensional representing a zero-mode; then the integral vanishes, as does the determinant. Of course, the same is true for even-dimensional A if one of the values ω_i vanishes.

   Another useful result is, that one can define a Grassmann-valued delta-function:

$$\delta(\xi - \xi') = -\delta(\xi' - \xi) = \xi - \xi', \quad (178)$$

with the properties

$$\int d\xi\, \delta(\xi - \xi') = 1, \qquad \int d\xi\, \delta(\xi - \xi') f(\xi) = f(\xi'). \quad (179)$$

The proof follows simply by writing out the integrants and using the fundamental rule of integration (168).

2.2   Classical BRST Transformations

Consider again a general dynamical system subject to a set of constraints $G_\alpha = 0$, as defined in (41) or (60). We take the algebra of constraints to be first-class, as in (69):

$$\{G_\alpha, G_\beta\} = P_{\alpha\beta}(G), \qquad \{G_\alpha, H\} = Z_\alpha(G). \quad (180)$$

Here $P(G)$ and $Z(G)$ are polynomial expressions in the constraints, such that $P(0) = Z(0) = 0$; in particular this implies that the constant terms vanish: $c_{\alpha\beta} = 0$.

   The BRST construction starts with the introduction of canonical pairs of Grassmann degrees of freedom (c^α, b_β), one for each constraint G_α, with Poisson brackets

$$\{c^\alpha, b_\beta\} = \{b_\beta, c^\alpha\} = -i\delta_\beta^\alpha, \quad (181)$$

These anti-commuting variables are known as ghosts; the complete Poisson brackets on the extended phase space are given by

$$\{A, B\} = \frac{\partial A}{\partial q^i}\frac{\partial B}{\partial p_i} - \frac{\partial A}{\partial p_i}\frac{\partial B}{\partial q^i} + i(-1)^A\left(\frac{\partial A}{\partial c^\alpha}\frac{\partial B}{\partial b_\alpha} + \frac{\partial A}{\partial b_\alpha}\frac{\partial B}{\partial c^\alpha}\right), \quad (182)$$

where $(-1)^A$ denotes the Grassmann parity of A: $+1$ if A is Grassmann-even (commuting) and -1 if A is Grassmann-odd (anti-commuting).

With the help of these ghost degrees of freedom, one defines the BRST charge Ω, which has Grassmann parity $(-1)^\Omega = -1$, as

$$\Omega = c^\alpha (G_\alpha + M_\alpha), \tag{183}$$

where M_α is Grassmann-even and of the form

$$M_\alpha = \sum_{n \geq 1} \frac{i^n}{2n!} c^{\alpha_1}...c^{\alpha_n} M_{\alpha\alpha_1...\alpha_n}^{\beta_1...\beta_n} b_{\beta_1}...b_{\beta_n}$$

$$= \frac{i}{2} c^{\alpha_1} M_{\alpha\alpha_1}^{\beta_1} b_{\beta_1} - \frac{1}{4} c^{\alpha_1} c^{\alpha_2} M_{\alpha\alpha_1\alpha_2}^{\beta_1\beta_2} b_{\beta_1} b_{\beta_2} + ... \tag{184}$$

The quantities $M_{\alpha\alpha_1...\alpha_p}^{\beta_1...\beta_p}$ are functions of the classical phase-space variables via the constraints G_α, and are defined such that

$$\{\Omega, \Omega\} = 0. \tag{185}$$

As Ω is Grassmann-odd, this is a non-trivial property, from which the BRST charge can be constructed inductively:

$$\{\Omega, \Omega\} = c^\alpha c^\beta \left(P_{\alpha\beta} + M_{\alpha\beta}^\gamma G_\gamma \right)$$

$$+ i c^\alpha c^\beta c^\gamma \left(\{G_\alpha, M_{\beta\gamma}^\delta\} - M_{\alpha\beta}^\varepsilon M_{\gamma\varepsilon}^\delta + M_{\alpha\beta\gamma}^{\delta\varepsilon} G_\varepsilon \right) b_\delta + ... \tag{186}$$

This vanishes if and only if

$$M_{\alpha\beta}^\gamma G_\gamma = -P_{\alpha\beta},$$

$$M_{\alpha\beta\gamma}^{\delta\varepsilon} G_\varepsilon = \left\{ M_{[\alpha\beta}^\delta, G_{\gamma]} \right\} + M_{[\alpha\beta}^\varepsilon M_{\gamma]\varepsilon}^\delta, \tag{187}$$

$$...$$

Observe, that the first relation can only be satisfied under the condition $c_{\alpha\beta} = 0$, with the solution

$$M_{\alpha\beta}^\gamma = f_{\alpha\beta}^\gamma + \frac{1}{2} g_{\alpha\beta}^{\gamma\delta} G_\delta + ... \tag{188}$$

The same condition guarantees that the second relation can be solved: the bracket on the right-hand side is

$$\{M_{\alpha\beta}^\delta, G_\gamma\} = \frac{\partial M_{\alpha\beta}^\delta}{\partial G_\varepsilon} P_{\varepsilon\gamma} = \frac{1}{2} g_{\alpha\beta}^{\delta\varepsilon} f_{\varepsilon\gamma}^\sigma G_\sigma + ... \tag{189}$$

whilst the Jacobi identity (67) implies that

$$f_{[\alpha\beta}^\varepsilon f_{\gamma]\varepsilon}^\delta = 0, \tag{190}$$

and therefore $M_{[\alpha\beta}{}^{\varepsilon} M_{\gamma]\varepsilon}{}^{\delta} = \mathcal{O}[G_{\sigma}]$. This allows to determine $M_{\alpha\beta\gamma}{}^{\delta\varepsilon}$. Any higher-order terms can be calculated similarly. In practice $P_{\alpha\beta}$ and M_{α} usually contain only a small number of terms.

Next, we observe that we can extend the classical hamiltonian $H = H_0$ with ghost terms such that

$$H_c = H_0 + \sum_{n \geq 1} \frac{i^n}{n!} c^{\alpha_1} \ldots c^{\alpha_n} h_{\alpha_1 \ldots \alpha_n}^{(n)\,\beta_1 \ldots \beta_n}(G)\, b_{\beta_1} \ldots b_{\beta_n}, \qquad \{\Omega, H_c\} = 0. \quad (191)$$

Observe that on the physical hypersurface in the phase space this hamiltonian coincides with the original classical hamiltonian modulo terms which do not affect the time-evolution of the classical phase-space variables (q, p). We illustrate the procedure by constructing the first term:

$$\{\Omega, H_c\} = \{c^{\alpha} G_{\alpha}, H_0\} + \frac{i}{2}\left\{c^{\alpha} G_{\alpha}, c^{\gamma} h_{\gamma}^{(1)\,\beta} b_{\beta}\right\} + \frac{i}{2}\left\{c^{\alpha_1} c^{\alpha_2} M_{\alpha_1 \alpha_2}^{\beta} b_{\beta}, H_0\right\} + \ldots$$

$$= c^{\alpha}\left(Z_{\alpha} - h_{\alpha}^{(1)\,\beta} G_{\beta}\right) + \ldots$$

$$(192)$$

Hence the bracket vanishes if the hamiltonian is extended by ghost terms such that

$$h_{\alpha}^{(1)\,\beta}(G)\, G_{\beta} = Z_{\alpha}(G), \qquad \ldots \qquad (193)$$

This equation is guaranteed to have a solution by the condition $Z(0) = 0$.

As the BRST charge commutes with the ghost-extended hamiltonian, we can use it to generate ghost-dependent symmetry transformations of the classical phase-space variables: the BRST transformations

$$\delta_{\Omega}\, q^i = -\{\Omega, q^i\} = \frac{\partial\Omega}{\partial p_i} = c^{\alpha}\frac{\partial G_{\alpha}}{\partial p_i} + \text{ghost extensions},$$

$$(194)$$

$$\delta_{\Omega}\, p_i = -\{\Omega, p_i\} = -\frac{\partial\Omega}{\partial q_i} = c^{\alpha}\frac{\partial G_{\alpha}}{\partial q^i} + \text{ghost extensions}.$$

These BRST transformations are just the gauge transformations with the parameters ϵ^{α} replaced by the ghost variables c^{α}, plus (possibly) some ghost-dependent extension.

Similarly, one can define BRST transformations of the ghosts:

$$\delta_{\Omega}\, c^{\alpha} = -\{\Omega, c^{\alpha}\} = i\frac{\partial\Omega}{\partial b_{\alpha}} = -\frac{1}{2}\, c^{\beta} c^{\gamma} M_{\beta\gamma}{}^{\alpha} + \ldots,$$

$$(195)$$

$$\delta_{\Omega}\, b_{\alpha} = -\{\Omega, b_{\alpha}\} = i\frac{\partial\Omega}{\partial c^{\alpha}} = iG_{\alpha} - c^{\beta} M_{\alpha\beta}{}^{\gamma} b_{\gamma} + \ldots$$

An important property of these transformations is their nilpotence:

$$\delta_{\Omega}^2 = 0. \qquad (196)$$

This follows most directly from the Jacobi identity for the Poisson brackets of the BRST charge with any phase-space function A:

$$\delta_\Omega^2 A = \{\Omega, \{\Omega, A\}\} = -\frac{1}{2}\{A, \{\Omega, \Omega\}\} = 0. \tag{197}$$

Thus, the BRST variation δ_Ω behaves like an exterior derivative. Next we observe that gauge invariant physical quantities F have the properties

$$\{F, c^\alpha\} = i\frac{\partial F}{\partial b_\alpha} = 0, \qquad \{F, b_\alpha\} = i\frac{\partial F}{\partial c^\alpha} = 0, \qquad \{F, G_\alpha\} = \delta_\alpha F = 0. \tag{198}$$

As a result, such physical quantities must be BRST invariant:

$$\delta_\Omega F = -\{\Omega, F\} = 0. \tag{199}$$

In the terminology of algebraic geometry, such a function F is called BRST closed. Now because of the nilpotence, there are trivial solutions to this condition, of the form

$$F_0 = \delta_\Omega F_1 = -\{\Omega, F_1\}. \tag{200}$$

These solutions are called BRST exact; they always depend on the ghosts (c^α, b_α), and cannot be physically relevant. We conclude, that true physical quantities must be BRST closed, but not BRST exact. Such non-trivial solutions of the BRST condition (199) define the BRST cohomology, which is the set

$$\mathcal{H}(\delta_\Omega) = \frac{\text{Ker}(\delta_\Omega)}{\text{Im}(\delta_\Omega)}. \tag{201}$$

We will make this more precise later on.

2.3 Examples

As an application of the above construction, we now present the classical BRST charges and transformations for the gauge systems discussed in Sect. 1.

The Relativistic Particle. We consider the gauge-fixed version of the relativistic particle. Taking $c = 1$, the only constraint is

$$H_0 = \frac{1}{2m}(p^2 + m^2) = 0, \tag{202}$$

and hence in this case $P_{\alpha\beta} = 0$. We only introduce one pair of ghost variables, and define

$$\Omega = \frac{c}{2m}(p^2 + m^2). \tag{203}$$

It is trivially nilpotent, and the BRST transformations of the phase space variables read

$$\delta_\Omega x^\mu = \{x^\mu, \Omega\} = \frac{cp^\mu}{m}, \quad \delta_\Omega p_\mu = \{p_\mu, \Omega\} = 0,$$

$$\delta_\Omega c = -\{c, \Omega\} = 0, \quad \delta_\Omega b = -\{b, \Omega\} = \frac{i}{2m}(p^2 + m^2) \approx 0. \tag{204}$$

The b-ghost transforms into the constraint, hence it vanishes on the physical hypersurface in the phase space. It is straightforward to verify that $\delta_\Omega^2 = 0$.

Electrodynamics. In the gauge fixed Maxwell's electrodynamics there is again only a single constraint, and a single pair of ghost fields to be introduced. We define the BRST charge

$$\Omega = \int d^3x \, c\boldsymbol{\nabla} \cdot \boldsymbol{E}. \tag{205}$$

The classical BRST transformations are just ghost-dependent gauge transformations:

$$\delta_\Omega \boldsymbol{A} = \{\boldsymbol{A}, \Omega\} = \boldsymbol{\nabla} c, \quad \delta_\Omega \boldsymbol{E} = \{\boldsymbol{E}, \Omega\} = 0,$$

$$\delta_\Omega c = -\{c, \Omega\} = 0, \quad \delta_\Omega b = -\{b, \Omega\} = i\boldsymbol{\nabla} \cdot \boldsymbol{E} \approx 0. \tag{206}$$

Yang–Mills Theory. One of the simplest non-trivial systems of constraints is that of Yang–Mills theory, in which the constraints define a local Lie algebra (137). The BRST charge becomes

$$\Omega = \int d^3x \left(c^a G_a - \frac{ig}{2} c^a c^b f_{ab}{}^c b_c \right), \tag{207}$$

with $G_a = (\boldsymbol{D} \cdot \boldsymbol{E})_a$. It is now non-trivial that the bracket of Ω with itself vanishes; it is true because of the closure of the Lie algebra, and the Jacobi identity for the structure constants.

The classical BRST transformations of the fields become

$$\delta_\Omega A^a = \{A^a, \Omega\} = (\boldsymbol{D}c)^a, \quad \delta_\Omega E_a = \{E_a, \Omega\} = g f_{ab}{}^c c^b E_c,$$

$$\delta_\Omega c^a = -\{c^a, \Omega\} = \frac{g}{2} f_{bc}{}^a c^b c^c, \quad \delta_\Omega b_a = -\{b_a, \Omega\} = i G_a + g f_{ab}{}^c c^b b_c. \tag{208}$$

Again, it can be checked by explicit calculation that $\delta_\Omega^2 = 0$ for all variations (208). It follows, that

$$\delta_\Omega G_a = g f_{ab}{}^c c^b G_c,$$

and as a result $\delta_\Omega^2 b_a = 0$.

The Relativistic String. Finally, we discuss the free relativistic string. We take the reduced constraints (154), satisfying the algebra (155), (156). The BRST charge takes the form

$$\Omega = \int d\sigma \left[\frac{1}{2} c^0 \left(\Pi^2 + [\partial_\sigma X]^2 \right) + c^1 \Pi \cdot \partial_\sigma X \right.$$
$$\left. - i \left(c^1 \partial_\sigma c^0 + c^0 \partial_\sigma c^1 \right) b_0 - i \left(c^0 \partial_\sigma c^0 + c^1 \partial_\sigma c^1 \right) b_1 \right]. \tag{209}$$

The BRST transformations generated by the Poisson brackets of this charge read

$$\delta_\Omega X^\mu = \{X^\mu, \Omega\} = c^0 \Pi^\mu + c^1 \partial_\sigma X^\mu \approx c^a \partial_a X^\mu,$$

$$\delta_\Omega \Pi_\mu = \{\Pi_\mu, \Omega\} = \partial_\sigma \left(c^0 \partial_\sigma X_\mu + c^1 \Pi_\mu \right) \approx \partial_\sigma \left(\varepsilon^{ab} c_a \partial_b X^\mu \right),$$

$$\delta_\Omega c^0 = - \{c^0, \Omega\} = c^1 \partial_\sigma c^0 + c^0 \partial_\sigma c^1,$$

$$\delta_\Omega c^1 = - \{c^0, \Omega\} = c^0 \partial_\sigma c^0 + c^1 \partial_\sigma c^1,$$

$$\delta_\Omega b_0 = - \{b_0, \Omega\} = \tfrac{i}{2} \left(\Pi^2 + [\partial_\sigma X]^2 \right) + c^1 \partial_\sigma b_0 + c^0 \partial_\sigma b_1 + 2 \partial_\sigma c^1 \, b_0 + 2 \, \partial_\sigma c^0 \, b_1,$$

$$\delta_\Omega b_1 = - \{b_1, \Omega\} = i \, \Pi \cdot \partial_\sigma X + c^0 \partial_\sigma b_0 + c^1 \partial_\sigma b_1 + 2 \partial_\sigma c^0 \, b_0 + 2 \, \partial_\sigma c^1 \, b_1. \tag{210}$$

A tedious calculation shows that these transformations are indeed nilpotent: $\delta_\Omega^2 = 0$.

2.4   Quantum BRST Cohomology

The construction of a quantum theory for constrained systems poses the following problem: to have a local and/or covariant description of the quantum system, it is advantageous to work in an extended Hilbert space of states, with unphysical components, like gauge and ghost degrees of freedom. Therefore we need first of all a way to characterize physical states within this extended Hilbert space and then a way to construct a unitary evolution operator, which does not mix physical and unphysical components. In this section we show that the BRST construction can solve both of these problems.

We begin with a quantum system subject to constraints G_α; we impose these constraints on the physical states:

$$G_\alpha |\Psi\rangle = 0, \tag{211}$$

implying that physical states are gauge invariant. In the quantum theory the generators of constraints are operators, which satisfy the commutation relations (80):

$$-i [G_\alpha, G_\beta] = P_{\alpha\beta}(G), \tag{212}$$

where we omit the hat on operators for ease of notation.

Next, we introduce corresponding ghost field operators (c_α, b_β) with equal-time anti-commutation relations

$$[c^\alpha, b_\beta]_+ = c^\alpha b_\beta + \beta_\beta c^\alpha = \delta^\alpha_\beta. \tag{213}$$

(For simplicity, the time-dependence in the notation has been suppressed). In the ghost-extended Hilbert space we now construct a BRST operator

$$\Omega = c^\alpha \left(G_\alpha + \sum_{n \geq 1} \frac{i^n}{2n!} c^{\alpha_1} ... c^{\alpha_n} M^{\beta_1 ... \beta_n}_{\alpha \alpha_1 ... \alpha_n} b_{\beta_1} ... b_{\beta_n} \right), \tag{214}$$

which is required to satisfy the anti-commutation relation

$$[\Omega, \Omega]_+ = 2\Omega^2 = 0. \tag{215}$$

In words, the BRST operator is nilpotent. Working out the square of the BRST operator, we get

$$\Omega^2 = \frac{i}{2} c^\alpha c^\beta \left(-i [G_\alpha, G_\beta] + M^\gamma_{\alpha\beta} G_\gamma \right)$$
$$- \frac{1}{2} c^\alpha c^\beta c^\gamma \left(-i [G_\alpha, M^\delta_{\beta\gamma}] + M^\varepsilon_{\alpha\beta} M^\delta_{\gamma\varepsilon} + M^{\delta\varepsilon}_{\alpha\beta\gamma} G_\varepsilon \right) b_\delta + ... \tag{216}$$

As a consequence, the coefficients M_α are defined as the solutions of the set of equations

$$i [G_\alpha, G_\beta] = -P_{\alpha\beta} = M^\gamma_{\alpha\beta} G_\gamma,$$

$$i \left[G_{[\alpha}, M^\delta_{\beta\gamma]} \right] + M^\varepsilon_{[\alpha\beta} M^\delta_{\gamma]\varepsilon} = M^{\delta\varepsilon}_{\alpha\beta\gamma} G_\varepsilon \tag{217}$$

...

These are operator versions of the classical equations (187). As in the classical case, their solution requires the absence of a central charge: $c_{\alpha\beta} = 0$.

Observe, that the Jacobi identity for the generators G_α implies some restrictions on the higher terms in the expansion of Ω:

$$0 = [G_\alpha, [G_\beta, G_\gamma]] + (\text{terms cyclic in } [\alpha\beta\gamma]) = -3i \left[G_{[\alpha}, M^\delta_{\beta\gamma]} G_\delta \right]$$
$$= -3 \left(i \left[G_{[\alpha}, M^\delta_{\beta\gamma]} \right] + M^\varepsilon_{[\alpha\beta} M^\delta_{\alpha]\varepsilon} \right) G_\delta = -\frac{3i}{2} M^{\delta\varepsilon}_{\alpha\beta\gamma} M^\sigma_{\delta\varepsilon} G_\sigma. \tag{218}$$

The equality on the first line follows from the first equation (217), the last equality from the second one.

To describe the states in the extended Hilbert space, we introduce a ghost-state module, a basis for the ghost states consisting of monomials in the ghost operators c^α:

$$|[\alpha_1 \alpha_2 ... \alpha_p]\rangle_{gh} = \frac{1}{p!} c^{\alpha_1} c^{\alpha_2} ... c^{\alpha_p} |0\rangle_{gh}, \tag{219}$$

with $|0\rangle_{\text{gh}}$ the ghost vacuum state annihilated by all b_β. By construction these states are completely anti-symmetric in the indices $[\alpha_1\alpha_2...\alpha_p]$, i.e. the ghosts satisfy Fermi-Dirac statistics, even though they do not carry spin. This confirms their unphysical nature. As a result of this choice of basis, we can decompose an arbitrary state in components with different ghost number (= rank of the ghost polynomial):

$$|\Psi\rangle = |\Psi^{(0)}\rangle + c^\alpha|\Psi_\alpha^{(1)}\rangle + \frac{1}{2}\,c^\alpha c^\beta|\Psi_{\alpha\beta}^{(2)}\rangle + ... \tag{220}$$

where the states $|\Psi_{\alpha_1...\alpha_n}^{(n)}\rangle$ corresponding to ghost number n are of the form $|\psi_{\alpha_1...\alpha_n}^{(n)}(q)\rangle \times |0\rangle_{\text{gh}}$, with $|\psi_{\alpha_1...\alpha_n}^{(n)}(q)\rangle$ states of zero-ghost number, depending only on the degrees of freedom of the constrained (gauge) system; therefore we have

$$b_\beta|\Psi_{\alpha_1...\alpha_n}^{(n)}\rangle = 0. \tag{221}$$

To do the ghost-counting, it is convenient to introduce the ghost-number operator

$$N_{\text{g}} = \sum_\alpha c^\alpha b_\alpha, \qquad [N_{\text{g}}, c^\alpha] = c^\alpha, \qquad [N_{\text{g}}, b_\alpha] = -b_\alpha, \tag{222}$$

where as usual the summation over α has to be interpreted in a generalized sense (it includes integration over space when appropriate). It follows, that the BRST operator has ghost number $+1$:

$$[N_{\text{g}}, \Omega] = \Omega. \tag{223}$$

Now consider a BRST-invariant state:

$$\Omega|\Psi\rangle = 0. \tag{224}$$

Substitution of the ghost-expansions of Ω and $|\Psi\rangle$ gives

$$\Omega|\Psi\rangle = c^\alpha G_\alpha|\Psi^{(0)}\rangle + \frac{1}{2}\,c^\alpha c^\beta\left(G_\alpha|\Psi_\beta^{(1)}\rangle - G_\beta|\Psi_\alpha^{(1)}\rangle + iM_{\alpha\beta}{}^\gamma|\Psi_\gamma^{(1)}\rangle\right)$$

$$+ \frac{1}{2}\,c^\alpha c^\beta c^\gamma\left(G_\alpha|\Psi_{\beta\gamma}^{(2)}\rangle - iM_{\alpha\beta}{}^\delta|\Psi_{\gamma\delta}^{(2)}\rangle + \frac{1}{2}\,M_{\alpha\beta\gamma}{}^{\delta\varepsilon}|\Psi_{\delta\varepsilon}^{(2)}\rangle\right) + ... \tag{225}$$

Its vanishing then implies

$$G_\alpha|\Psi^{(0)}\rangle = 0,$$

$$G_\alpha|\Psi_\beta^{(1)}\rangle - G_\beta|\Psi_\alpha^{(1)}\rangle + iM_{\alpha\beta}{}^\gamma|\Psi_\gamma^{(1)}\rangle = 0,$$

$$G_{[\alpha}|\Psi_{\beta\gamma]}^{(2)}\rangle - iM_{[\alpha\beta}{}^\delta|\Psi_{\gamma]\delta}^{(2)}\rangle + \frac{1}{2}\,M_{\alpha\beta\gamma}{}^{\delta\varepsilon}|\Psi_{\delta\varepsilon}^{(2)}\rangle = 0, \tag{226}$$

...

These conditions admit solutions of the form

$$|\Psi_\alpha^{(1)}\rangle = G_\alpha|\chi^{(0)}\rangle,$$

$$|\Psi_{\alpha\beta}^{(2)}\rangle = G_\alpha|\chi_\beta^{(1)}\rangle - G_\beta|\chi_\alpha^{(1)}\rangle + iM_{\alpha\beta}^\gamma|\chi_\gamma^{(1)}\rangle, \tag{227}$$

...

where the states $|\chi^{(n)}\rangle$ have zero ghost number: $b_\alpha|\chi^{(n)}\rangle = 0$. Substitution of these expressions into (220) gives

$$|\Psi\rangle = |\Psi^{(0)}\rangle + c^\alpha G_\alpha|\chi^{(0)}\rangle + c^\alpha c^\beta G_\alpha|\chi_\beta^{(1)}\rangle + \frac{i}{2}c^\alpha c^\beta M_{\alpha\beta}^\gamma|\chi_\gamma^{(1)}\rangle + \ldots$$

$$= |\Psi^{(0)}\rangle + \Omega\left(|\chi^{(0)}\rangle + c^\alpha|\chi_\alpha^{(1)}\rangle + \ldots\right) \tag{228}$$

$$= |\Psi^{(0)}\rangle + \Omega|\chi\rangle.$$

The second term is trivially BRST invariant because of the nilpotence of the BRST operator: $\Omega^2 = 0$. Assuming that Ω is hermitean, it follows, that $|\Psi\rangle$ is normalized if and only if $|\Psi^{(0)}\rangle$ is:

$$\langle\Psi|\Psi\rangle = \langle\Psi^{(0)}|\Psi^{(0)}\rangle + 2\operatorname{Re}\langle\chi|\Omega|\Psi^{(0)}\rangle + \langle\chi|\Omega^2|\chi\rangle = \langle\Psi^{(0)}|\Psi^{(0)}\rangle. \tag{229}$$

We conclude, that the class of normalizable BRST-invariant states includes the set of states which can be decomposed into a normalizable gauge-invariant state $|\Psi^{(0)}\rangle$ at ghost number zero, plus a trivially invariant zero-norm state $\Omega|\chi\rangle$. These states are members of the BRST cohomology, the classes of states which are BRST invariant (BRST closed) modulo states in the image of Ω (BRST-exact states):

$$\mathcal{H}(\Omega) = \frac{\operatorname{Ker}\Omega}{\operatorname{Im}\Omega}. \tag{230}$$

2.5   BRST-Hodge Decomposition of States

We have shown by explicit construction, that physical states can be identified with the BRST-cohomology classes of which the lowest, non-trivial, component has zero ghost-number. However, our analysis does not show to what extent these solutions are unique. In this section we present a general discussion of BRST cohomology to establish conditions for the existence of a direct correspondence between physical states and BRST cohomology classes [24].

We assume that the BRST operator is self-adjoint with respect to the physical inner product. As an immediate consequence, the ghost-extended Hilbert space of states contains zero-norm states. Let

$$|\Lambda\rangle = \Omega|\chi\rangle. \tag{231}$$

These states are all orthogonal to each other, including themselves, and thus they have zero-norm indeed:

$$\langle \Lambda' | \Lambda \rangle = \langle \chi' | \Omega^2 | \chi \rangle = 0 \quad \Rightarrow \quad \langle \Lambda | \Lambda \rangle = 0. \tag{232}$$

Moreover, these states are orthogonal to all normalizable BRST-invariant states:

$$\Omega | \Psi \rangle = 0 \quad \Rightarrow \quad \langle \Lambda | \Psi \rangle = 0. \tag{233}$$

Clearly, the BRST-exact states cannot be physical. On the other hand, BRST-closed states are defined only modulo BRST-exact states. We prove, that if on the extended Hilbert space $\mathcal{H}_{\text{ext}}$ there exists a non-degenerate inner product (*not* the physical inner product), which is also non-degenerate when restricted to the subspace Im Ω of BRST-exact states, then all physical states must be members of the BRST cohomology.

A non-degenerate inner product $(\,,\,)$ on $\mathcal{H}_{\text{ext}}$ is an inner product with the following property:

$$(\phi, \chi) = 0 \quad \forall \phi \quad \Leftrightarrow \quad \chi = 0. \tag{234}$$

If the restriction of this inner product to Im Ω is non-degenerate as well, then

$$(\Omega\phi, \Omega\chi) = 0 \quad \forall \phi \quad \Leftrightarrow \quad \Omega\chi = 0. \tag{235}$$

As there are no non-trivial zero-norm states with respect to this inner product, the BRST operator cannot be self-adjoint; its adjoint, denoted by $^*\Omega$, then defines a second nilpotent operator:

$$(\Omega\phi, \chi) = (\phi, {}^*\Omega\chi) \quad \Rightarrow \quad (\Omega^2\phi, \chi) = (\phi, {}^*\Omega^2\chi) = 0, \quad \forall\phi. \tag{236}$$

The non-degeneracy of the inner product implies that $^*\Omega^2 = 0$. The adjoint $^*\Omega$ is called the co-BRST operator. Note, that from (235) one infers

$$(\phi, {}^*\Omega\,\Omega\chi) = 0, \quad \forall\phi, \quad \Leftrightarrow \quad {}^*\Omega\,\Omega\chi = 0 \quad \Leftrightarrow \quad \Omega\chi = 0. \tag{237}$$

It follows immediately, that any BRST-closed vector $\Omega\psi = 0$ is determined uniquely by requiring it to be co-closed as well. Indeed, let $^*\Omega\psi = 0$; then

$$^*\Omega(\psi + \Omega\chi) = 0 \quad \Leftrightarrow \quad {}^*\Omega\,\Omega\chi = 0 \quad \Leftrightarrow \quad \Omega\chi = 0. \tag{238}$$

Thus, if we regard the BRST transformations as gauge transformations on states in the extended Hilbert space generated by Ω, then $^*\Omega$ represents a gauge-fixing operator determining a single particular state out of the complete BRST orbit. States which are both closed and co-closed are called (BRST) *harmonic*.

Denoting the subspace of harmonic states by $\mathcal{H}_{\text{harm}}$, we can now prove the following theorem: the extended Hilbert space $\mathcal{H}_{\text{ext}}$ can be decomposed exactly into three subspaces (Fig. 1):

$$\mathcal{H}_{\text{ext}} = \mathcal{H}_{\text{harm}} + \text{Im}\,\Omega + \text{Im}\,{}^*\Omega. \tag{239}$$

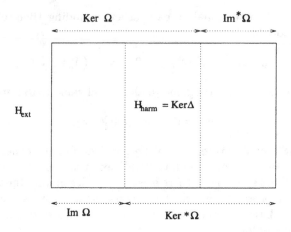

Fig. 1. Decomposition of the extended Hilbert space

Equivalently, any vector in $\mathcal{H}_{\text{ext}}$ can be decomposed as

$$\psi = \omega + \Omega\chi + {}^*\Omega\phi, \qquad \text{where} \quad \Omega\omega = {}^*\Omega\omega = 0. \qquad (240)$$

We sketch the proof. Denote the space of zero modes of the BRST operator (the BRST-closed vectors) by Ker Ω, and the zero modes of the co-BRST operator (co-closed vectors) by Ker ${}^*\Omega$. Then

$$\psi \in \text{Ker}\,\Omega \quad \Leftrightarrow \quad (\Omega\psi, \phi) = 0 \quad \forall\phi \quad \Leftrightarrow \quad (\psi, {}^*\Omega\phi) = 0 \quad \forall\phi. \qquad (241)$$

With ψ being orthogonal to all vectors in Im ${}^*\Omega$, it follows that

$$\text{Ker}\,\Omega = (\text{Im}\,{}^*\Omega)^{\perp}, \qquad (242)$$

the orthoplement of Im ${}^*\Omega$. Similarly we prove

$$\text{Ker}\,{}^*\Omega = (\text{Im}\,\Omega)^{\perp}. \qquad (243)$$

Therefore, any vector which is not in Im Ω *and* not in Im ${}^*\Omega$ must belong to the orthoplement of both, i.e. to Ker ${}^*\Omega$ and Ker Ω simultaneously; such a vector is therefore harmonic.

Now as the BRST-operator and the co-BRST operator are both nilpotent,

$$\text{Im}\,\Omega \subset \text{Ker}\,\Omega = (\text{Im}\,{}^*\Omega)^{\perp}, \qquad \text{Im}\,{}^*\Omega \subset \text{Ker}\,{}^*\Omega = (\text{Im}\,\Omega)^{\perp}. \qquad (244)$$

Therefore Im Ω and Im ${}^*\Omega$ have no elements in common (recall that the null-vector is not in the space of states). Obviously, they also have no elements in common with their own orthoplements (because of the non-degeneracy of the inner product), and in particular with $\mathcal{H}_{\text{harm}}$, which is the set of common states in both orthoplements. This proves the theorem.

We can define a BRST-laplacian $\varDelta_{\mathrm{BRST}}$ as the semi positive definite self-adjoint operator

$$\varDelta_{\mathrm{BRST}} = (\varOmega + {}^*\varOmega)^2 = {}^*\varOmega\,\varOmega + \varOmega\,{}^*\varOmega, \qquad (245)$$

which commutes with both $\varOmega$ and ${}^*\varOmega$. Consider its zero-modes ω:

$$\varDelta_{\mathrm{BRST}}\,\omega = 0 \quad\Leftrightarrow\quad {}^*\varOmega\,\varOmega\,\omega + \varOmega\,{}^*\varOmega\,\omega = 0. \qquad (246)$$

The left-hand side of the last expression is a sum of a vector in Im $\varOmega$ and one in Im ${}^*\varOmega$; as these subspaces are orthogonal with respect to the non-degenerate inner product, it follows that

$$ {}^*\varOmega\,\varOmega\,\omega = 0 \quad\wedge\quad \varOmega\,{}^*\varOmega\,\omega = 0, \qquad (247)$$

separately. This in turn implies $\varOmega\omega = 0$ and ${}^*\varOmega\omega = 0$, and ω must be a harmonic state:

$$\varDelta_{\mathrm{BRST}}\,\omega = 0 \quad\Leftrightarrow\quad \omega \in \mathcal{H}_{\mathrm{harm}}; \qquad (248)$$

hence Ker $\varDelta_{\mathrm{BRST}} = \mathcal{H}_{\mathrm{harm}}$. The BRST-Hodge decomposition theorem can therefore be expressed as

$$\mathcal{H}_{\mathrm{ext}} = \mathrm{Ker}\,\varDelta_{\mathrm{BRST}} + \mathrm{Im}\,\varOmega + \mathrm{Im}^*\varOmega. \qquad (249)$$

The BRST-laplacian allows us to discuss the representation theory of BRST-transformations. First of all, the BRST-laplacian commutes with the BRST- and co-BRST operators $\varOmega$ and ${}^*\varOmega$:

$$[\varDelta_{\mathrm{BRST}}, \varOmega] = 0, \qquad [\varDelta_{\mathrm{BRST}}, {}^*\varOmega] = 0. \qquad (250)$$

As a result, BRST-multiplets can be characterized by the eigenvalues of $\varDelta_{\mathrm{BRST}}$: the action of $\varOmega$ or ${}^*\varOmega$ does not change this eigenvalue. Basically we must then distinguish between zero-modes and non-zero modes of the BRST-laplacian. The zero-modes, the harmonic states, are BRST-singlets:

$$\varOmega|\omega\rangle = 0, \qquad {}^*\varOmega|\omega\rangle = 0.$$

In contrast, the non-zero modes occur in pairs of BRST- and co-BRST-exact states:

$$\varDelta_{\mathrm{BRST}}|\phi_\pm\rangle = \lambda^2|\phi_\pm\rangle \quad\Rightarrow\quad \varOmega|\phi_+\rangle = \lambda\,|\phi_-\rangle, \quad {}^*\varOmega|\phi_-\rangle = \lambda\,|\phi_+\rangle. \qquad (251)$$

Equation (232) guarantees that $|\phi_\pm\rangle$ have zero (physical) norm; we can however rescale these states such that

$$\langle\phi_-|\phi_+\rangle = \langle\phi_+|\phi_-\rangle = 1. \qquad (252)$$

It follows, that the linear combinations

$$|\chi_\pm\rangle = \frac{1}{\sqrt{2}}\left(|\phi_+\rangle \pm |\phi_-\rangle\right) \qquad (253)$$

define a pair of positive- and negative-norm states:

$$\langle \chi_\pm | \chi_\pm \rangle = \pm 1, \qquad \langle \chi_\mp | \chi_\pm \rangle = 0. \tag{254}$$

They are eigenstates of the operator $\Omega + {}^*\Omega$ with eigenvalues $(\lambda, -\lambda)$:

$$(\Omega + {}^*\Omega)|\chi_\pm\rangle = \pm \lambda |\chi_\pm\rangle. \tag{255}$$

As physical states must have positive norm, all BRST-doublets must be unphysical, and only BRST-singlets (harmonic) states can represent physical states. Conversely, if all harmonic states are to be physical, only the components of the BRST-doublets are allowed to have non-positive norm. Observe, however, that this condition can be violated if the inner product (,) becomes degenerate on the subspace Im Ω; in that case the harmonic gauge does not remove all freedom to make BRST-transformations and zero-norm states can survive in the subspace of harmonic states.

2.6   BRST Operator Cohomology

The BRST construction replaces a complete set of constraints, imposed by the generators of gauge transformations, by a single condition: BRST invariance. However, the normalizable solutions of the BRST condition (224):

$$\Omega |\Psi\rangle = 0, \qquad \langle \Psi | \Psi \rangle = 1,$$

are not unique: from any solution one can construct an infinite set of other solutions

$$|\Psi'\rangle = |\Psi\rangle + \Omega|\chi\rangle, \qquad \langle \Psi' | \Psi' \rangle = 1, \tag{256}$$

provided the BRST operator is self-adjoint with respect to the physical inner product. Under the conditions discussed in Sect. 2.5, the normalizable part of the state vector is unique. Hence, the transformed state is not physically different from the original one, and we actually identify a single physical state with the complete class of solutions (256). As observed before, in this respect the quantum theory in the extended Hilbert space behaves much like an abelian gauge theory, with the BRST transformations acting as gauge transformations.

Keeping this in mind, it is clearly necessary that the action of dynamical observables of the theory on physical states is invariant under BRST transformations: an observable $\mathcal{O}$ maps physical states to physical states; therefore if $|\Psi\rangle$ is a physical state, then

$$\Omega \mathcal{O} |\Psi\rangle = [\Omega, \mathcal{O}] |\Psi\rangle = 0. \tag{257}$$

Again, the solution of this condition for any given observable is not unique: for an observable with ghost number $N_g = 0$, and any operator Φ with ghost number $N_g = -1$,

$$\mathcal{O}' = \mathcal{O} + [\Omega, \Phi]_+ \tag{258}$$

also satisfies condition (257). The proof follows directly from the Jacobi identity:

$$[\Omega, [\Omega, \Phi]_+] = [\Omega^2, \Phi] = 0. \tag{259}$$

This holds in particular for the hamiltonian; indeed, the time-evolution of states in the unphysical sector (the gauge and ghost fields) is not determined a priori, and can be chosen by an appropriate BRST extension of the hamiltonian:

$$H_{\text{ext}} = H_{\text{phys}} + [\Omega, \Phi]_+. \tag{260}$$

Here H_{phys} is the hamiltonian of the physical degrees of freedom. The BRST-exact extension $[\Omega, \Phi]_+$ acts only on the unphysical sector, and can be used to define the dynamics of the gauge- and ghost degrees of freedom.

2.7 Lie-Algebra Cohomology

We illustrate the BRST construction with a simple example: a system of constraints defining an ordinary n-dimensional compact Lie algebra [25]. The Lie algebra is taken to be a direct sum of semi-simple and abelian $u(1)$ algebras, of the form

$$[G_a, G_b] = if_{ab}{}^c G_c, \qquad (a, b, c) = 1, ..., n, \tag{261}$$

where the generators G_a are hermitean, and the $f_{ab}{}^c = -f_{ba}{}^c$ are real structure constants. We assume the generators normalized such that the Killing metric is unity:

$$-\frac{1}{2} f_{ac}{}^d f_{bd}{}^c = \delta_{ab}. \tag{262}$$

Then $f_{abc} = f_{ab}{}^d \delta_{dc}$ is completely anti-symmetric. We introduce ghost operators (c^a, b_b) with canonical anti-commutation relations (213):

$$[c^a, b_b]_+ = \delta_b^a, \qquad [c^a, c^b]_+ = [b_a, b_b]_+ = 0.$$

This implies, that in the 'co-ordinate representation', in which the ghosts c^a are represented by Grassmann variables, the b_a can be represented by a Grassmann derivative:

$$b_a = \frac{\partial}{\partial c^a}. \tag{263}$$

The nilpotent BRST operator takes the simple form

$$\Omega = c^a G_a - \frac{i}{2} c^a c^b f_{ab}{}^c b_c, \qquad \Omega^2 = 0. \tag{264}$$

We define a ghost-extended state space with elements

$$\psi[c] = \sum_{k=0}^{n} \frac{1}{k!} c^{a_1} ... c^{a_k} \psi^{(k)}_{a_1...a_k}. \tag{265}$$

The coefficients $\psi^{(k)}_{a_1..a_k}$ of ghost number k carry completely anti-symmetric product representations of the Lie algebra.

On the state space we introduce an indefinite inner product, with respect to which the ghosts c^a and b_a are self-adjoint; this is realized by the Berezin integral over the ghost variables

$$\langle \phi, \psi \rangle = \int [dc^n ... dc^1] \, \phi^\dagger \, \psi = \frac{1}{n!} \, \varepsilon^{a_1 ... a_n} \sum_{k=0}^{n} \binom{n}{k} \, \phi^{(n-k)\,*}_{a_{n-k}...a_1} \, \psi^{(k)}_{a_{n-k+1}...a_n}. \quad (266)$$

In components, the action of the ghosts is given by

$$(c^a \psi)^{(k)}_{a_1...a_k} = \delta^a_{a_1} \psi^{(k-1)}_{a_2 a_3...a_k} - \delta^a_{a_2} \psi^{(k-1)}_{a_1 a_3...a_k} + ... + (-1)^{k-1} \delta^a_{a_k} \psi^{(k-1)}_{a_1 a_2...a_{k-1}}, \quad (267)$$

and similarly

$$(b_a \psi)^{(k)}_{a_1...a_k} = \left(\frac{\partial \psi}{\partial c^a} \right)^{(k)}_{a_1...a_k} = \psi^{(k+1)}_{a a_1...a_k}. \quad (268)$$

It is now easy to check that the ghost operators are self-adjoint with respect to the inner product (266):

$$\langle \phi, c^a \psi \rangle = \langle c^a \phi, \psi \rangle, \qquad \langle \phi, b_a \psi \rangle = \langle b_a \phi, \psi \rangle. \quad (269)$$

It follows directly that the BRST operator (264) is self-adjoint as well:

$$\langle \phi, \Omega \psi \rangle = \langle \Omega \phi, \psi \rangle. \quad (270)$$

Now we can introduce a second inner product, which is positive definite and therefore manifestly non-degenerate:

$$(\phi, \psi) = \sum_{k=0}^{n} \frac{1}{k!} \left(\phi^{(k)\,*} \right)^{a_1...a_k} \psi^{(k)}_{a_1...a_k}. \quad (271)$$

It is related to the first indefinite inner product by Hodge duality: define the Hodge $*$-operator by

$$*\psi^{(k)\,a_1...a_k} = \frac{1}{(n-k)!} \, \varepsilon^{a_1...a_k a_{k+1}...a_n} \, \psi^{(n-k)}_{a_{k+1}...a_n}. \quad (272)$$

Furthermore, define the ghost permutation operator $\mathcal{P}$ as the operator which reverses the order of the ghosts in $\psi[c]$; equivalently:

$$(\mathcal{P}\psi)^{(k)}_{a_1...a_k} = \psi^{(k)}_{a_k...a_1}. \quad (273)$$

Then the two inner products are related by

$$(\phi, \psi) = \langle \mathcal{P}^* \phi, \psi \rangle. \quad (274)$$

An important property of the non-degenerate inner product is, that the ghosts c^a and b_a are adjoint to one another:

$$(\phi, c^a \psi) = (b_a \phi, \psi). \quad (275)$$

Then the adjoint of the BRST operator is given by the co-BRST operator

$$^*\Omega = b_a G^a - \frac{i}{2} c^c f^{ab}_{\ \ c} b_a b_b. \tag{276}$$

Here raising and lowering indices on the generators and structure constants is done with the help of the Killing metric (δ_{ab} in our normalization). It is easy to check that $^*\Omega^2 = 0$, as expected.

The harmonic states are both BRST- and co-BRST-closed: $\Omega\psi = {}^*\Omega\psi = 0$. They are zero-modes of the BRST-laplacian:

$$\Delta_{\mathrm{BRST}} = {}^*\Omega\,\Omega + \Omega^*\Omega = ({}^*\Omega + \Omega)^2, \tag{277}$$

as follows from the observation that

$$(\psi, \Delta_{\mathrm{BRST}}\,\psi) = (\Omega\psi, \Omega\psi) + ({}^*\Omega\psi, {}^*\Omega\psi) = 0 \quad \Leftrightarrow \quad \Omega\psi = {}^*\Omega\psi = 0. \tag{278}$$

For the case at hand, these conditions become

$$G_a\psi = 0, \qquad \Sigma_a\psi = 0, \tag{279}$$

where Σ_a is defined as

$$\Sigma_a = \Sigma_a^\dagger = -if_{ab}^{\ \ c} c^b b_c. \tag{280}$$

From the Jacobi identity, it is quite easy to verify that Σ_a defines a representation of the Lie algebra:

$$[\Sigma_a, \Sigma_b] = if_{ab}^{\ \ c}\Sigma_c, \qquad [G_a, \Sigma_b] = 0. \tag{281}$$

The conditions (279) are proven as follows. Substitute the explicit expressions for Ω and $^*\Omega$ into (277) for Δ_{BRST}. After some algebra one then finds

$$\Delta_{\mathrm{BRST}} = G^2 + G\cdot\Sigma + \frac{1}{2}\,\Sigma^2 = \frac{1}{2}\,G^2 + \frac{1}{2}\,(G+\Sigma)^2. \tag{282}$$

This being a sum of squares, any zero mode must satisfy (279). Q.E.D.

Looking for solutions, we observe that in components the second condition reads

$$(\Sigma_a\psi)^{(k)}_{a_1...a_k} = -if_{a[a_1}^{\ \ \ b}\psi^{(k)}_{a_2...a_k]b} = 0. \tag{283}$$

It acts trivially on states of ghost number $k = 0$; hence bona fide solutions are the gauge-invariant states of zero ghost number:

$$\psi = \psi^{(0)}, \qquad G_a\psi^{(0)} = 0. \tag{284}$$

However, other solutions with non-zero ghost number exist. A general solution is for example

$$\psi = \frac{1}{3!}\,f_{abc}\,c^a c^b c^c\,\chi, \qquad G_a\chi = 0. \tag{285}$$

The 3-ghost state $\psi^{(3)}_{abc} = f_{abc}\chi$ indeed satisfies (283) as a result of the Jacobi identity. The states χ are obviously in one-to-one correspondence with the states $\psi^{(0)}$. Hence, in general there exist several copies of the space of physical states in the BRST cohomology, at different ghost number. We infer that in addition to requiring physical states to belong to the BRST cohomology, it is also necessary to fix the ghost number for the definition of physical states to be unique.

3   Action Formalism

The canonical construction of the BRST cohomology we have described, can be given a basis in the action formulation, either in lagrangean or hamiltonian form. The latter one relates most directly to the canonical bracket formulation. It is then straightforward to switch to a gauge-fixed lagrangean formulation. Once we have the lagrangean formulation, a covariant approach to gauge fixing and quantization can be developed. In this section these constructions are presented and the relations between various formulations are discussed.

3.1   BRST Invariance from Hamilton's Principle

We have observed in Sect. 2.6, that the effective hamiltonian in the ghost-extended phase space is defined only modulo BRST-exact terms:

$$H_{\text{eff}} = H_c + i\{\Omega, \Psi\} = H_c - i\delta_\Omega \Psi, \tag{286}$$

where Ψ is a function of the phase space variables with ghost number $N_g(\Psi) = -1$. Moreover, the ghosts (c, b) are canonically conjugate:

$$\{c^\alpha, b_\beta\} = -i\delta_\beta^\alpha.$$

Thus, we are led to construct a pseudo-classical action of the form

$$S_{\text{eff}} = \int dt \left(p_i \dot{q}^i + ib_\alpha \dot{c}^\alpha - H_{\text{eff}}\right). \tag{287}$$

That this is indeed the correct action for our purposes follows from the ghost equations of motion obtained from this action, reading

$$\dot{c}^\alpha = -i\frac{\partial H_{\text{eff}}}{\partial b_\alpha}. \qquad \dot{b}_\alpha = -i\frac{\partial H_{\text{eff}}}{\partial c^\alpha}. \tag{288}$$

These equations are in full agreement with the definition of the extended Poisson brackets (182):

$$\dot{c}^\alpha = -\{H_{\text{eff}}, c^\alpha\}, \qquad \dot{b}_\alpha = -\{H_{\text{eff}}, b_\alpha\}. \tag{289}$$

As H_c is BRST invariant, H_{eff} is BRST invariant as well: the BRST variations are nilpotent and therefore $\delta_\Omega^2 \Phi = 0$. It is then easy to show that the action S_{eff} is BRST-symmetric and that the conserved Noether charge is the BRST charge as defined previously:

$$\delta_\Omega S_{\text{eff}} = \int dt \left[\left(\delta_\Omega p_i \dot{q}^i - \delta_\Omega q^i \dot{p}_i + i\delta_\Omega b_\alpha \dot{c}^\alpha + i\delta_\Omega c^\alpha \dot{b}_\alpha - \delta_\Omega H_{\text{eff}}\right)\right.$$

$$\left. + \frac{d}{dt}(p_i \delta_\Omega q^i - ib_\alpha \delta_\Omega c^\alpha)\right] \tag{290}$$

$$= \int dt \frac{d}{dt}\left(p_i \delta_\Omega q^i - ib_\alpha \delta_\Omega c^\alpha - \Omega\right).$$

To obtain the last equality, we have used (194) and (195), which can be summarized

$$\delta_\Omega q^i = \frac{\partial \Omega}{\partial p_i}, \qquad \delta_\Omega p_i = -\frac{\partial \Omega}{\partial q^i},$$

$$\delta_\Omega c^\alpha = i\frac{\partial \Omega}{\partial b_\alpha}, \qquad \delta_\Omega b_\alpha = i\frac{\partial \Omega}{\partial c^\alpha}.$$

The therefore action is invariant up to a total time derivative. By comparison with (59), we conclude that Ω is the conserved Noether charge.

3.2   Examples

The Relativistic Particle. A simple example of the procedure presented above is the relativistic particle. The canonical hamiltonian H_0 is constrained to vanish itself. As a result, the effective hamiltonian is a pure BRST term:

$$H_{\text{eff}} = i\{\Omega, \Psi\}. \tag{291}$$

A simple choice for the gauge fermion is $\Psi = b$, which has the correct ghost number $N_g = -1$. With this choice and the BRST generator Ω of (203), the effective hamiltonian is

$$H_{\text{eff}} = i\left\{\frac{c}{2m}(p^2 + m^2), b\right\} = \frac{1}{2m}(p^2 + m^2). \tag{292}$$

Then the effective action becomes

$$S_{\text{eff}} = \int d\tau \left(p \cdot \dot{x} + ib\dot{c} - \frac{1}{2m}(p^2 + m^2)\right). \tag{293}$$

This action is invariant under the BRST transformations (204) :

$$\delta_\Omega x^\mu = \{x^\mu, \Omega\} = \frac{cp^\mu}{m}, \, \delta_\Omega p_\mu = \{p_\mu, \Omega\} = 0,$$

$$\delta_\Omega c = -\{c, \Omega\} = 0, \qquad \delta_\Omega b = -\{b, \Omega\} = \frac{i}{2m}(p^2 + m^2),$$

up to a total proper-time derivative:

$$\delta_\Omega S_{\text{eff}} = \int d\tau \frac{d}{d\tau}\left[c\left(\frac{p^2 - m^2}{2m}\right)\right]. \tag{294}$$

Implementing the Noether construction, the conserved charge resulting from the BRST transformations is

$$\Omega = p \cdot \delta_\Omega x + ib\,\delta_\Omega c - \frac{c}{2m}(p^2 - m^2) = \frac{c}{2m}(p^2 + m^2). \tag{295}$$

Thus, we have reobtained the BRST charge from the action (293) and the transformations (204) confirming that together with the BRST-cohomology principle, they correctly describe the dynamics of the relativistic particle.

From the hamiltonian formulation (293) it is straightforward to construct a lagrangean one by using the hamilton equation $p^\mu = m\dot{x}^\mu$ to eliminate the momenta as independent variables; the result is

$$S_{\text{eff}} \simeq \int d\tau \left(\frac{m}{2} (\dot{x}^2 - 1) + ib\dot{c} \right). \tag{296}$$

Maxwell–Yang–Mills Theory. The BRST generator of the Maxwell–Yang–Mills theory in the temporal gauge has been given in (207):

$$\Omega = \int d^3x \left(c^a G_a - \frac{ig}{2} f_{ab}{}^c c^a c^b b_c \right),$$

with $G_a = (\boldsymbol{D} \cdot \boldsymbol{E})_a$. The BRST-invariant effective hamiltonian takes the form

$$H_{\text{eff}} = \frac{1}{2} \left(\boldsymbol{E}_a^2 + \boldsymbol{B}_a^2 \right) + i \{ \Omega, \Psi \}. \tag{297}$$

Then, a simple choice of the gauge fermion, $\Psi = \lambda^a b_a$, with some constants λ_a gives the effective action

$$S_{\text{eff}} = \int d^4x \left[-\boldsymbol{E} \cdot \frac{\partial \boldsymbol{A}}{\partial t} + ib_a \dot{c}^a - \frac{1}{2} \left(\boldsymbol{E}_a^2 + \boldsymbol{B}_a^2 \right) - \lambda^a (\boldsymbol{D} \cdot \boldsymbol{E})_a + ig\lambda^a f_{ab}{}^c c^b b_c \right]. \tag{298}$$

The choice $\lambda^a = 0$ would in effect turn the ghosts into free fields. However, if we eliminate the electric fields $\boldsymbol{E}^a$ as independent degrees of freedom by the substitution $E_i^a = F_{i0}^a = \partial_i A_0^a - \partial_0 A_i^a - g f_{bc}{}^a A_i^b A_0^c$ and recalling the classical hamiltonian (127), we observe that we might actually interpret λ^a as a constant scalar potential, $A_0^a = \lambda^a$, in a BRST-extended relativistic action

$$S_{\text{eff}} = \int d^4x \left[-\frac{1}{4} (F_{\mu\nu}^a)^2 + ib_a (D_0 c)^a \right]_{A_0^a = \lambda^a}, \tag{299}$$

where $(D_0 c)^a = \partial_0 c^a - g f_{bc}{}^a A_0^b c^c$. The action is invariant under the classical BRST transformations (208):

$$\delta_\Omega \boldsymbol{A}^a = (\boldsymbol{D}c)^a, \qquad \delta_\Omega \boldsymbol{E}_a = g f_{ab}{}^c c^b \boldsymbol{E}_c,$$

$$\delta_\Omega c^a = \frac{g}{2} f_{bc}{}^a c^b c^c, \ \delta_\Omega b_a = i G_a + g f_{ab}{}^c c^b b_c,$$

with the above BRST generator (207) as the conserved Noether charge. All of the above applies to Maxwell electrodynamics as well, except that in an abelian theory there is only a single vector field, and all structure constants vanish: $f_{ab}{}^c = 0$.

3.3 Lagrangean BRST Formalism

From the hamiltonian formulation of BRST-invariant dynamical systems it is straightforward to develop an equivalent lagrangean formalism, by eliminating the momenta p_i as independent degrees of freedom. This proceeds as usual by solving Hamilton's equation

$$\dot{q}^i = \frac{\partial H}{\partial p_i},$$

for the momenta in terms of the velocities, and performing the inverse Legendre transformation. We have already seen how this works for the examples of the relativistic particle and the Maxwell–Yang–Mills theory. As the lagrangean is a scalar function under space-time transformations, it is better suited for the development of a manifestly covariant formulation of gauge-fixed BRST-extended dynamics of theories with local symmetries, including Maxwell–Yang–Mills theory and the relativistic particle as well as string theory and general relativity.

The procedure follows quite naturally the steps outlined in the previous Sects. 3.1 and 3.2:

a. Start from a gauge-invariant lagrangean $L_0(q, \dot{q})$.

b. For each gauge degree of freedom (each gauge parameter), introduce a ghost variable c^a; by definition these ghost variables carry ghost number $N_g[c^a] = +1$. Construct BRST transformations $\delta_\Omega X$ for the extended configuration-space variables $X = (q^i, c^a)$, satisfying the requirement that they leave L_0 invariant (possibly modulo a total derivative), and are nilpotent: $\delta_\Omega^2 X = 0$.

c. Add a trivially BRST-invariant set of terms to the action, of the form $\delta_\Omega \Psi$ for some anti-commuting function Ψ (the gauge fermion).

The last step is to result in an effective lagrangean L_{eff} with net ghost number $N_g[L_{\text{eff}}] = 0$. To achieve this, the gauge fermion must have ghost number $N_g[\Psi] = -1$. However, so far we only have introduced dynamical variables with non-negative ghost number: $N_g[q^i, c^a] = (0, +1)$. To solve this problem we introduce anti-commuting anti-ghosts b_a, with ghost number $N_g[b_a] = -1$. The BRST-transforms of these variables must then be commuting objects α_a, with ghost number $N_g[\alpha] = 0$. In order for the BRST-transformations to be nilpotent, we require

$$\delta_\Omega \, b_a = i\alpha_a, \qquad \delta_\Omega \, \alpha_a = 0, \tag{300}$$

which indeed satisfy $\delta_\Omega^2 = 0$ trivially. The examples of the previous section illustrate this procedure.

The Relativistic Particle. The starting point for the description of the relativistic particle was the reparametrization-invariant action (8). We identify the integrand as the lagrangean L_0. Next we introduce the Grassmann-odd ghost variable $c(\lambda)$, and define the BRST transformations

$$\delta_\Omega \, x^\mu = c \frac{dx^\mu}{d\lambda}, \qquad \delta_\Omega \, e = \frac{d(ce)}{d\lambda}, \qquad \delta_\Omega \, c = c \frac{dc}{d\lambda}. \tag{301}$$

As $c^2 = 0$, these transformations are nilpotent indeed. In addition, introduce the anti-ghost representation (b, α) with the transformation rules (300). We can now construct a gauge fermion. We make the choice

$$\Psi(b, e) = b(e - 1) \quad \Rightarrow \quad \delta_\Omega \Psi = i\alpha(e - 1) - b \frac{d(ce)}{d\lambda}. \tag{302}$$

As a result, the effective lagrangean (in natural units) becomes

$$L_{\text{eff}} = L_0 - i\delta_\Omega\Psi = \frac{m}{2e}\frac{dx_\mu}{d\lambda}\frac{dx^\mu}{d\lambda} - \frac{em}{2} + \alpha(e-1) + ib\,\frac{d(ce)}{d\lambda}. \tag{303}$$

Observing that the variable α plays the role of a Lagrange multiplier, fixing the einbein to its canonical value $e = 1$ such that $d\lambda = d\tau$, this lagrangean is seen to reproduce the action (296):

$$S_{\text{eff}} = \int d\tau L_{\text{eff}} \simeq \int d\tau \left(\frac{m}{2}\,(\dot{x}^2 - 1) + ib\,\dot{c}\right).$$

Maxwell–Yang–Mills Theory. The covariant classical action of the Maxwell–Yang–Mills theory was presented in (121):

$$S_0 = -\frac{1}{4}\int d^4x\,\left(F_{\mu\nu}^a\right)^2.$$

Introducing the ghost fields c^a, we can define nilpotent BRST transformations

$$\delta_\Omega A_\mu^a = (D_\mu c)^a\,, \qquad \delta_\Omega c^a = \frac{g}{2}\,f_{bc}{}^a c^b c^c. \tag{304}$$

Next we add the anti-ghost BRST multiplets (b_a, α_a), with the transformation rules (300). Choose the gauge fermion

$$\Psi(A_0^a, b_a) = b_a(A_0^a - \lambda^a) \quad \Rightarrow \quad \delta_\Omega\Psi = i\alpha_a(A_0^a - \lambda^a) - b_a(D_0 c)^a, \tag{305}$$

where λ^a are some constants (possibly zero). Adding this to the classical action gives

$$S_{\text{eff}} = \int d^4x\,\left[-\frac{1}{4}\,(F_{\mu\nu}^a)^2 + \alpha_a(A_0^a - \lambda^a) + ib_a(D_0 c)^a\right]. \tag{306}$$

Again, the fields α_a act as Lagrange multipliers, fixing the electric potentials to the constant values λ^a. After substitution of these values, the action reduces to the form (299).

We have thus demonstrated that the lagrangean and canonical procedures lead to equivalent results; however, we stress that in both cases the procedure involves the choice of a gauge fermion Ψ, restricted by the requirement that it has ghost number $N_g[\Psi] = -1$.

The advantage of the lagrangean formalism is, that it is easier to formulate the theory with different choices of the gauge fermion. In particular, it is possible to make choices of gauge which manifestly respect the Lorentz-invariance of Minkowski space. This is not an issue for the study of the relativistic particle, but it is an issue in the case of Maxwell–Yang–Mills theory, which we have constructed so far only in the temporal gauge $A_0^a = $ constant.

We now show how to construct a covariant gauge-fixed and BRST-invariant effective lagrangean for Maxwell–Yang–Mills theory, using the same procedure. In stead of (305), we choose the gauge fermion

$$\Psi = b_a\left(\partial \cdot A^a - \frac{\lambda}{2}\,\alpha^a\right) \quad \Rightarrow \quad \delta_\Omega\Psi = i\alpha_a\,\partial \cdot A^a - \frac{i\lambda}{2}\,\alpha_a^2 - b_a\,\partial\cdot(Dc)^a. \tag{307}$$

Here the parameter λ is a arbitrary real number, which can be used to obtain a convenient form of the propagator in perturbation theory. The effective action obtained with this choice of gauge-fixing fermion is, after a partial integration:

$$S_{\text{eff}} = \int d^4x \left[-\frac{1}{4} (F^a_{\mu\nu})^2 + \alpha_a \, \partial \cdot A^a - \frac{\lambda}{2} \alpha_a^2 - i\partial b_a \cdot (Dc)^a \right]. \tag{308}$$

As we have introduced quadratic terms in the bosonic variables α_a, they now behave more like auxiliary fields, rather than Lagrange multipliers. Their variational equations lead to the result

$$\alpha^a = \frac{1}{\lambda} \partial \cdot A^a. \tag{309}$$

Eliminating the auxiliary fields by this equation, the effective action becomes

$$S_{\text{eff}} = \int d^4x \left[-\frac{1}{4} (F^a_{\mu\nu})^2 + \frac{1}{2\lambda} (\partial \cdot A^a)^2 - i\partial b_a \cdot (Dc)^a \right]. \tag{310}$$

This is the standard form of the Yang–Mills action used in covariant perturbation theory. Observe, that the elimination of the auxiliary field α_a also changes the BRST-transformation of the anti-ghost b_a to:

$$\delta_\Omega b^a = \frac{i}{\lambda} \partial \cdot A^a \quad \Rightarrow \quad \delta_\Omega^2 b^a = \frac{i}{\lambda} \partial \cdot (Dc)^a \simeq 0. \tag{311}$$

The transformation is now nilpotent only after using the ghost field equation.

The BRST-Noether charge can be computed from the action (310) by the standard procedure, and leads to the expression

$$\Omega = \int d^3x \left(\pi^\mu_a (D_\mu c)^a - \frac{ig}{2} f_{ab}{}^c c^a c^b \gamma_c \right), \tag{312}$$

where π^μ_a is the canonical momentum of the vector potential A^a_μ, and (β^a, γ_a) denote the canonical momenta of the ghost fields (b_a, c^a):

$$\pi^i_a = \frac{\partial \mathcal{L}_{\text{eff}}}{\partial \dot{A}^a_i} = -F^{0i}_a = -E^i_a, \ \pi^0_a = \frac{\partial \mathcal{L}_{\text{eff}}}{\partial \dot{A}^a_0} = -\frac{1}{\lambda} \partial \cdot A_a,$$

$$\beta^a = i\frac{\partial \mathcal{L}_{\text{eff}}}{\partial \dot{b}_a} = -(D_0 c)^a, \quad \gamma_a = i\frac{\partial \mathcal{L}_{\text{eff}}}{\partial \dot{c}^a} = \partial_0 b_a. \tag{313}$$

Each ghost field (b_a, c^a) now has its own conjugate momentum, because the ghost terms in the action (310) are quadratic in derivatives, rather than linear as before. Note also, that a factor i has been absorbed in the ghost momenta to make them real; this leads to the standard Poisson brackets

$$\{c^a(\boldsymbol{x};t), \gamma_b(\boldsymbol{y};t)\} = -i\delta^a_b \delta^3(\boldsymbol{x} - \boldsymbol{y}), \qquad \{b_a(\boldsymbol{x};t), \beta^b(\boldsymbol{y};t)\} = -i\delta^b_a \delta^3(\boldsymbol{x} - \boldsymbol{y}). \tag{314}$$

As our calculation shows, all explicit dependence on (b_a, β^a) has dropped out of the expression (312) for the BRST charge.

The parameter λ is still a free parameter, and in actual calculations it is often useful to check partial gauge-independence of physical results, like cross sections, by establishing that they do not depend on this parameter. What needs to be shown more generally is, that physical results do not depend on the choice of gauge fermion. This follows formally from the BRST cohomology being independent of the choice of gauge fermion. Indeed, from the expression (312) for Ω we observe that it is of the same form as the one we have used previously in the temporal gauge, even though now π_a^0 no longer vanishes identically. In the quantum theory this implies, that the BRST-cohomology classes at ghost number zero correspond to gauge-invariant states, in which

$$(\boldsymbol{D} \cdot \boldsymbol{E})^a = 0, \qquad \partial \cdot A^a = 0. \tag{315}$$

The second equation implies, that the time-evolution of the 0-component of the vector potential is fixed completely by the initial conditions and the evolution of the spatial components $\boldsymbol{A}^a$. In particular, $A_0^a = \lambda^a = $ constant is a consistent solution if by a gauge transformation we take the spatial components to satisfy $\boldsymbol{\nabla} \cdot \boldsymbol{A}^a = 0$.

In actual computations, especially in perturbation theory, the matter is more subtle however: the theory needs to be renormalized, and this implies that the action and BRST-transformation rules have to be adjusted to the introduction of counter terms. To prove the gauge independence of the renormalized theory it must be shown, that the renormalized action still possesses a BRST-invariance, and the cohomology classes at ghost-number zero satisfy the renormalized conditions (315). In four-dimensional space-time this can indeed be done for the pure Maxwell–Yang–Mills theory, as there exists a manifestly BRST-invariant regularization scheme (dimensional regularization) in which the theory defined by the action (310) is renormalizable by power counting. The result can be extended to gauge theories interacting with scalars and spin-1/2 fermions, except for the case in which the Yang–Mills fields interact with chiral fermions in anomalous representations of the gauge group.

3.4   The Master Equation

Consider a BRST-invariant action $S_{\text{eff}}[\Phi^A] = S_0 + \int dt\,(i\delta_\Omega\Psi)$, where the variables $\Phi^A = (q^i, c^a, b_a, \alpha_a)$ parametrize the extended configuration space of the system, and Ψ is the gauge fermion, which is Grassmann-odd and has ghost number $N_g[\Psi] = -1$. Now by construction,

$$\delta_\Omega\Psi = \delta_\Omega\Phi^A\,\frac{\partial\Psi}{\partial\Phi^A}, \tag{316}$$

and therefore we can write the effective action also as

$$S_{\text{eff}}[\Phi^A] = S_0 + i\int dt\,\left[\delta_\Omega\Phi^A\,\Phi_A^*\right]_{\Phi_A^* = \frac{\partial\Psi}{\partial\Phi^A}}. \tag{317}$$

In this way of writing, one considers the action as a functional on a doubled configuration space, parametrized by variables (Φ^A, Φ^*_A) of which the first set Φ^A is called the *fields*, and the second set Φ^*_A is called the *anti-fields*. In the generalized action

$$S^*[\Phi^A, \Phi^*_A] = S_0 + i \int dt\, \delta_\Omega \Phi^A\, \Phi^*_A, \tag{318}$$

the anti-fields play the role of sources for the BRST-variations of the fields Φ^A; the effective action S_{eff} is the restriction to the hypersurface $\Sigma[\Psi]$: $\Phi^*_A = \partial\Psi/\partial\Phi^A$. We observe, that by construction the antifields have Grassmann parity opposite to that of the corresponding fields, and ghost number $N_g[\Phi^*_A] = -(N_g[\Phi^A]+1)$.

In the doubled configuration space the BRST variations of the fields can be written as

$$i\delta_\Omega \Phi^A = (-1)^A \frac{\delta S^*}{\delta \Phi^*_A}, \tag{319}$$

where $(-1)^A$ is the Grassmann parity of the field Φ^A, whilst $-(-1)^A = (-1)^{A+1}$ is the Grassmann parity of the anti-field Φ^*_A. We now define the *anti-bracket* of two functionals $F(\Phi^A, \Phi^*_A)$ and $G(\Phi^A, \Phi^*_A)$ on the large configuration space by

$$(F, G) = (-1)^{F+G+FG} (G, F) = (-1)^{A(F+1)} \left(\frac{\delta F}{\delta \Phi^A} \frac{\delta G}{\delta \Phi^*_A} + (-1)^F \frac{\delta F}{\delta \Phi^*_A} \frac{\delta G}{\delta \Phi^A} \right). \tag{320}$$

These brackets are symmetric in F and G if both are Grassmann-even (bosonic), and anti-symmetric in all other cases. Sometimes one introduces the notion of *right derivative*:

$$\frac{F \overleftarrow{\delta}}{\delta \Phi^A} \equiv (-1)^{A(F+1)} \frac{\delta F}{\delta \Phi^A}. \tag{321}$$

Then the anti-brackets take the simple form

$$(F, G) = \frac{F \overleftarrow{\delta}}{\delta \Phi^A} \frac{\overrightarrow{\delta} G}{\delta \Phi^*_A} - \frac{F \overleftarrow{\delta}}{\delta \Phi^*_A} \frac{\overrightarrow{\delta} G}{\delta \Phi^A}, \tag{322}$$

where the derivatives with a right arrow denote the standard *left* derivatives. In terms of the anti-brackets, the BRST transformations (319) can be written in the form

$$i\delta_\Omega \Phi^A = (S^*, \Phi^A). \tag{323}$$

In analogy, we can define

$$i\delta_\Omega \Phi^*_A = (S^*, \Phi^*_A) = (-1)^A \frac{\delta S^*}{\delta \Phi^A}. \tag{324}$$

Then the BRST transformation of any functional $Y(\Phi^A, \Phi^*_A)$ is given by

$$i\delta_\Omega Y = (S^*, Y). \tag{325}$$

In particular, the BRST invariance of the action S^* can be expressed as

$$(S^*, S^*) = 0. \tag{326}$$

This equation is known as the *master equation*. Next we observe, that on the physical hypersurface $\Sigma[\Psi]$ the BRST transformations of the antifields are given by the classical field equations; indeed, introducing an anti-commuting parameter μ for infinitesimal BRST transformations

$$i\mu \, \delta_\Omega \Phi_A^* = \frac{\delta S^*}{\delta \Phi^A} \mu \xrightarrow{\Sigma[\Psi]} \frac{\delta S_{\text{eff}}}{\delta \Phi^A} \mu \simeq 0, \tag{327}$$

where the last equality holds only for solutions of the classical field equations. Because of this result, it is customary to redefine the BRST transformations of the antifields such that they vanish:

$$\delta_\Omega \Phi_A^* = 0, \tag{328}$$

instead of (324). As the BRST transformations are nilpotent, this is consistent with the identification $\Phi_A^* = \partial \Psi / \partial \Phi^A$ in the action; indeed, it now follows that

$$\delta_\Omega \left(\delta_\Omega \Phi^A \, \Phi_A^* \right) = 0, \tag{329}$$

which holds before the identification as a result of (328), and after the identification because it reduces to $\delta_\Omega^2 \Psi = 0$. Note, that the condition for BRST invariance of the action now becomes

$$i\delta_\Omega S^* = \frac{1}{2} \left(S^*, S^* \right) = 0, \tag{330}$$

which still implies the master equation (326).

3.5   Path-Integral Quantization

The construction of BRST-invariant actions $S_{\text{eff}} = S^*[\Phi_A^* = \partial \Psi / \partial \Phi^A]$ and the anti-bracket formalism is especially useful in the context of path-integral quantization. The path integral provides a representation of the matrix elements of the evolution operator in the configuration space:

$$\langle q_f, T/2 | e^{-iTH} | q_i, -T/2 \rangle = \int_{q_i}^{q_f} Dq(t) \, e^{i \int_{-T/2}^{T/2} L(q, \dot{q}) dt}. \tag{331}$$

In field theory one usually considers the vacuum-to-vacuum amplitude in the presence of sources, which is a generating functional for time-ordered vacuum Green's functions:

$$Z[J] = \int D\Phi \, e^{iS[\Phi] + i \int J\Phi}, \tag{332}$$

such that

$$\langle 0 | T(\Phi_1 ... \Phi_k) | 0 \rangle = \frac{\delta^k Z[J]}{\delta J_1 ... \delta J_k} \bigg|_{J=0}. \tag{333}$$

The corresponding generating functional $W[J]$ for the connected Green's functions is related to $Z[J]$ by

$$Z[J] = e^{i\,W[J]}. \tag{334}$$

For theories with gauge invariances, the evolution operator is constructed from the BRST-invariant hamiltonian; then the action to be used is the in the path integral (332) is the BRST invariant action:

$$Z[J] = e^{i\,W[J]} = \int D\Phi^A \; e^{i\,S^*[\Phi^A,\Phi^*_A]+i\int J_A\Phi^A}\Big|_{\Phi^*_A=\partial\Psi/\partial\Phi^A}, \tag{335}$$

where the sources J_A for the fields are supposed to be BRST invariant themselves. For the complete generating functional to be BRST invariant, it is not sufficient that only the action S^* is BRST invariant, as guaranteed by the master equation (326): the functional integration measure must be BRST invariant as well. Under an infinitesimal BRST transformation $\mu\delta_\Omega\Phi^A$ the measure changes by a graded jacobian (superdeterminant) [18,19]

$$\mathcal{J} = \mathrm{SDet}\left(\delta^A_B + \mu(-1)^B \frac{\delta(\delta_\Omega\Phi^A)}{\delta\Phi^B}\right) \approx 1 + \mu\,\mathrm{Tr}\frac{\delta(\delta_\Omega\Phi^A)}{\delta\Phi^B}. \tag{336}$$

We now define

$$\frac{\delta(i\delta_\Omega\Phi^A)}{\delta\Phi^A} = (-1)^A \frac{\delta^2 S^*}{\delta\Phi^A\delta\Phi^*_A} \equiv \bar{\Delta}S^*. \tag{337}$$

The operator $\bar{\Delta}$ is a laplacian on the field/anti-field configuration space, which for an arbitrary functional $Y(\Phi^A,\Phi^*_A)$ is defined by

$$\bar{\Delta}Y = (-1)^{A(1+Y)} \frac{\delta^2 Y}{\delta\Phi^A\delta\Phi^*_A}. \tag{338}$$

The condition of invariance of the measure requires the BRST jacobian (336) to be unity:

$$\mathcal{J} = 1 - i\mu\,\bar{\Delta}S^* = 1, \tag{339}$$

which reduces to the vanishing of the laplacian of S^*:

$$\bar{\Delta}S^* = 0. \tag{340}$$

The two conditions (326) and (340) imply the BRST invariance of the path integral (335). Actually, a somewhat more general situation is possible, in which neither the action nor the functional measure are invariant independently, only the combined functional integral. Let the action generating the BRST transformations be denoted by $W^*[\Phi^A,\Phi^*_A]$:

$$i\delta_\Omega\Phi^A = (W^*,\Phi^A), \qquad i\delta_\Omega\Phi^*_A = 0. \tag{341}$$

As a result the graded jacobian for a transformation with parameter μ is

$$\mathrm{SDet}\left(\delta^A_B + \mu(-1)^B \frac{\delta(\delta_\Omega\Phi^A)}{\delta\Phi^B}\right) \approx 1 - i\mu\,\bar{\Delta}W^*. \tag{342}$$

Then the functional W^* itself needs to satisfy the generalized master equation

$$\frac{1}{2}\left(W^*, W^*\right) = i\bar{\Delta}W^*, \tag{343}$$

for the path-integral to be BRST invariant. This equation can be neatly summarized in the form

$$\bar{\Delta}\,e^{iW^*} = 0. \tag{344}$$

Solutions of this equation restricted to the hypersurface $\Phi_A^* = \partial\Psi/\partial\Phi^A$ are acceptable actions for the construction of BRST-invariant path integrals.

4   Applications of BRST Methods

In the final section of these lecture notes, we turn to some applications of BRST-methods other than the perturbative quantization of gauge theories. We deal with two topics; the first is the construction of BRST field theories, presented in the context of the scalar point particle. This is the simplest case [34]; for more complicated ones, like the superparticle [35,36] or the string [35,37,32], we refer to the literature.

The second application concerns the classification of anomalies in gauge theories of the Yang–Mills type. Much progress has been made in this field in recent years [40], of which a summary is presented here.

4.1   BRST Field Theory

The examples of the relativistic particle and string show that in theories with local reparametrization invariance the hamiltonian is one of the generators of gauge symmetries, and as such is constrained to vanish. The same phenomenon also occurs in general relativity, leading to the well-known Wheeler-deWitt equation. In such cases, the *full* dynamics of the system is actually contained in the BRST cohomology. This opens up the possibility for constructing quantum field theories for particles [32–34], or strings [32,35,37], in a BRST formulation, in which the usual BRST operator becomes the kinetic operator for the fields. This formulation has some formal similarities with the Dirac equation for spin-1/2 fields.

As our starting point we consider the BRST-operator for the relativistic quantum scalar particle, which for free particles, after some rescaling, reads

$$\Omega = c(p^2 + m^2), \qquad \Omega^2 = 0. \tag{345}$$

It acts on fields $\Psi(x, c) = \psi_0(x) + c\psi_1(x)$, with the result

$$\Omega\Psi(x, c) = c(p^2 + m^2)\,\psi_0(x). \tag{346}$$

As in the case of Lie-algebra cohomology (271), we introduce the non-degenerate (positive definite) inner product

$$(\Phi, \Psi) = \int d^d x \,\left(\phi_0^*\psi_0 + \phi_1^*\psi_1\right). \tag{347}$$

With respect to this inner product the ghosts (b, c) are mutually adjoint:

$$(\Phi, c\Psi) = (b\Phi, \Psi) \quad \leftrightarrow \quad b = c^\dagger. \tag{348}$$

Then, the BRST operator Ω is not self-adjoint but rather

$$\Omega^\dagger = b(p^2 + m^2), \qquad \Omega^{\dagger 2} = 0. \tag{349}$$

Quite generally, we can construct actions for quantum scalar fields coupled to external sources J of the form

$$S_G[J] = \frac{1}{2}\,(\Psi, G\,\Omega\Psi) - (\Psi, J)\,, \tag{350}$$

where the operator G is chosen such that

$$G\Omega = (G\Omega)^\dagger = \Omega^\dagger G^\dagger. \tag{351}$$

This guarantees that the action is real. From the action we then derive the field equation

$$G\Omega\Psi = \Omega^\dagger G^\dagger \Psi = J. \tag{352}$$

Its consistency requires the co-BRST invariance of the source:

$$\Omega^\dagger J = 0. \tag{353}$$

This reflects the invariance of the action and the field equation under BRST transformations

$$\Psi \to \Psi' = \Psi + \Omega\chi. \tag{354}$$

In order to solve the field equation we therefore have to impose a gauge condition, selecting a particular element of the equivalence class of solutions (354).

A particularly convenient condition is

$$\Omega G^\dagger \Psi = 0. \tag{355}$$

In this gauge, the field equation can be rewritten in the form

$$\Delta G^\dagger \Psi = \left(\Omega^\dagger \Omega + \Omega\Omega^\dagger\right) G^\dagger \Psi = \Omega\,J. \tag{356}$$

Here Δ is the BRST laplacean, which can be inverted using a standard analytic continuation in the complex plane, to give

$$G^\dagger \Psi = \frac{1}{\Delta}\,\Omega\,J. \tag{357}$$

We interpret the operator $\Delta^{-1}\Omega$ on the right-hand side as the (tree-level) propagator of the field.

We now implement the general scheme (350)–(357) by choosing the inner product (347), and $G = b$. Then

$$G\Omega = bc(p^2 + m^2) = \Omega^\dagger G^\dagger, \tag{358}$$

and therefore

$$\frac{1}{2}\left(\Psi, G\,\Omega\Psi\right) = \frac{1}{2}\int d^d x\,\psi_0^*(p^2 + m^2)\psi_0, \tag{359}$$

which is the standard action for a free scalar field[4].

The laplacean for the BRST operators (346) and (349) is

$$\Delta = \Omega\,\Omega^\dagger + \Omega^\dagger\Omega = (p^2 + m^2)^2, \tag{360}$$

which is manifestly non-negative, but might give rise to propagators with double poles, or negative residues, indicating the appearance of ghost states. However, in the expression (357) for the propagator, one of the poles is canceled by the zero of the BRST operator; in the present context the equation reads

$$c\psi_0 = \frac{1}{(p^2 + m^2)^2}\,c(p^2 + m^2)\,J_0. \tag{361}$$

This leads to the desired result

$$\psi_0 = \frac{1}{p^2 + m^2}\,J_0, \tag{362}$$

and we recover the standard scalar field theory indeed. It is not very difficult to extend the theory to particles in external gravitational or electromagnetic fields[5], or to spinning particles [38].

However, a different and more difficult problem is the inclusion of self interactions. This question has been addressed mostly in the case of string theory [32]. As it is expected to depend on spin, no unique prescription has been constructed for the point particle so far.

4.2   Anomalies and BRST Cohomology

In the preceding sections we have seen how local gauge symmetries are encoded in the BRST-transformations. First, the BRST-transformations of the classical variables correspond to ghost-dependent gauge transformations. Second, the closure of the algebra of the gauge transformations (and the Poisson brackets or commutators of the constraints), as well as the corresponding Jacobi-identities, are part of the condition that the BRST transformations are nilpotent.

It is important to stress, as we observed earlier, that the closure of the classical gauge algebra does not necessarily guarantee the closure of the gauge algebra in the quantum theory, because it may be spoiled by anomalies. Equivalently, in the presence of anomalies there is no nilpotent quantum BRST operator, and no local action satisfying the master equation (344). A particular case in point is that of a Yang–Mills field coupled to chiral fermions, as in the electro-weak standard model. In the following we consider chiral gauge theories in some detail.

---

[4] Of course, there is no loss of generality here if we restrict the coefficients ψ_a to be real.

[5] See the discussion in [34], which uses however a less elegant implementation of the action.

The action of chiral fermions coupled to an abelian or non-abelian gauge field reads

$$S_F[A] = \int d^4x\, \bar{\psi}_L \slashed{D}\psi_L. \tag{363}$$

Here $D_\mu\psi_L = \partial_\mu\psi_L - gA_\mu^a T_a\psi_L$ with T_a being the generators of the gauge group in the representation according to which the spinors ψ_L transform. In the path-integral formulation of quantum field theory the fermions make the following contribution to the effective action for the gauge fields:

$$e^{iW[A]} = \int D\bar{\psi}_L D\psi_L\, e^{iS_F[A]}. \tag{364}$$

An infinitesimal local gauge transformation with parameter Λ^a changes the effective action $W[A]$ by

$$\delta(\Lambda)W[A] = \int d^4x\, (D_\mu\Lambda)^a\, \frac{\delta W[A]}{\delta A_\mu^a} = \int d^4x\, \Lambda^a \left(\partial_\mu \frac{\delta}{\delta A_\mu^a} - gf_{ab}{}^c A_\mu^b \frac{\delta}{\delta A_\mu^c} \right) W[A], \tag{365}$$

assuming boundary terms to vanish. By construction, the fermion action $S_F[A]$ itself is gauge invariant, but this is generally not true for the fermionic functional integration measure. If the measure is not invariant:

$$\delta(\Lambda)W[A] = \int d^4x\, \Lambda^a \Gamma_a[A] \neq 0, \tag{366}$$

$$\Gamma_a[A] = \mathcal{D}_a W[A] \equiv \left(\partial_\mu \frac{\delta}{\delta A_\mu^a} - gf_{ab}{}^c A_\mu^b \frac{\delta}{\delta A_\mu^c} \right) W[A].$$

Even though the action $W[A]$ may not be invariant, its variation should still be covariant and satisfy the condition

$$\mathcal{D}_a\Gamma_b[A] - \mathcal{D}_b\Gamma_a[A] = [\mathcal{D}_a, \mathcal{D}_b]\, W[A] = gf_{ab}{}^c \mathcal{D}_c W[A] = gf_{ab}{}^c \Gamma_c[A]. \tag{367}$$

This consistency condition was first derived by Wess and Zumino [41], and its solutions determine the functional form of the anomalous variation $\Gamma_a[A]$ of the effective action $W[A]$. It can be derived from the BRST cohomology of the gauge theory [39,44,40].

To make the connection, observe that the Wess–Zumino consistency condition (367) can be rewritten after contraction with ghosts as follows:

$$0 = \int d^4x\, c^a c^b\, (\mathcal{D}_a\Gamma_b[A] - \mathcal{D}_b\Gamma_a[A] - gf_{ab}{}^c \Gamma_c[A])$$

$$= 2\int d^4x\, c^a c^b \left(\mathcal{D}_a\Gamma_b - \frac{g}{2} f_{ab}{}^c \Gamma_c \right) = -2\,\delta_\Omega \int d^4x\, c^a \Gamma_a, \tag{368}$$

provided we can ignore boundary terms. The integrand is a 4-form of ghost number $+1$:

$$I_4^1 = d^4x\, c^a \Gamma_a[A] = \frac{1}{4!}\, \varepsilon_{\mu\nu\kappa\lambda}\, dx^\mu \wedge dx^\nu \wedge dx^\kappa \wedge dx^\lambda\, c^a \Gamma_a[A]. \tag{369}$$

The Wess–Zumino consistency condition (368) then implies that non-trivial solutions of this condition must be of the form

$$\delta_\Omega I_4^1 = dI_3^2, \tag{370}$$

where I_3^2 is a 3-form of ghost number $+2$, vanishing on any boundary of the space-time $\mathcal{M}$.

Now we make a very interesting and useful observation: the BRST construction can be mapped to a standard cohomology problem on a principle fibre bundle with local structure $\mathcal{M} \times G$, where $\mathcal{M}$ is the space-time and G is the gauge group viewed as a manifold [42]. First note that the gauge field is a function of both the co-ordinates x^μ on the space-time manifold $\mathcal{M}$ and the parameters Λ^a on the group manifold G. We denote the combined set of these co-ordinates by $\xi = (x, \Lambda)$. To make the dependence on space-time and gauge group explicit, we introduce the Lie-algebra valued 1-form

$$A(x) = dx^\mu A_\mu^a(x) T_a, \tag{371}$$

with a generator T_a of the gauge group, and the gauge field $A_\mu^a(x)$ at the point x in the space-time manifold $\mathcal{M}$. Starting from A, all gauge-equivalent configurations are obtained by local gauge transformations, generated by group elements $a(\xi)$ according to

$$\mathcal{A}(\xi) = -\frac{1}{g} a^{-1}(\xi)\, da(\xi) + a^{-1}(\xi)\, A(x)\, a(\xi), \qquad A(x) = \mathcal{A}(x, 0) \tag{372}$$

where d is the ordinary differential operator on the space-time manifold $\mathcal{M}$:

$$da(x, \Lambda) = dx^\mu \frac{\partial a}{\partial x^\mu}(x, \Lambda). \tag{373}$$

Furthermore, the parametrization of the group is chosen such that $a(x, 0) = 1$, the identity element. Then, if $a(\xi)$ is close to the identity:

$$a(\xi) = e^{g\Lambda(x)\cdot T} \approx 1 + g\,\Lambda^a(x) T_a + \mathcal{O}(g^2\Lambda^2), \tag{374}$$

and (372) represents the infinitesimally transformed gauge field 1-form (124). In the following we interpret $\mathcal{A}(\xi)$ as a particular 1-form living on the fibre bundle with local structure $\mathcal{M} \times G$.

A general one-form $\mathbf{N}$ on the bundle can be decomposed as

$$\mathbf{N}(\xi) = d\xi^i N_i = dx^\mu N_\mu + d\Lambda^a N_a. \tag{375}$$

Correspondingly, we introduce the differential operators

$$d = dx^\mu \frac{\partial}{\partial x^\mu}, \qquad s = d\Lambda^a \frac{\partial}{\partial \Lambda^a}, \qquad \mathbf{d} = d + s, \tag{376}$$

with the properties

$$d^2 = 0, \qquad s^2 = 0, \qquad \mathbf{d}^2 = ds + sd = 0. \tag{377}$$

Next define the left-invariant 1-forms on the group $C(\xi)$ by

$$C = a^{-1} \, sa, \qquad c(x) = C(x, 0). \tag{378}$$

By construction, using $sa^{-1} = -a^{-1} sa \, a^{-1}$, these forms satisfy

$$sC = -C^2. \tag{379}$$

The action of the group differential s on the one-form $\mathcal{A}$ is

$$s\mathcal{A} = \frac{1}{g} DC = \frac{1}{g} (dC - g[\mathcal{A}, C]_+). \tag{380}$$

Finally, the field strength $\mathcal{F}(\xi)$ for the gauge field $\mathcal{A}$ is defined as the 2-form

$$\mathcal{F} = d\mathcal{A} - g\mathcal{A}^2 = a^{-1} F \, a, \qquad F(x) = \mathcal{F}(x, 0). \tag{381}$$

The action of s on $\mathcal{F}$ is given by

$$s\mathcal{F} = [\mathcal{F}, C]. \tag{382}$$

Clearly, the above equations are in one-to-one correspondence with the BRST transformations of the Yang–Mills fields, described by the Lie-algebra valued one-form $A = dx^\mu A^a_\mu T_a$, and the ghosts described by the Lie-algebra valued Grassmann variable $c = c^a T_a$, upon the identification $-gs|_{\Lambda=0} \to \delta_\Omega$:

$$-gs\mathcal{A}|_{\Lambda=0} \to \delta_\Omega A = -dx^\mu (D_\mu c)^a T_a = -Dc,$$

$$-gsC|_{\Lambda=0} \to \delta_\Omega c = \frac{g}{2} f^c_{ab} c^a c^b T_c = \frac{g}{2} c^a c^b [T_a, T_b] = gc^2. \tag{383}$$

$$-gs\mathcal{F}|_{\Lambda=0} \to \delta_\Omega F = -\frac{g}{2} dx^\mu \wedge dx^\nu f^c_{ab} F^a_{\mu\nu} c^b T_c = -g[F, c],$$

provided we take the BRST variational derivative δ_Ω and the ghosts c to anti-commute with the differential operator d:

$$d\delta_\Omega + \delta_\Omega d = 0, \qquad dc + cd = dx^\mu(\partial_\mu c). \tag{384}$$

Returning to the Wess–Zumino consistency condition (370), we now see that it can be restated as a cohomology problem on the principle fibre bundle on which the 1-form $\mathcal{A}$ lives. This is achieved by mapping the 4-form of ghost number $+1$ to a particular 5-form on the bundle, which is a local 4-form on $\mathcal{M}$ and a 1-form on G; similarly one maps the 3-form of ghost number $+2$ to another 5-form which is a local 3-form on $\mathcal{M}$ and a 2-form on G:

$$I^1_4 \to \omega^1_4, \qquad I^2_3 \to \omega^2_3, \tag{385}$$

where the two 5-forms must be related by

$$-gs\omega^1_4 = d\omega^2_3. \tag{386}$$

We now show how to solve this equation as part of a whole chain of equations known as the *descent equations*. The starting point is a set of invariant polynomials known as the Chern characters of order n. They are constructed in terms of the field-strength 2-form:

$$F = dA - gA^2 = \frac{1}{2} dx^\mu \wedge dx^\nu F^a_{\mu\nu} T_a, \tag{387}$$

which satisfies the Bianchi identity

$$DF = dF - g[A, F] = 0. \tag{388}$$

The two-form F transforms covariantly under gauge transformations (372):

$$F \to a^{-1} F a = \mathcal{F}. \tag{389}$$

It follows that the Chern character of order n, defined by

$$Ch_n[A] = \text{Tr } F^n = \text{Tr } \mathcal{F}^n, \tag{390}$$

is an invariant $2n$-form: $Ch_n[A] = Ch_n[\mathcal{A}]$. It is also closed, as a result of the Bianchi identity:

$$d\, Ch_n[A] = n\text{Tr}\,[(DF)F^{n-1}] = 0. \tag{391}$$

The solution of this equation is given by the exact $(2n-1)$-forms:

$$Ch_n[A] = d\omega^0_{2n-1}[A]. \tag{392}$$

Note that the exact $2n$-form on the right-hand side lies entirely in the local space-time part $\mathcal{M}$ of the bundle because this is manifestly true for the left-hand side.

Proof of the result (392) is to be given; for the time being we take it for granted and continue our argument. First, we define a generalized connection on the bundle by

$$\mathbf{A}(\xi) \equiv -\frac{1}{g} a^{-1}(\xi)\, \mathbf{d}a(\xi) + a^{-1}(\xi)A(x)a(\xi) = -\frac{1}{g} C(\xi) + \mathcal{A}(\xi). \tag{393}$$

It follows that the corresponding field strength on the bundle is

$$\mathbf{F} = \mathbf{dA} - g\mathbf{A}^2 = (d + s)\left(\mathcal{A} - \frac{1}{g}C\right) - g\left(\mathcal{A} - \frac{1}{g}C\right)^2 \tag{394}$$

$$= d\mathcal{A} - g\mathcal{A}^2 = \mathcal{F}.$$

To go from the first to the second line we have used (380). This result is sometimes referred to as *the Russian formula* [43]. The result implies that the components of the generalized field strength in the directions of the group manifold all vanish.

It is now obvious that

$$Ch_n[\mathbf{A}] = \mathrm{Tr}\,\mathbf{F}^n = Ch_n[A]; \tag{395}$$

moreover, $\mathbf{F}$ satisfies the Bianchi identity

$$\mathbf{DF} = \mathbf{dF} - g[\mathbf{A}, \mathbf{F}] = 0. \tag{396}$$

Again, this leads us to infer that

$$\mathbf{d}Ch_n[\mathbf{A}] = 0 \quad \Rightarrow \quad Ch_n[\mathbf{A}] = \mathbf{d}\omega^0_{2n-1}[\mathbf{A}] = d\omega^0_{2n-1}[A], \tag{397}$$

where the last equality follows from (395) and (392). The middle step, which states that the $(2n-1)$-form of which $Ch_n[\mathbf{A}]$ is the total exterior derivative has the same functional form in terms of $\mathbf{A}$, as the one of which it is the exterior space-time derivative has in terms of A, will be justified shortly.

We first conclude the derivation of the chain of descent equations, which follow from the last result by expansion in terms of C:

$$d\omega^0_{2n-1}[A] = (d+s)\,\omega^0_{2n-1}[\mathcal{A} - C/g]$$

$$= (d+s)\left(\omega^0_{2n-1}[\mathcal{A}] + \frac{1}{g}\,\omega^1_{2n-2}[\mathcal{A}, C] + \dots + \frac{1}{g^{2n-1}}\,\omega^{2n-1}_0[\mathcal{A}, C]\right). \tag{398}$$

Comparing terms of the same degree, we find

$$d\omega^0_{2n-1}[A] = d\omega^0_{2n-1}[\mathcal{A}],$$

$$-gs\omega^0_{2n-1}[\mathcal{A}] = d\omega^1_{2n-1}[\mathcal{A}, C],$$

$$-gs\omega^1_{2n-2}[\mathcal{A}, C] = d\omega^2_{2n-3}[\mathcal{A}, C], \tag{399}$$

$$\dots$$

$$-gs\omega^{2n-1}_0[\mathcal{A}, C] = 0.$$

Obviously, this result carries over to the BRST differentials: with $I^0_n[A] = \omega^0_n[A]$, one obtains

$$\delta_\Omega I^k_m[A, c] = dI^{k+1}_{m-1}[A, c], \qquad m + k = 2n - 1, \quad k = 0, 1, 2, \dots, 2n - 1. \tag{400}$$

The first line just states the gauge independence of the Chern character. Taking $n = 3$, we find that the third line is the Wess–Zumino consistency condition (386):

$$\delta_\Omega\, I^1_4[A, c] = dI^2_3[A, c].$$

Proofs and Solutions. We now show how to derive the result (392); this will provide us at the same time with the tools to solve the Wess–Zumino consistency condition. Consider an arbitrary gauge field configuration described by the Lie-algebra valued 1-form A. From this we define a whole family of gauge fields

$$A_t = tA, \qquad t \in [0, 1]. \tag{401}$$

It follows, that

$$F_t \equiv F[A_t] = tdA - gt^2 A^2 = tF[A] - g(t^2 - t)A^2. \tag{402}$$

This field strength 2-form satisfies the appropriate Bianchi identity:

$$D_t F_t = dF_t - g[A_t, F] = 0. \tag{403}$$

In addition, one easily derives

$$\frac{dF_t}{dt} = dA - [A_t, A]_+ = D_t A, \tag{404}$$

where the anti-commutator of the 1-forms implies a *commutator* of the Lie-algebra elements. Now we can compute the Chern character

$$Ch_n[A] = \int_0^1 dt \frac{d}{dt} \mathrm{Tr} F_t^n = n \int_0^1 dt \, \mathrm{Tr} \left((D_t A) F_t^{n-1} \right)$$

$$= nd \int_0^1 dt \, \mathrm{Tr} \left(A F_t^{n-1} \right). \tag{405}$$

In this derivation we have used both (404) and the Bianchi identity (403).

It is now straightforward to compute the forms w_5^0 and w_4^1. First, taking $n = 3$ in the result (405) gives $Ch_3[A] = dw_5^0$ with

$$I_5^0[A] = w_5^0[A] = 3 \int_0^1 dt \, \mathrm{Tr} \left((D_t A) F_t^2 \right) = \mathrm{Tr} \left(AF^2 + \frac{g}{2} A^3 F + \frac{g^2}{10} A^5 \right). \tag{406}$$

Next, using (383), the BRST differential of this expression gives $\delta_\Omega I_5^0 = dI_4^1$, with

$$I_4^1[A, c] = -\mathrm{Tr} \left(c \left[F^2 + \frac{g}{2} \left(A^2 F + AFA + FA^2 \right) + \frac{g^2}{2} A^4 \right] \right). \tag{407}$$

This expression determines the anomaly up to a constant of normalization $\mathcal{N}$:

$$\Gamma_a[A] = \mathcal{N} \mathrm{Tr} \left(T_a \left[F^2 + \frac{g}{2} \left(A^2 F + AFA + FA^2 \right) + \frac{g^2}{2} A^4 \right] \right). \tag{408}$$

Of course, the component form depends on the gauge group; for example, for $SU(2) \simeq SO(3)$ it vanishes identically, which is true for any orthogonal group $SO(N)$; in contrast the anomaly does not vanish identically for $SU(N)$, for any $N \geq 3$. In that case it has to be anulled by cancellation between the contributions of chiral fermions in different representations of the gauge group G.

Appendix. Conventions

In these lecture notes the following conventions are used. Whenever two objects carrying a same index are multiplied (as in $a_i b_i$ or in $u_\mu v^\mu$) the index is a dummy index and is to be summed over its entire range, unless explicitly stated otherwise (summation convention). Symmetrization of objects enclosed is denoted by braces $\{...\}$, anti-symmetrization by square brackets $[...]$; the total weight of such (anti-)symmetrizations is always unity.

In these notes we deal both with classical and quantum hamiltonian systems. To avoid confusion, we use braces $\{\,,\}$ to denote classical Poisson brackets, brackets $[\,,]$ to denote commutators and suffixed brackets $[\,,]_+$ to denote anti-commutators.

The Minkowski metric $\eta_{\mu\nu}$ has signature $(-1, +1, ..., +1)$, the first co-ordinate in a pseudo-cartesian co-ordinate system x^0 being time-like. Arrows above symbols $(\boldsymbol{x})$ denote purely spatial vectors (most often 3-dimensional).

Unless stated otherwise, we use natural units in which $c = \hbar = 1$. Therefore we usually do not write these dimensional constants explicitly. However, in a few places where their role as universal constants is not a priori obvious they are included in the equations.

References

1. C. Becchi, A. Rouet and R. Stora, Ann. Phys. 98 (1976), 28
2. I. V. Tyutin, Lebedev preprint FIAN 39 (1975), unpublished
3. R. P. Feynman, Acta Phys. Pol. 26 (1963), 697
4. B. DeWitt, Phys. Rev. 162 (1967), 1195
5. L. D. Faddeev and V. N. Popov, Phys. Lett. B25 (1967), 29
6. T. Kugo and I. Ojima, Supp. Progr. Theor. Phys. 66 (1979), 1
7. M. Henneaux, Phys. Rep. 126 (1985), 1
8. L. Baulieu, Phys. Rep. 129 (1985), 1
9. B. Kostant and S. Sternberg, Ann. Phys. (NY) 176 (1987), 49
10. R. Marnelius, in: *Geometrical and Algebraic Aspects of Non-Linear Field Theory*, eds. S. De Filippo, M. Marinaro, G. Marmo and G. Vilasi (North Holland, 1989)
11. J. W. van Holten, in: *Functional Integration, Geometry and Strings*, eds. Z. Haba and J. Sobczyk (Birkhäuser, 1989), 388
12. D. M. Gitman and I. V. Tyutin, *Quantization of fields with constraints* (Springer, 1990)
13. S. K. Blau, Ann. Phys. 205 (1991), 392
14. M. Henneaux and C. Teitelboim, *Quantization of Gauge Systems* (Princeton Univ. Press, 1992)
15. H. Goldstein, *Classical Mechanics* (Addison-Wesley, 1950)
16. L. Fadeev and R. Jackiw, Phys. Rev. Lett. 60 (1988), 1692
    R. Jackiw, in: *Constraint Theory and Quantization Methods*, eds. F. Colomo, L. Lusanna and G. Marmo (World Scientific, 1994), 163
17. B. de Wit and J. W. van Holten, Phys. Lett. B79 (1978), 389
    J. W. van Holten, *On the construction of supergravity theories* (PhD. thesis; Leiden Univ., 1980)
18. F. A. Berezin, *The method of second quantization* (Academic Press, NY; 1966)

19. B. S. DeWitt, *Supermanifolds* (Cambridge University Press; 1984)
20. M. Spiegelglas, Nucl. Phys. B283 (1986), 205
21. S. A. Frolov and A. A. Slavnov, Phys. Lett. B218 (1989), 461
22. S. Hwang and R. Marnelius, Nucl. Phys. B320 (1989), 476
23. A. V. Razumov and G. N. Rybkin, Nucl. Phys. B332 (1990), 209
24. W. Kalau and J. W. van Holten, Nucl. Phys. B361 (1991), 233
25. J. W. van Holten, Phys. Rev. Lett. 64 (1990), 2863; Nucl. Phys. B339 (1990), 158
26. J. Fisch and M. Henneaux, in: *Constraints Theory and Relativistic Dynamics*, eds. G. Longhi and L. Lusanna (World Scientific, Singapore; 1987), 57
27. J. Zinn-Justin, Lect. Notes Phys. **37**, 1 (Springer-Verlag Berlin Heidelberg 1975)
28. I. A. Batalin and G. A. Vilkovisky, Phys. Lett. B102 (1981), 27; Phys. Rev. D28 (1983), 2567
29. E. Witten, Mod. Phys. Lett. A5 (1990), 487
30. A. Schwarz, Comm. Math. Phys. 155 (1993), 249
31. S. Aoyama and S. Vandoren, Mod. Phys. Lett. A8 (1993), 3773
32. W. Siegel, *Introduction to String Field Theory*, (World Scientific, Singapore, 1988)
33. H. Hüffel, Phys. Lett. B241 (1990), 369
    Int. J. Mod. Phys. A6 (1991), 4985
34. J. W. van Holten, J. Mod. Phys. A7 (1992), 7119
35. W. Siegel, Phys. Lett. B151 (1985), 391, 396
36. S. Aoyama, J. Kowalski-Glikman, L. Lukierski and J. W. van Holten, Phys. Lett. B216 (1989), 133
37. E. Witten, Nucl. Phys. B268 (1986), 79
    Nucl. Phys. B276 (1986), 291
38. J. W. van Holten, Nucl. Phys. B457 (1995), 375
39. B. Zumino, in *Relativity, Groups and Topology II* (Les Houches 1983), eds. B. S. DeWitt and R. Stora (North Holland, Amsterdam, 1984)
40. G. Barnich, F. Brandt and M. Henneaux, Phys. Rep. 338 (2000), 439
41. J. Wess and B. Zumino, Phys. Lett. B37 (1971), 95
42. L. Bonora and P. Cotta-Ramusino, Comm. Math. Phys. 87, (1983), 589
43. R. Stora, in: *New Developments in Quantum Field Theory and Statistical Mechanics*, eds. H. Levy and P. Mitter (Plenum, 1977)
44. B. Zumino, Y-S. Wu and A. Zee, Nucl. Phys. B239 (1984), 477

Chiral Anomalies and Topology

J. Zinn-Justin

Dapnia, CEA/Saclay, Département de la Direction des Sciences de la Matière, and Institut de Mathématiques de Jussieu–Chevaleret, Université de Paris VII

Abstract. When a field theory has a symmetry, global or local like in gauge theories, in the tree or classical approximation formal manipulations lead to believe that the symmetry can also be implemented in the full quantum theory, provided one uses the proper quantization rules. While this is often true, it is not a general property and, therefore, requires a proof because simple formal manipulations ignore the unavoidable divergences of perturbation theory. The existence of invariant regularizations allows solving the problem in most cases but the combination of gauge symmetry and chiral fermions leads to subtle issues. Depending on the specific group and field content, *anomalies* are found: obstructions to the quantization of chiral gauge symmetries. Because anomalies take the form of local polynomials in the fields, are linked to local group transformations, but vanish for global (rigid) transformations, one discovers that they have a *topological* character. In these notes we review various perturbative and non-perturbative regularization techniques, and show that they leave room for possible anomalies when both gauge fields and chiral fermions are present. We determine the form of anomalies in simple examples. We relate anomalies to the index of the Dirac operator in a gauge background. We exhibit gauge instantons that contribute to the anomaly in the example of the $CP(N-1)$ models and $SU(2)$ gauge theories. We briefly mention a few physical consequences. For many years the problem of anomalies had been discussed only within the framework of perturbation theory. New non-perturbative solutions based on lattice regularization have recently been proposed. We describe the so-called overlap and domain wall fermion formulations.

1 Symmetries, Regularization, Anomalies

Divergences. Symmetries of the classical lagrangian or tree approximation do not always translate into symmetries of the corresponding complete quantum theory. Indeed, local quantum field theories are affected by UV divergences that invalidate simple algebraic proofs.

The origin of UV divergences in field theory is double. First, a field contains an infinite number of degrees of freedom. The corresponding divergences are directly related to the renormalization group and reflect the property that, even in renormalizable quantum field theories, degrees of freedom remain coupled on all scales.

However, another of type of divergences can appear, which is related to the order between quantum operators and the transition between classical and quantum hamiltonians. Such divergences are already present in ordinary quantum mechanics in perturbation theory, for instance, in the quantization of the geodesic

J. Zinn-Justin, Chiral Anomalies and Topology, Lect. Notes Phys. **659**, 167–236 (2005)
http://www.springerlink.com/

motion of a particle on a manifold (like a sphere). Even in the case of forces linear in the velocities (like a coupling to a magnetic field), finite ambiguities are found. In local quantum field theories the problem is even more severe. For example, the commutator of a scalar field operator $\hat{\phi}$ and its conjugate momentum $\hat{\pi}$, in the Schrödinger picture (in d space–time dimension), takes the form

$$[\hat{\phi}(x), \hat{\pi}(y)] = i\hbar\, \delta^{d-1}(x - y).$$

Hamiltonians contain products of fields and conjugate momenta as soon as derivative couplings are involved (in covariant theories), or when fermions are present. Because in a local theory all operators are taken at the same point, products of this nature lead to divergences, except in quantum mechanics ($d = 1$ with our conventions). These divergences reflect the property that the knowledge of the classical theory is not sufficient, in general, to determine the quantized theory completely.

Regularization. Regularization is a useful intermediate step in the renormalization program that consists in modifying the initial theory at short distance, large momentum or otherwise to render perturbation theory finite. Note that from the point of view of Particle Physics, all these modifications affect in some essential way the physical properties of the theory and, thus, can only be considered as intermediate steps in the removal of divergences.

When a regularization can be found which preserves the symmetry of the initial classical action, a symmetric quantum field theory can be constructed.

Momentum cut-off regularization schemes, based on modifying propagators at large momenta, are specifically designed to cut the infinite number of degrees of freedom. With some care, these methods will preserve formal symmetries of the un-renormalized theory that correspond to global (space-independent) linear group transformations. Problems may, however, arise when the symmetries correspond to non-linear or local transformations, like in the examples of non-linear σ models or gauge theories, due to the unavoidable presence of derivative couplings. It is easy to verify that in this case regularizations that only cut momenta do not in general provide a complete regularization.

The addition of regulator fields has, in general, the same effect as modifying propagators but offers a few new possibilities, in particular, when regulator fields have the wrong spin–statistics connection. Fermion loops in a gauge background can be regularized by such a method.

Other methods have to be explored. In many examples dimensional regularization solves the problem because then the commutator between field and conjugated momentum taken at the same point vanishes. However, in the case of chiral fermions dimensional regularization fails because chiral symmetries are specific to even space–time dimensions.

Of particular interest is the method of lattice regularization, because it can be used, beyond perturbation theory, either to discuss the existence of a quantum field theory, or to determine physical properties of field theories by non-perturbative numerical techniques. One verifies that such a regularization indeed specifies an order between quantum operators. Therefore, it solves the ordering problem in non-linear σ-models or non-abelian gauge theories. However, again it fails in the presence of chiral fermions: the manifestation of this difficulty takes the form of a doubling of the fermion degrees of freedom. Until recently, this had prevented a straightforward numerical study of chiral theories.

Anomalies. That no conventional regularization scheme can be found in the case of gauge theories with chiral fermions is not surprising since we know theories with anomalies, that is theories in which a local symmetry of the tree or classical approximation cannot be implemented in the full quantum theory. This may create obstructions to the construction of chiral gauge theories because exact gauge symmetry, and thus the absence of anomalies, is essential for the physical consistency of a gauge theory.

Note that we study in these lectures only local anomalies, which can be determined by perturbative calculations; peculiar global non-perturbative anomalies have also been exhibited.

The anomalies discussed in these lectures are local quantities because they are consequences of short distance singularities. They are responses to local (space-dependent) group transformations but vanish for a class of space-independent transformations. This gives them a topological character that is further confirmed by their relations with the index of the Dirac operator in a gauge background.

The recently discovered solutions of the Ginsparg–Wilson relation and the methods of overlap and domain wall fermions seem to provide an unconventional solution to the problem of lattice regularization in gauge theories involving chiral fermions. They evade the fermion doubling problem because chiral transformations are no longer strictly local on the lattice (though remain local from the point of view of the continuum limit), and relate the problem of anomalies with the invariance of the fermion measure. The absence of anomalies can then be verified directly on the lattice, and this seems to confirm that the theories that had been discovered anomaly-free in perturbation theory are also anomaly-free in the non-perturbative lattice construction. Therefore, the specific problem of lattice fermions was in essence technical rather than reflecting an inconsistency of chiral gauge theories beyond perturbation theory, as one may have feared.

Finally, since these new regularization schemes have a natural implementation in five dimensions in the form of domain wall fermions, this again opens the door to speculations about additional space dimensions.

We first discuss the advantages and shortcomings, from the point of view of symmetries, of three regularization schemes, momentum cut-off, dimensional, lattice regularizations. We show that they leave room for possible anomalies when both gauge fields and chiral fermions are present.

We then recall the origin and the form of anomalies, beginning with the simplest example of the so-called abelian anomaly, that is the anomaly in the conservation of the abelian axial current in gauge theories. We relate anomalies to the index of a covariant Dirac operator in the background of a gauge field.

In the two-dimensional $CP(N-1)$ models and in four-dimensional non-abelian gauge theories, we exhibit gauge instantons. We show that they can be classified in terms of a topological charge, the space integral of the chiral anomaly. The existence of gauge field configurations that contribute to the anomaly has direct physical implications, like possible strong CP violation and the solution to the $U(1)$ problem.

We examine the form of the anomaly for a general axial current, and infer conditions for gauge theories that couple differently to fermion chiral components to be anomaly-free. A few physical applications are also briefly mentioned.

Finally, the formalism of overlap fermions on the lattice and the role of the Ginsparg–Wilson relation are explained. The alternative construction of domain wall fermions is explained, starting from the basic mechanism of zero-modes in supersymmetric quantum mechanics.

Conventions. Throughout these notes we work in *euclidean space* (with imaginary or euclidean time), and this also implies a formalism of *euclidean fermions*. For details see

J. Zinn-Justin, *Quantum Field Theory and Critical Phenomena*, Clarendon Press (Oxford 1989, fourth ed. 2002).

2 Momentum Cut-Off Regularization

We first discuss methods that work in the continuum (compared to lattice methods) and at fixed dimension (unlike dimensional regularization). The idea is to modify field propagators beyond a large momentum cut-off to render all Feynman diagrams convergent. The regularization must satisfy one important condition: the inverse of the regularized propagator must remain a *smooth* function of the momentum **p**. Indeed, singularities in momentum variables generate, after Fourier transformation, contributions to the large distance behaviour of the propagator, and regularization should modify the theory only at short distance.

Note, however, that such modifications result in unphysical properties of the quantum field theory at cut-off scale. They can be considered as intermediate steps in the renormalization program (physical properties would be recovered in the large cut-off limit). Alternatively, in modern thinking, the necessity of a regularization often indicates that quantum field theories cannot rendered consistent on all distance scales, and have eventually to be embedded in a more complete non field theory framework.

2.1 Matter Fields: Propagator Modification

Scalar Fields. A simple modification of the propagator improves the convergence of Feynman diagrams at large momentum. For example in the case of the action of the self-coupled scalar field,

$$S(\phi) = \int \mathrm{d}^d x \left[\frac{1}{2}\phi(x)(-\nabla_x^2 + m^2)\phi(x) + V_{\mathrm{I}}\big(\phi(x)\big) \right], \tag{1}$$

the propagator in Fourier space $1/(m^2 + p^2)$ can be replaced by

$$\Delta_{\rm B}(p) = \left(\frac{1}{p^2 + m^2}\right)_{\rm reg.}$$

with

$$\Delta_{\rm B}^{-1}(p) = (p^2 + m^2) \prod_{i=1}^{n}(1 + p^2/M_i^2). \tag{2}$$

The masses M_i are proportional to the momentum cut-off Λ,

$$M_i = \alpha_i \Lambda, \quad \alpha_i > 0.$$

If the degree n is chosen large enough, all diagrams become convergent. In the formal large cut-off limit $\Lambda \to \infty$, at parameters α fixed, the initial propagator is recovered. This is the spirit of momentum cut-off or Pauli–Villars's regularization.

Note that such a propagator cannot be derived from a hermitian hamiltonian. Indeed, hermiticity of the hamiltonian implies that if the propagator is, as above, a rational function, it must be a sum of poles in p^2 with positive residues (as a sum over intermediate states of the two-point function shows) and thus cannot decrease faster than $1/p^2$.

While this modification can be implemented also in Minkowski space because the regularized propagators decrease in all complex p^2 directions (except real negative), in euclidean time more general modifications are possible. Schwinger's proper time representation suggests

$$\Delta_{\rm B}(p) = \int_0^\infty {\rm d}t\, \rho(t\Lambda^2) {\rm e}^{-t(p^2 + m^2)}, \tag{3}$$

in which the function $\rho(t)$ is positive (to ensure that $\Delta_{\rm B}(p)$ does not vanish and thus is invertible) and satisfies the condition

$$|1 - \rho(t)| < C{\rm e}^{-\sigma t} \ (\sigma > 0) \text{ for } t \to +\infty.$$

By choosing a function $\rho(t)$ that decreases fast enough for $t \to 0$, the behaviour of the propagator can be arbitrarily improved. If $\rho(t) = O(t^n)$, the behaviour (2) is recovered. Another example is

$$\rho(t) = \theta(t - 1),$$

$\theta(t)$ being the step function, which leads to an exponential decrease:

$$\Delta_{\rm B}(p) = \frac{{\rm e}^{-(p^2 + m^2)/\Lambda^2}}{p^2 + m^2}. \tag{4}$$

As the example shows, it is thus possible to find in this more general class propagators without unphysical singularities, but they do not follow from a hamiltonian formalism because continuation to real time becomes impossible.

Spin 1/2 Fermions. For spin 1/2 fermions similar methods are applicable. To the free Dirac action,

$$\mathcal{S}_{\mathrm{F0}} = \int \mathrm{d}^d x \, \bar{\psi}(x)(\partial\!\!\!/ + m)\psi(x) \,,$$

corresponds in Fourier representation the propagator $1/(m + i p\!\!\!/)$. It can be replaced by the regularized propagator $\Delta_{\mathrm{F}}(p)$ where

$$\Delta_{\mathrm{F}}^{-1}(p) = (m + i p\!\!\!/) \prod_{i=1}^{n}(1 + p^2/M_i^2). \tag{5}$$

Note that we use the standard notation $p\!\!\!/ \equiv p_\mu \gamma_\mu$, with euclidean fermion conventions, analytic continuation to imaginary or euclidean time of the usual Minkowski fermions, and hermitian matrices γ_μ.

Remarks. Momentum cut-off regularizations have several advantages: one can work at fixed dimension and in the continuum. However, two potential weaknesses have to be stressed:

(i) The generating functional of correlation functions, obtained by adding to the action (1) a source term for the fields:

$$\mathcal{S}(\phi) \mapsto \mathcal{S}(\phi) - \int \mathrm{d}^d x \, J(x)\phi(x),$$

can be written as

$$\mathcal{Z}(J) = \det{}^{1/2}(\Delta_{\mathrm{B}}) \exp\left[-\mathcal{V}_{\mathrm{I}}\left(\delta/\delta J\right)\right] \exp\left(\frac{1}{2} \int \mathrm{d}^d x \, \mathrm{d}^d y \, J(x)\Delta_{\mathrm{B}}(x-y)J(y)\right),$$
$$\tag{6}$$

where the determinant is generated by the gaussian integration, and

$$\mathcal{V}_{\mathrm{I}}(\phi) \equiv \int \mathrm{d}^d x \, V_{\mathrm{I}}(\phi(x)).$$

None of the momentum cut-off regularizations described so far can deal with the determinant. As long as the determinant is a divergent constant that cancels in normalized correlation functions, this is not a problem, but in the case of a determinant in the background of an external field (which generates a set of one-loop diagrams) this may become a serious issue.

(ii) This problem is related to another one: even in simple quantum mechanics, some models have divergences or ambiguities due to problem of the order between quantum operators in products of position and momentum variables. A class of Feynman diagrams then cannot be regularized by this method. Quantum field theories where this problem occurs include models with non-linearly realized (like in the non-linear σ model) or gauge symmetries.

Global Linear Symmetries. To implement symmetries of the classical action in the quantum theory, we need a regularization scheme that preserves the symmetry. This requires some care but can always be achieved for linear global symmetries, that is symmetries that correspond to transformations of the fields of the form

$$\phi_R(x) = R\,\phi(x)\,,$$

where R is a constant matrix. The main reason is that in the quantum hamiltonian field operators and conjugate momenta are not mixed by the transformation and, therefore, the order of operators is to some extent irrelevant. To take an example directly relevant here, a theory with massless fermions may, in four dimensions, have a chiral symmetry

$$\psi_\theta(x) = e^{i\theta\gamma_5}\psi(x), \qquad \bar\psi_\theta(x) = \bar\psi(x)e^{i\theta\gamma_5}\,.$$

The substitution (5)(for $m = 0$) preserves chiral symmetry. Note the importance here of being able to work at fixed dimension four because chiral symmetry is defined only in even dimensions. In particular, the invariance of the integration measure $[d\bar\psi(x)d\psi(x)]$ relies on the property that $\operatorname{tr}\gamma_5 = 0$.

2.2   Regulator Fields

Regularization in the form (2) or (5) has another equivalent formulation based on the introduction of regulator fields. Note, again, that some of the regulator fields have unphysical properties; for instance, they violate the spin–statistics connection. The regularized quantum field theory is physically consistent only for momenta much smaller than the masses of the regulator fields.

Scalar Fields. In the case of scalar fields, to regularize the action (1) for the scalar field ϕ, one introduces additional dynamical fields $\phi_r, r = 1, \ldots, r_{\max}$, and considers the modified action

$$S_{\text{reg.}}(\phi, \phi_r) = \int d^d x \left[\frac{1}{2}\phi\left(-\nabla^2 + m^2\right)\phi + \sum_r \frac{1}{2z_r}\phi_r\left(-\nabla^2 + M_r^2\right)\phi_r \right.$$
$$\left. + V_I(\phi + \textstyle\sum_r \phi_r) \right]. \tag{7}$$

With the action 7 any internal ϕ propagator is replaced by the sum of the ϕ propagator and all the ϕ_r propagators $z_r/(p^2 + M_r^2)$. For an appropriate choice of the constants z_r, after integration over the regulator fields, the form (2) is recovered. Note that the condition of cancellation of the $1/p^2$ contribution at large momentum implies

$$1 + \sum_r z_r = 0\,.$$

Therefore, not all z_r can be positive and, thus, the fields ϕ_r, corresponding to the negative values, necessarily are unphysical. In particular, in the integral over these fields, one must integrate over imaginary values.

Fermions. The fermion inverse propagator (5) can be written as

$$\Delta_F^{-1}(p) = (m + i\slashed{p}) \prod_{r=1}^{r_{max}} (1 + i\slashed{p}/M_r)(1 - i\slashed{p}/M_r).$$

This indicates that, again, the same form can be obtained by a set of regulator fields $\{\bar{\psi}_{r\pm}, \psi_{r\pm}\}$. One replaces the kinetic part of the action by

$$\int d^d x \, \bar{\psi}(x)(\slashed{\partial} + m)\psi(x) \mapsto \int d^d x \, \bar{\psi}(x)(\slashed{\partial} + m)\psi(x)$$
$$+ \sum_{\epsilon=\pm,r} \frac{1}{z_{r\epsilon}} \int d^d x \, \bar{\psi}_{r\epsilon}(x)(\slashed{\partial} + \epsilon M_r)\psi_{r\epsilon}(x).$$

Moreover, in the interaction term the fields ψ and $\bar{\psi}$ are replaced by the sums

$$\psi \mapsto \psi + \sum_{r,\epsilon} \psi_{r\epsilon} , \qquad \bar{\psi} \mapsto \bar{\psi} + \sum_{r,\epsilon} \bar{\psi}_{r\epsilon} .$$

For a proper choice of the constants z_r, after integration over the regulator fields, the form (5) is recovered.

For $m = 0$, the propagator (5) is chiral invariant. Chiral transformations change the sign of mass terms. Here, chiral symmetry can be maintained only if, in addition to normal chiral transformations, $\psi_{r,+}$ and ψ_{-r} are exchanged (which implies $z_{r+} = z_{r-}$). Thus, chiral symmetry is preserved by the regularization, even though the regulators are massive, by fermion doubling. The fermions ψ_+ and ψ_- are chiral partners. For a pair $\psi \equiv (\psi_+, \psi_-)$, $\bar{\psi} \equiv (\bar{\psi}_+, \bar{\psi}_-)$, the action can be written as

$$\int d^d x \, \bar{\psi}(x) \, (\slashed{\partial} \otimes \mathbf{1} + M \mathbf{1} \otimes \sigma_3) \, \psi(x),$$

where the first matrix $\mathbf{1}$ and the Pauli matrix σ_3 act in $\pm$ space. The spinors then transform like

$$\psi_\theta(x) = e^{i\theta\gamma_5 \otimes \sigma_1} \psi(x), \qquad \bar{\psi}_\theta(x) = \bar{\psi}(x) e^{i\theta\gamma_5 \otimes \sigma_1} ,$$

because σ_1 anticommutes with σ_3.

2.3   Abelian Gauge Theory

The problem of matter in presence of a gauge field can be decomposed into two steps, first matter in an external gauge field, and then the integration over the gauge field. For gauge fields, we choose a covariant gauge, in such a way that power counting is the same as for scalar fields.

Charged Fermions in a Gauge Background. The new problem that arises in presence of a gauge field is that only covariant derivatives are allowed because gauge invariance is essential for the physical consistency of the theory. The regularized action in a gauge background now reads

$$S(\bar\psi, \psi, A) = \int d^dx\, \bar\psi(x)\, (m + \not{D}) \prod_r \left(1 - \not{D}^2/M_r^2\right) \psi(x),$$

where D_μ is the covariant derivative

$$D_\mu = \partial_\mu + ieA_\mu.$$

Note that up to this point the regularization, unlike dimensional or lattice regularizations, preserves a possible chiral symmetry for $m = 0$.

The higher order covariant derivatives, however, generate new, more singular, gauge interactions and it is no longer clear whether the theory can be rendered finite.

Fermion correlation functions in the gauge background are generated by

$$\mathcal{Z}(\bar\eta, \eta; A) = \int \left[d\psi(x)d\bar\psi(x)\right]$$
$$\times \exp\left[-S(\bar\psi, \psi, A) + \int d^dx\, \left(\bar\eta(x)\psi(x) + \bar\psi(x)\eta(x)\right)\right], \quad (8)$$

where $\bar\eta, \eta$ are Grassmann sources. Integrating over fermions explicitly, one obtains

$$\mathcal{Z}(\bar\eta, \eta; A) = \mathcal{Z}_0(A) \exp\left[-\int d^dx\, d^dy\, \bar\eta(y)\Delta_{\rm F}(A; y, x)\eta(x)\right],$$
$$\mathcal{Z}_0(A) \quad = \mathcal{N} \det\left[(m + \not{D}) \prod_r \left(1 - \not{D}^2/M_r^2\right)\right],$$

where $\mathcal{N}$ is a gauge field-independent normalization ensuring $\mathcal{Z}_0(0) = 1$ and $\Delta_{\rm F}(A; y, x)$ the fermion propagator in an external gauge field.

Diagrams constructed from $\Delta_{\rm F}(A; y, x)$ belong to loops with gauge field propagators and, therefore, can be rendered finite if the gauge field propagator can be improved, a condition that we check below. The other problem involves the determinant, which generates closed fermion loops in a gauge background. Using $\ln \det = \operatorname{tr} \ln$, one finds

$$\ln \mathcal{Z}_0(A) = \operatorname{tr} \ln (m + \not{D}) + \sum_r \operatorname{tr} \ln \left(1 - \not{D}^2/M_r^2\right) - (A = 0),$$

or, using the anticommutation of γ_5 with $\not{D}$,

$$\det(\not{D} + m) = \det \gamma_5 (\not{D} + m)\gamma_5 = \det(m - \not{D}),$$

$$\ln \mathcal{Z}_0(A) = \tfrac{1}{2} \operatorname{tr} \ln \left(m^2 - \not{D}^2\right) + \sum_r \operatorname{tr} \ln \left(1 - \not{D}^2/M_r^2\right) - (A = 0).$$

One sees that the regularization has no effect, from the point of view of power counting, on the determinant because all contributions add. The determinant generates one-loop diagrams of the form of closed fermion loops with external gauge fields, which therefore require an additional regularization.

As an illustration, Fig. 1 displays on the first line two Feynman diagrams involving only $\Delta_F(A; y, x)$, and on the second line two diagrams involving the determinant.

The Fermion Determinant. Finally, the fermion determinant can be regularized by adding to the action a boson regulator field with fermion spin (unphysical since violating the spin–statisitics connection) and, therefore, a propagator similar to Δ_F but with different masses:

$$\mathcal{S}_B(\bar{\phi}, \phi; A) = \int d^d x \, \bar{\phi}(x) \left(M_0^B + \slashed{D}\right) \prod_{r=1} \left(1 - \slashed{D}^2/(M_r^B)^2\right) \phi(x).$$

The integration over the boson ghost fields $\bar{\phi}, \phi$ adds to $\ln \mathcal{Z}_0$ the quantity

$$\delta \ln \mathcal{Z}_0(A) = -\tfrac{1}{2} \operatorname{tr} \ln \left((M_0^B)^2 - \slashed{D}^2\right) - \sum_{r=1} \operatorname{tr} \ln \left(1 - \slashed{D}^2/(M_r^B)^2\right) - (A = 0).$$

Expanding the sum $\ln \mathcal{Z}_0 + \delta \ln \mathcal{Z}_0$ in inverse powers of $\slashed{D}$, one adjusts the masses to cancel as many powers of $\slashed{D}$ as possible. However, the unpaired initial fermion mass m is the source of a problem. The corresponding determinant can only be regularized with an unpaired boson M_0^B. In the chiral limit $m = 0$, two options are available: either one gives a chiral charge to the boson field and the mass M_0^B breaks chiral symmetry, or one leaves it invariant in a chiral transformation. In the latter case one finds the determinant of the transformed operator

$$e^{i\theta(x)\gamma_5} \slashed{D} e^{i\theta(x)\gamma_5} \left(\slashed{D} + M_0^B\right)^{-1}.$$

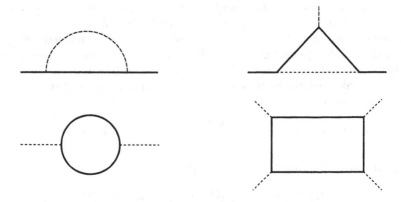

Fig. 1. Gauge–fermion diagrams (the fermions and gauge fields correspond to continuous and dotted lines, respectively).

For $\theta(x)$ constant $e^{i\theta\gamma_5}\not{D} = \not{D}e^{-i\theta\gamma_5}$ and the θ-dependence cancels. Otherwise a non-trivial contribution remains. The method thus suggests possible difficulties with space-dependent chiral transformations.

Actually, since the problem reduces to the study of a determinant in an external background, one can study it directly, as we will starting with in Sect. 4. One examines whether it is possible to define some regularized form in a way consistent with chiral symmetry. When this is possible, one then inserts the one-loop renormalized diagrams in the general diagrams regularized by the preceding cut-off methods.

The Boson Determinant in a Gauge Background. The boson determinant can be regularized by introducing a massive spinless charged fermion (again unphysical since violating the spin–statisitics connection). Alternatively, it can be expressed in terms of the statistical operator using Schwinger's representation ($\operatorname{tr}\ln = \ln\det$)

$$\ln\det H - \ln\det H_0 = \operatorname{tr}\int_0^\infty \frac{dt}{t}\left[e^{-tH_0} - e^{-tH}\right],$$

where the operator H is analogous to a non-relativistic hamiltonian in a magnetic field,

$$H = -D_\mu D_\mu + m^2, \quad H_0 = -\nabla^2 + m^2.$$

UV divergences then arise from the small t integration. The integral over time can thus be regularized by cutting it for t small, integrating for example over $t \geq 1/\Lambda^2$.

The Gauge Field Propagator. For the free gauge action in a covariant gauge, ordinary derivatives can be used because in an abelian theory the gauge field is neutral. The tensor $F_{\mu\nu}$ is gauge invariant and the action for the scalar combination $\partial_\mu A_\mu$ is arbitrary. Therefore, the large momentum behaviour of the gauge field propagator can be arbitrarily improved by the substitution

$$F_{\mu\nu}F_{\mu\nu} \mapsto F_{\mu\nu}P(\nabla^2/\Lambda^2)F_{\mu\nu},$$
$$(\partial_\mu A_\mu)^2 \mapsto \partial_\mu A_\mu P(\nabla^2/\Lambda^2)\partial_\mu A_\mu.$$

2.4   Non-Abelian Gauge Theories

Compared with the abelian case, the new features of the non-abelian gauge action are the presence of gauge field self-interactions and ghost terms. For future purpose we define our notation. We introduce the covariant derivative, as acting on a matter field,

$$\mathbf{D}_\mu = \partial_\mu + \mathbf{A}_\mu(x), \tag{9}$$

where $\mathbf{A}_\mu$ is an anti-hermitian matrix, and the curvature tensor

$$\mathbf{F}_{\mu\nu} = [\mathbf{D}_\mu, \mathbf{D}_\nu] = \partial_\mu\mathbf{A}_\nu - \partial_\nu\mathbf{A}_\mu + [\mathbf{A}_\mu, \mathbf{A}_\nu]. \tag{10}$$

The pure gauge action then is

$$S(\mathbf{A}_\mu) = -\frac{1}{4g^2} \int d^d x \ \mathrm{tr}\, \mathbf{F}_{\mu\nu}(x) \mathbf{F}_{\mu\nu}(x).$$

In the covariant gauge

$$S_{\text{gauge}}(\mathbf{A}_\mu) = -\frac{1}{2\xi} \int d^d x \ \mathrm{tr}(\partial_\mu \mathbf{A}_\mu)^2,$$

the ghost field action takes the form

$$S_{\text{ghost}}(\mathbf{A}_\mu, \bar{\mathbf{C}}, \mathbf{C}) = -\int d^d x \ \mathrm{tr}\, \bar{\mathbf{C}}\, \partial_\mu \left(\partial_\mu \mathbf{C} + [\mathbf{A}_\mu, \mathbf{C}] \right).$$

The ghost fields thus have a simple δ_{ab}/p^2 propagator and canonical dimension one in four dimensions.

The problem of regularization in non-abelian gauge theories shares several features both with the abelian case and with the non-linear σ-model. The regularized gauge action takes the form

$$S_{\text{reg.}}(\mathbf{A}_\mu) = -\int d^d x \ \mathrm{tr}\, \mathbf{F}_{\mu\nu} P\left(\mathbf{D}^2 / \Lambda^2 \right) \mathbf{F}_{\mu\nu},$$

in which P is a polynomial of arbitrary degree. In the same way, the gauge function $\partial_\mu \mathbf{A}_\mu$ is changed into

$$\partial_\mu \mathbf{A}_\mu \longmapsto Q\left(\partial^2 / \Lambda^2 \right) \partial_\mu \mathbf{A}_\mu,$$

in which Q is a polynomial of the same degree as P. As a consequence, both the gauge field propagator and the ghost propagator can be arbitrarily improved. However, as in the abelian case, the covariant derivatives generate new interactions that are more singular. It is easy to verify that the power counting of one-loop diagrams is unchanged while higher order diagrams can be made convergent by taking the degrees of P and Q large enough: Regularization by higher derivatives takes care of all diagrams except, as in non-linear σ models, some one-loop diagrams (and thus subdiagrams).

As with charged matter, the one-loop diagrams have to be examined separately. However, for fermion matter it is still possible as, in the abelian case, to add a set of regulator fields, massive fermions and bosons with fermion spin. In the chiral situation, the problem of the compatibility between the gauge symmetry and the quantization is reduced to an explicit verification of the Ward–Takahashi (WT) identities for the one-loop diagrams. Note that the preservation of gauge symmetry is necessary for the cancellation of unphysical states in physical amplitudes and, thus, essential to ensure the physical relevance of the quantum field theory.

3   Other Regularization Schemes

The other regularization schemes we now discuss, have the common property that they modify in some essential way the structure of space–time: dimensional regularization because it relies on defining Feynman diagrams for non-integer dimensions, lattice regularization because continuum space is replaced by a discrete lattice.

3.1 Dimensional Regularization

Dimensional regularization involves continuation of Feynman diagrams in the parameter d (d is the space dimension) to arbitrary complex values and, therefore, seems to have no meaning outside perturbation theory. However, this regularization very often leads to the simplest perturbative calculations.

In addition, it solves the problem of commutation of quantum operators in local field theories. Indeed commutators, for example in the case of a scalar field, take the form (in the Schrödinger picture)

$$[\hat{\phi}(x), \hat{\pi}(y)] = i\hbar\, \delta^{d-1}(x-y) = i\hbar(2\pi)^{1-d} \int \mathrm{d}^{d-1}p\, \mathrm{e}^{ip(x-y)},$$

where $\hat{\pi}(x)$ is the momentum conjugate to the field $\hat{\phi}(x)$. As we have already stressed, in a local theory all operators are taken at the same point and, therefore, a commutation in the product $\hat{\phi}(x)\hat{\pi}(x)$ generates a divergent contribution (for $d > 1$) proportional to

$$\delta^{d-1}(0) = (2\pi)^{1-d} \int \mathrm{d}^{d-1}p.$$

The rules of dimensional regularization imply the consistency of the change of variables $p \mapsto \lambda p$ and thus

$$\int \mathrm{d}^d p = \lambda^d \int \mathrm{d}^d p \;\Rightarrow\; \int \mathrm{d}^d p = 0,$$

in contrast to momentum regularization where it is proportional to a power of the cut-off. Therefore, the order between operators becomes irrelevant because the commutator vanishes. Dimensional regularization thus is applicable to geometric models where these problems of quantization occur, like non-linear σ models or gauge theories.

Its use, however, requires some care in massless theories. For instance, in a massless theory in two dimensions, integrals of the form $\int \mathrm{d}^d k/k^2$ are met. They also vanish in dimensional regularization for the same reason. However, here they correspond to an unwanted cancellation between UV and IR logarithmic divergences.

More important here, it is not applicable when some essential property of the field theory is specific to the initial dimension. An example is provided by theories containing fermions in which Parity symmetry is violated.

Fermions. For fermions transforming under the fundamental representation of the spin group $\mathrm{Spin}(d)$, the strategy is the same. The evaluation of diagrams with fermions can be reduced to the calculation of traces of γ matrices. Therefore, only one additional prescription for the trace of the unit matrix is needed. There is no natural continuation since odd and even dimensions behave differently. Since no algebraic manipulation depends on the explicit value of the trace, any smooth continuation in the neighbourhood of the relevant dimension is satisfactory. A convenient choice is to take the trace constant. In even dimensions, as long as only γ_μ matrices are involved, no other problem arises. However, no dimensional continuation that preserves all properties of γ_{d+1}, which is the product of all other γ matrices, can be found. This leads to serious difficulties if γ_{d+1} in the calculation of Feynman diagrams has to be replaced by its explicit expression in terms of the other γ matrices. For example, in four dimensions γ_5 is related to the other γ matrices by

$$4!\,\gamma_5 = -\epsilon_{\mu_1\ldots\mu_4}\gamma_{\mu_1}\cdots\gamma_{\mu_4}\,, \qquad (11)$$

where $\epsilon_{\mu_1\cdots\mu_4}$ is the complete antisymmetric tensor with $\epsilon_{1234} = 1$. Therefore, problems arise in the case of gauge theories with chiral fermions, because the special properties of γ_5 are involved as we recall below. This difficulty is the source of chiral anomalies.

Since perturbation theory involves the calculation of traces, one possibility is to define γ_5 near four dimensions by

$$\gamma_5 = E_{\mu_1\ldots\mu_4}\gamma_{\mu_1}\cdots\gamma_{\mu_4}\,, \qquad (12)$$

where $E_{\mu\nu\rho\sigma}$ is a completely antisymmetric tensor, which reduces to $-\epsilon_{\mu\nu\rho\sigma}/4!$ in four dimensions. It is easy to then verify that, with this definition, γ_5 anticommutes with the other γ_μ matrices only in four dimensions. If, for example, one evaluates the product $\gamma_\nu\gamma_5\gamma_\nu$ in d dimensions, replacing γ_5 by (12) and using systematically the anticommutation relations $\gamma_\mu\gamma_\nu + \gamma_\nu\gamma_\mu = 2\delta_{\mu\nu}$, one finds

$$\gamma_\nu\gamma_5\gamma_\nu = (d-8)\gamma_5\,.$$

Anticommuting properties of the γ_5 would have led to a factor $-d$, instead. This additional contribution, proportional to $d-4$, if it is multiplied by a factor $1/(d-4)$ consequence of UV divergences in one-loop diagrams, will lead to a finite difference with the formal result.

The other option would be to keep the anticommuting property of γ_5 but the preceding example shows that this is contradictory with a form (12). Actually, one verifies that the only consistent prescription for generic dimensions then is that the traces of γ_5 with any product of γ_μ matrices vanishes and, thus, this prescription is useless.

Finally, an alternative possibility consists in breaking $O(d)$ symmetry and keeping the four γ matrices of $d = 4$.

3.2   Lattice Regularization

We have explained that Pauli–Villars's regularization does not provide a complete regularization for field theories in which the geometric properties generate interactions like models where fields belong to homogeneous spaces (e.g. the non-linear σ-model) or non-abelian gauge theories. In these theories some divergences are related to the problem of quantization and order of operators, which already appears in simple quantum mechanics. Other regularization methods are then needed. In many cases lattice regularization may be used.

Lattice Field Theory. To each site x of a lattice are attached field variables corresponding to fields in the continuum. To the action S in the continuum corresponds a lattice action, the energy of lattice field configurations in the language of classical statistical physics. The functional integral becomes a sum over configurations and the regularized partition function is the partition function of a lattice model.

All expressions in these notes will refer implicitly to a hypercubic lattice and we denote the lattice spacing by a.

The advantages of lattice regularization are:

(i) Lattice regularization indeed corresponds to a specific choice of quantization.

(ii) It is the only established regularization that for gauge theories and other geometric models has a meaning outside perturbation theory. For instance the regularized functional integral can be calculated by numerical methods, like stochastic methods (Monte-Carlo type simulations) or strong coupling expansions.

(iii) It preserves most global and local symmetries with the exception of the space $O(d)$ symmetry, which is replaced by a hypercubic symmetry (but this turns out not to be a major difficulty), and fermion chirality, which turns out to be a more serious problem, as we will show.

The main disadvantage is that it leads to rather complicated perturbative calculations.

3.3 Boson Field Theories

Scalar Fields. To the action (1) for a scalar field ϕ in the continuum corresponds a lattice action, which is obtained in the following way: The euclidean lagrangian density becomes a function of lattice variables $\phi(x)$, where x now is a lattice site. Locality can be implemented by considering lattice lagrangians that depend only on a site and its neighbours (though this is a too strong requirement; lattice interactions decreasing exponentially with distance are also local). Derivatives $\partial_\mu \phi$ of the continuum are replaced by finite differences, for example:

$$\partial_\mu \phi \mapsto \nabla_\mu^{\text{lat.}} \phi = [\phi(x + a n_\mu) - \phi(x)]/a, \tag{13}$$

where a is the lattice spacing and n_μ the unit vector in the μ direction. The lattice action then is the sum over lattice sites.

With the choice (13), the propagator $\Delta_a(p)$ for the Fourier components of a massive scalar field is given by

$$\Delta_a^{-1}(p) = m^2 + \frac{2}{a^2} \sum_{\mu=1}^{d} (1 - \cos(ap_\mu)).$$

It is a periodic function of the components p_μ of the momentum vector with period $2\pi/a$. In the small lattice spacing limit, the continuum propagator is recovered:

$$\Delta_a^{-1}(p) = m^2 + p^2 - \tfrac{1}{12} \sum_\mu a^2 p_\mu^4 + O\left(p_\mu^6\right).$$

In particular, hypercubic symmetry implies $O(d)$ symmetry at order p^2.

Gauge Theories. Lattice regularization defines unambiguously a quantum theory. Therefore, once one has realized that gauge fields should be replaced by link variables corresponding to parallel transport along links of the lattice, one can regularize a gauge theory.

The link variables $\mathbf{U}_{xy}$ are group elements associated with the links joining the sites x and y on the lattice. The regularized form of $\int \mathrm{d}x\, F_{\mu\nu}^2$ is a sum of products of link variables along closed curves on the lattice. On a hypercubic lattice, the smallest curve is a square leading to the well-known plaquette action (each square forming a plaquette). The typical gauge invariant lattice action corresponding to the continuum action of a gauge field coupled to scalar bosons then has the form

$$S(\mathbf{U}, \phi^*, \phi) = \beta \sum_{\substack{\text{all} \\ \text{plaquettes}}} \operatorname{tr} \mathbf{U}_{xy} \mathbf{U}_{yz} \mathbf{U}_{zt} \mathbf{U}_{tx} + \kappa \sum_{\substack{\text{all} \\ \text{links}}} \phi_x^* \mathbf{U}_{xy} \phi_y + \sum_{\substack{\text{all} \\ \text{sites}}} V(\phi_x^* \phi_x),$$

(14)

where x, y,... denotes lattice sites, and β and κ are coupling constants. The action (14) is invariant under independent group transformations on each lattice site, lattice equivalents of gauge transformations in the continuum theory. The measure of integration over the gauge variables is the group invariant measure on each site. Note that on the lattice and in a finite volume, the gauge invariant action leads to a well-defined partition function because the gauge group (finite product of compact groups) is compact. However, in the continuum or infinite volume limits the compact character of the group is lost. Even on the lattice, regularized perturbation theory is defined only after gauge fixing.

Finally, we note that, on the lattice, the difficulties with the regularization do not come from the gauge field directly but involve the gauge field only through the integration over chiral fermions.

3.4   Fermions and the Doubling Problem

We now review a few problems specific to relativistic fermions on the lattice. We consider the free action for a Dirac fermion

$$S(\bar\psi, \psi) = \int \mathrm{d}^d x\, \bar\psi(x)\, (\partial\!\!\!/ + m)\, \psi(x).$$

A lattice regularization of the derivative $\partial_\mu \psi(x)$, which preserves chiral properties in the massless limit, is, for example, the symmetric combination

$$\nabla_\mu^{\text{lat.}} \psi(x) = [\psi(x + an_\mu) - \psi(x - an_\mu)]/2a.$$

In the boson case, there is no equivalent constraint and thus a possible choice is the expression 13.

The lattice Dirac operator for the Fourier components $\tilde{\psi}(p)$ of the field (inverse of the fermion propagator $\Delta_{\text{lat.}}(p)$) is

$$D_{\text{lat.}}(p) = m + i \sum_\mu \gamma_\mu \frac{\sin ap_\mu}{a}, \tag{15}$$

a periodic function of the components p_μ of the momentum vector. A problem then arises: the equations relevant to the small lattice spacing limit,

$$\sin(a\,p_\mu) = 0,$$

have each two solutions $p_\mu = 0$ and $p_\mu = \pi/a$ within one period, that is 2^d solutions within the Brillouin zone. Therefore, the propagator (15) propagates 2^d fermions. To remove this degeneracy, it is possible to add to the regularized action an additional scalar term δS involving second derivatives:

$$\delta S(\bar{\psi}, \psi) = \tfrac{1}{2} M \sum_{x,\mu} \left[2\bar{\psi}(x)\psi(x) - \bar{\psi}(x + an_\mu)\psi(x) - \bar{\psi}(x)\psi(x + an_\mu) \right]. \tag{16}$$

The modified Dirac operator for the Fourier components of the field reads

$$D_W(p) = m + M \sum_\mu (1 - \cos ap_\mu) + \frac{i}{a} \sum_\mu \gamma_\mu \sin ap_\mu. \tag{17}$$

The fermion propagator becomes

$$\Delta(p) = D_W^\dagger(p) \left(D_W(p) D_W^\dagger(p) \right)^{-1}$$

with

$$D_W(p) D_W^\dagger(p) = \left[m + M \sum_\mu (1 - \cos ap_\mu) \right]^2 + \frac{1}{a^2} \sum_\mu \sin^2 ap_\mu.$$

Therefore, the degeneracy between the different states has been lifted. For each component p_μ that takes the value π/a the mass is increased by $2M$. If M is of order $1/a$ the spurious states are eliminated in the continuum limit. This is the recipe of Wilson's fermions.

However, a problem arises if one wants to construct a theory with massless fermions and chiral symmetry. Chiral symmetry implies that the Dirac operator $D(p)$ anticommutes with γ_5:

$$\{D(p), \gamma_5\} = 0,$$

and, therefore, both the mass term and the term (16) are excluded. It remains possible to add various counter-terms and try to adjust them to recover chiral symmetry in the continuum limit. But there is no *a priori* guarantee that this is indeed possible and, moreover, calculations are plagued by fine tuning problems and cancellations of unnecessary UV divergences.

One could also think about modifying the fermion propagator by adding terms connecting fermions separated by more than one lattice spacing. But it has been proven that this does not solve the doubling problem. (Formal solutions can be exhibited but they violate locality that implies that $D(p)$ should be a smooth periodic function.) In fact, this doubling of the number of fermion degrees of freedom is directly related to the problem of anomalies.

Since the most naive form of the propagator yields 2^d fermion states, one tries in practical calculations to reduce this number to a smaller multiple of two, using for instance the idea of staggered fermions introduced by Kogut and Susskind.

However, the general picture has recently changed with the discovery of the properties of overlap fermions and solutions of the Ginsparg–Wilson relation or domain wall fermions, a topic we postpone and we will study in Sect. 7.

4   The Abelian Anomaly

We have pointed out that none of the standard regularization methods can deal in a straightforward way with one-loop diagrams in the case of gauge fields coupled to chiral fermions. We now show that indeed chiral symmetric gauge theories, involving gauge fields coupled to massless fermions, can be found where the axial current is not conserved. The divergence of the axial current in a chiral quantum field theory, when it does not vanish, is called an *anomaly*. Anomalies in particular lead to obstructions to the construction of gauge theories when the gauge field couples differently to the two fermion chiral components.

Several examples are physically important like the theory of weak electromagnetic interactions, the electromagnetic decay of the π_0 meson, or the $U(1)$ problem. We first discuss the abelian axial current, in four dimensions (the generalization to all even dimensions then is straightforward), and then the general non-abelian situation.

4.1   Abelian Axial Current and Abelian Vector Gauge Fields

The only possible source of anomalies are one-loop fermion diagrams in gauge theories when chiral properties are involved. This reduces the problem to the discussion of fermions in background gauge fields or, equivalently, to the properties of the determinant of the gauge covariant Dirac operator.

We thus consider a QED-like fermion action for massless Dirac fermions $\psi, \bar{\psi}$ in the background of an abelian gauge field A_μ of the form

$$\mathcal{S}(\bar{\psi}, \psi; A) = -\int \mathrm{d}^4x\, \bar{\psi}(x) \slashed{D} \psi(x), \quad \slashed{D} \equiv \slashed{\partial} + ie\slashed{A}(x), \tag{18}$$

and the corresponding functional integral

$$\mathcal{Z}(A_\mu) = \int [\mathrm{d}\psi \mathrm{d}\bar\psi] \exp\left[-\mathcal{S}(\psi,\bar\psi;A)\right] = \det \not{D}.$$

We can find regularizations that preserve gauge invariance, that is invariance under the transformations

$$\psi(x) = e^{i\Lambda(x)}\psi'(x), \quad \bar\psi(x) = e^{-i\Lambda(x)}\bar\psi'(x), \quad A_\mu(x) = -\frac{1}{e}\partial_\nu\Lambda(x) + A'_\mu(x),$$
(19)

and, since the fermions are massless, chiral symmetry. Therefore, we would naively expect the corresponding axial current to be conserved (symmetries are generally related to current conservation). However, the proof of current conservation involves space-dependent chiral transformations and, therefore, steps that cannot be regularized without breaking local chiral symmetry.

Under the space-dependent chiral transformation

$$\psi_\theta(x) = e^{i\theta(x)\gamma_5}\psi(x), \qquad \bar\psi_\theta(x) = \bar\psi(x)e^{i\theta(x)\gamma_5},$$
(20)

the action becomes

$$\mathcal{S}_\theta(\bar\psi,\psi;A) = -\int \mathrm{d}^4x\, \bar\psi_\theta(x)\not{D}\psi_\theta(x) = \mathcal{S}(\bar\psi,\psi;A) + \int \mathrm{d}^4x\, \partial_\mu\theta(x)J_\mu^5(x),$$

where $J_\mu^5(x)$, the coefficient of $\partial_\mu\theta$, is the axial current:

$$J_\mu^5(x) = i\bar\psi(x)\gamma_5\gamma_\mu\psi(x).$$

After the transformation 20, $\mathcal{Z}(A_\mu)$ becomes

$$\mathcal{Z}(A_\mu,\theta) = \det\left[e^{i\gamma_5\theta(x)}\not{D}e^{i\gamma_5\theta(x)}\right].$$

Note that $\ln[\mathcal{Z}(A_\mu,\theta)]$ is the generating functional of connected $\partial_\mu J_\mu^5$ correlation functions in an external field A_μ.

Since $e^{i\gamma_5\theta}$ has a determinant that is unity, one would naively conclude that $\mathcal{Z}(A_\mu,\theta) = \mathcal{Z}(A_\mu)$ and, therefore, that the current $J_\mu^5(x)$ is conserved. This is a conclusion we now check by an explicit calculation of the expectation value of $\partial_\mu J_\mu^5(x)$ in the case of the action 18.

Remarks.

(i) For any regularization that is consistent with the hermiticity of γ_5

$$|\mathcal{Z}(A_\mu,\theta)|^2 = \det\left[e^{i\gamma_5\theta(x)}\not{D}e^{i\gamma_5\theta(x)}\right]\det\left[e^{-i\gamma_5\theta(x)}\not{D}^\dagger e^{-i\gamma_5\theta(x)}\right] = \det\left(\not{D}\not{D}^\dagger\right),$$

and thus $|\mathcal{Z}(A_\mu,\theta)|$ is independent of θ. Therefore, an anomaly can appear only in the imaginary part of $\ln \mathcal{Z}$.

(ii) We have shown that one can find a regularization with regulator fields such that gauge invariance is maintained, and the determinant is independent of θ for $\theta(x)$ constant.

(iii) If the regularization is gauge invariant, $\mathcal{Z}(A_\mu, \theta)$ is also gauge invariant. Therefore, a possible anomaly will also be gauge invariant.

(iv) $\ln \mathcal{Z}(A_\mu, \theta)$ receives only connected, 1PI contributions. Short distance singularities coming from one-loop diagrams thus take the form of local polynomials in the fields and sources. Since a possible anomaly is a short distance effect (equivalently a large momentum effect), it must also take the form of a local polynomial of A_μ and $\partial_\mu \theta$ constrained by parity and power counting. The field A_μ and $\partial_\mu \theta$ have dimension 1 and no mass parameter is available. Thus,

$$\ln \mathcal{Z}(A_\mu, \theta) - \ln \mathcal{Z}(A_\mu, 0) = i \int \mathrm{d}^4 x \, \mathcal{L}(A, \partial \theta; x),$$

where $\mathcal{L}$ is the sum of monomials of dimension 4. At order θ only one is available:

$$\mathcal{L}(A, \partial \theta; x) \propto e^2 \epsilon_{\mu\nu\rho\sigma} \partial_\mu \theta(x) A_\nu(x) \partial_\rho A_\sigma(x),$$

where $\epsilon_{\mu\nu\rho\sigma}$ is the complete antisymmetric tensor with $\epsilon_{1234} = 1$. A simple integration by parts and anti-symmetrization shows that

$$\int \mathrm{d}^4 x \, \mathcal{L}(A, \partial \theta; x) \propto e^2 \epsilon_{\mu\nu\rho\sigma} \int \mathrm{d}^4 x \, F_{\mu\nu}(x) F_{\rho\sigma}(x) \theta(x),$$

where $F_{\mu\nu} = \partial_\mu A_\nu - \partial_\nu A_\mu$ is the electromagnetic tensor, an expression that is gauge invariant.

The coefficient of $\theta(x)$ is the expectation value in an external gauge field of $\partial_\mu J_\mu^5(x)$, the divergence of the axial current. It is determined up to a multiplicative constant:

$$\langle \partial_\lambda J_\lambda^5(x) \rangle \propto e^2 \epsilon_{\mu\nu\rho\sigma} \partial_\mu A_\nu(x) \partial_\rho A_\sigma(x) \propto e^2 \epsilon_{\mu\nu\rho\sigma} F_{\mu\nu}(x) F_{\rho\sigma}(x),$$

where we denote by $\langle \bullet \rangle$ expectation values with respect to the measure $e^{-S(\bar\psi, \psi; A)}$.

Since the possible anomaly is independent up to a multiplicative factor of the regularization, it must indeed be a gauge invariant local function of A_μ.

To find the multiplicative factor, which is the only regularization dependent feature, it is sufficient to calculate the coefficient of the term quadratic in A in the expansion of $\langle \partial_\lambda J_\lambda^5(x) \rangle$ in powers of A. We define the three-point function in momentum representation by

$$\Gamma_{\lambda\mu\nu}^{(3)}(k; p_1, p_2) = \frac{\delta}{\delta A_\mu(p_1)} \frac{\delta}{\delta A_\nu(p_2)} \langle J_\lambda^5(k) \rangle \Big|_{A=0}, \tag{21}$$

$$= \frac{\delta}{\delta A_\mu(p_1)} \frac{\delta}{\delta A_\nu(p_2)} i \, \mathrm{tr}\left[\gamma_5 \gamma_\lambda \slashed{D}^{-1}(k)\right]\Big|_{A=0}.$$

$\Gamma^{(3)}$ is the sum of the two Feynman diagrams of Fig. 2.

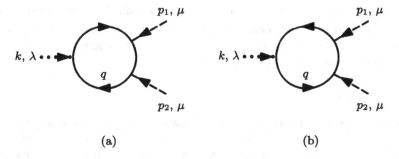

Fig. 2. Anomalous diagrams.

The contribution of diagram (a) is:

$$(a) \mapsto \frac{e^2}{(2\pi)^4} \operatorname{tr} \left[\int d^4q\, \gamma_5 \gamma_\lambda \, (\slashed{q} + \slashed{k})^{-1} \gamma_\mu \, (\slashed{q} - \slashed{p}_2)^{-1} \gamma_\nu \slashed{q}^{-1} \right], \qquad (22)$$

and the contribution of diagram (b) is obtained by exchanging $p_1, \gamma_\mu \leftrightarrow p_2, \gamma_\nu$.

Power counting tells us that the function $\Gamma^{(3)}$ may have a linear divergence that, due to the presence of the γ_5 factor, must be proportional to $\epsilon_{\lambda\mu\nu\rho}$, symmetric in the exchange $p_1, \gamma_\mu \leftrightarrow p_2, \gamma_\nu$, and thus proportional to

$$\epsilon_{\lambda\mu\nu\rho} \, (p_1 - p_2)_\rho \, . \qquad (23)$$

On the other hand, by commuting γ_5 in (22), we notice that $\Gamma^{(3)}$ is formally a symmetric function of the three sets of external arguments. A divergence, being proportional to (23), which is not symmetric, breaks the symmetry between external arguments. Therefore, a symmetric regularization, of the kind we adopt in the first calculation, leads to a finite result. The result is not ambiguous because a possible ambiguity again is proportional to (23).

Similarly, if the regularization is consistent with gauge invariance, the vector current is conserved:

$$p_{1\mu} \Gamma^{(3)}_{\lambda\mu\nu}(k; p_1, p_2) = 0 \, .$$

Applied to a possible divergent contribution, the equation implies

$$-p_{1\mu} p_{2\rho} \epsilon_{\lambda\mu\nu\rho} = 0 \, ,$$

which cannot be satisfied for arbitrary p_1, p_2. Therefore, the sum of the two diagrams is finite. Finite ambiguities must also have the form (23) and thus are also forbidden by gauge invariance. All regularizations consistent with gauge invariance must give the same answer.

Therefore, there are two possibilities:

(i) $k_\lambda \Gamma^{(3)}_{\lambda\mu\nu}(k; p_1, p_2)$ in a regularization respecting the symmetry between the three arguments vanishes. Then both $\Gamma^{(3)}$ is gauge invariant and the axial current is conserved.

(ii) $k_\lambda \Gamma^{(3)}_{\lambda\mu\nu}(k; p_1, p_2)$ in a symmetric regularization does not vanish. Then it is possible to add to $\Gamma^{(3)}$ a term proportional to (23) to restore gauge invariance but this term breaks the symmetry between external momenta: the axial current is not conserved and an anomaly is present.

4.2   Explicit Calculation

Momentum Regularization. The calculation can be done using one of the various gauge invariant regularizations, for example, Momentum cut-off regularization or dimensional regularization with γ_5 being defined as in dimension four and thus no longer anticommuting with other γ matrices. Instead, we choose a regularization that preserves the symmetry between the three external arguments and global chiral symmetry, but breaks gauge invariance. We modify the fermion propagator as

$$(\slashed{q})^{-1} \longmapsto (\slashed{q})^{-1}\rho(\varepsilon q^2),$$

where ε is the regularization parameter ($\varepsilon \to 0_+$), $\rho(z)$ is a positive differentiable function such that $\rho(0) = 1$, and decreasing fast enough for $z \to +\infty$, at least like $1/z$.

Then, as we have argued, current conservation and gauge invariance are compatible only if $k_\lambda \Gamma^{(3)}_{\lambda\mu\nu}(k; p_1, p_2)$ vanishes.

It is convenient to consider directly the contribution $C^{(2)}(k)$ of order A^2 to $\langle k_\lambda J^5_\lambda(k)\rangle$, which sums the two diagrams:

$$C^{(2)}(k) = e^2 \int_{p_1+p_2+k=0} d^4p_1\, d^4p_2\, A_\mu(p_1)A_\nu(p_2) \int \frac{d^4q}{(2\pi)^4}\rho\big(\varepsilon(q+k)^2\big)$$
$$\times \rho\big(\varepsilon(q-p_2)^2\big)\rho\big(\varepsilon q^2\big)\, \mathrm{tr}\left[\gamma_5\slashed{k}(\slashed{q}+\slashed{k})^{-1}\gamma_\mu(\slashed{q}-\slashed{p}_2)^{-1}\gamma_\nu\slashed{q}^{-1}\right],$$

because the calculation then suggests how the method generalizes to arbitrary even dimensions.

We first transform the expression, using the identity

$$\slashed{k}(\slashed{q}+\slashed{k})^{-1} = 1 - \slashed{q}(\slashed{q}+\slashed{k})^{-1}. \tag{24}$$

Then,

$$C^{(2)}(k) = C_1^{(2)}(k) + C_2^{(2)}(k)$$

with

$$C_1^{(2)}(k) = e^2 \int_{p_1+p_2+k=0} d^4p_1\, d^4p_2\, A_\mu(p_1)A_\nu(p_2) \int \frac{d^4q}{(2\pi)^4}\rho\big(\varepsilon(q+k)^2\big)$$
$$\times \rho\big(\varepsilon(q-p_2)^2\big)\rho\big(\varepsilon q^2\big)\, \mathrm{tr}\left[\gamma_5\gamma_\mu(\slashed{q}-\slashed{p}_2)^{-1}\gamma_\nu\slashed{q}^{-1}\right]$$

and

$$C_2^{(2)}(k) = -e^2 \int_{p_1+p_2+k=0} d^4p_1\, d^4p_2\, A_\mu(p_1)A_\nu(p_2) \int \frac{d^4q}{(2\pi)^4}\rho\big(\varepsilon(q+k)^2\big)$$
$$\times \rho\big(\varepsilon(q-p_2)^2\big)\rho\big(\varepsilon q^2\big)\, \mathrm{tr}\left[\gamma_5\slashed{q}(\slashed{q}+\slashed{k})^{-1}\gamma_\mu(\slashed{q}-\slashed{p}_2)^{-1}\gamma_\nu\slashed{q}^{-1}\right].$$

In $C_2^{(2)}(k)$ we use the cyclic property of the trace and the *commutation* of $\gamma_\nu \slashed{q}^{-1}$ and γ_5 to cancel the propagator $\slashed{q}^{-1}$ and obtain

$$C_2^{(2)}(k) = -e^2 \int_{p_1+p_2+k=0} d^4p_1\, d^4p_2\, A_\mu(p_1)A_\nu(p_2) \int \frac{d^4q}{(2\pi)^4} \rho\big(\varepsilon(q+k)^2\big)$$
$$\times \rho\big(\varepsilon(q-p_2)^2\big)\rho\big(\varepsilon q^2\big)\, \mathrm{tr}\,\big[\gamma_5\gamma_\nu(\slashed{q}+\slashed{k})^{-1}\gamma_\mu(\slashed{q}-\slashed{p}_2)^{-1}\big].$$

We then shift $q \mapsto q + p_2$ and interchange (p_1,μ) and (p_2,ν),

$$C_2^{(2)}(k) = -e^2 \int_{p_1+p_2+k=0} d^4p_1\, d^4p_2\, A_\mu(p_1)A_\nu(p_2) \int \frac{d^4q}{(2\pi)^4} \rho\big(\varepsilon(q-p_2)^2\big)$$
$$\times \rho\big(\varepsilon q^2\big)\rho\big(\varepsilon(q+p_1)^2\big)\, \mathrm{tr}\,\big[\gamma_5\gamma_\mu(\slashed{q}-\slashed{p}_2)^{-1}\gamma_\nu\slashed{q}^{-1}\big]. \tag{25}$$

We see that the two terms $C_1^{(2)}$ and $C_2^{(2)}$ would cancel in the absence of regulators. This would correspond to the formal proof of current conservation. However, without regularization the integrals diverge and these manipulations are not legitimate.

Instead, here we find a non-vanishing sum due to the difference in regulating factors:

$$C^{(2)}(k) = e^2 \int_{p_1+p_2+k=0} d^4p_1\, d^4p_2\, A_\mu(p_1)A_\nu(p_2) \int \frac{d^4q}{(2\pi)^4} \rho\big(\varepsilon(q-p_2)^2\big)\rho\big(\varepsilon q^2\big)$$
$$\times \mathrm{tr}\,\big[\gamma_5\gamma_\mu(\slashed{q}-\slashed{p}_2)^{-1}\gamma_\nu\slashed{q}^{-1}\big]\big[\rho\big(\varepsilon(q+k)^2\big) - \rho\big(\varepsilon(q+p_1)^2\big)\big].$$

After evaluation of the trace, $C^{(2)}$ becomes (using (11))

$$C^{(2)}(k) = -4e^2 \int_{p_1+p_2+k=0} d^4p_1\, d^4p_2\, A_\mu(p_1)A_\nu(p_2) \int \frac{d^4q}{(2\pi)^4} \rho\big(\varepsilon(q-p_2)^2\big)\rho\big(\varepsilon q^2\big)$$
$$\times \epsilon_{\mu\nu\rho\sigma} \frac{p_{2\rho}q_\sigma}{q^2(q-p_2)^2}\big[\rho\big(\varepsilon(q+k)^2\big) - \rho\big(\varepsilon(q+p_1)^2\big)\big].$$

Contributions coming from finite values of q cancel in the $\varepsilon \to 0$ limit. Due to the cut-off, the relevant values of q are of order $\varepsilon^{-1/2}$. Therefore, we rescale q accordingly, $q\varepsilon^{1/2} \mapsto q$, and find

$$C^{(2)}(k) = -4e^2 \int_{p_1+p_2+k=0} d^4p_1\, d^4p_2\, A_\mu(p_1)A_\nu(p_2) \int \frac{d^4q}{(2\pi)^4} \rho\big((q-p_2\sqrt{\varepsilon})^2\big)$$
$$\times \rho(q^2)\epsilon_{\mu\nu\rho\sigma} \frac{p_{2\rho}q_\sigma}{q^2(q-p_2\sqrt{\varepsilon})^2} \frac{\rho\big((q+k\sqrt{\varepsilon})^2\big) - \rho\big((q+p_1\sqrt{\varepsilon})^2\big)}{\sqrt{\varepsilon}}.$$

Taking the $\varepsilon \to 0$ limit, we obtain the finite result

$$C^{(2)}(k) = -4e^2\epsilon_{\mu\nu\rho\sigma} \int_{p_1+p_2+k=0} d^4p_1\, d^4p_2\, A_\mu(p_1)A_\nu(p_2)I_{\rho\sigma}(p_1,p_2)$$

with

$$I_{\rho\sigma}(p_1,p_2) \sim \int \frac{d^4q}{(2\pi)^4 q^4} p_{2\rho}q_\sigma \rho^2(q^2)\rho'(q^2)\,[2q_\lambda(k-p_1)_\lambda].$$

The identity

$$\int d^4q \, q_\alpha q_\beta f(q^2) = \tfrac{1}{4}\delta_{\alpha\beta} \int d^4q \, q^2 f(q^2)$$

transforms the integral into

$$I_{\rho\sigma}(p_1, p_2) \sim -\tfrac{1}{2} p_{2\rho}(2p_1 + p_2)_\sigma \int \frac{\varepsilon d^4 q}{(2\pi)^4 q^2} \rho^2(q^2)\rho'(q^2).$$

The remaining integral can be calculated explicitly (we recall $\rho(0) = 1$):

$$\int \frac{d^4q}{(2\pi)^4 q^2} \rho^2(q^2)\rho'(q^2) = \frac{1}{8\pi^2} \int_0^\infty q dq \, \rho^2(q^2)\rho'(q^2) = -\frac{1}{48\pi^2},$$

and yields a result independent of the function ρ. We finally obtain

$$\langle k_\lambda J_\lambda^5(k) \rangle = -\frac{e^2}{12\pi^2} \epsilon_{\mu\nu\rho\sigma} \int d^4p_1 \, d^4p_2 \, p_{1\mu} A_\nu(p_1) p_{2\rho} A_\sigma(p_2) \tag{26}$$

and, therefore, from the definition (21):

$$k_\lambda \Gamma_{\lambda\mu\nu}^{(3)}(k; p_1, p_2) = \frac{e^2}{6\pi^2} \epsilon_{\mu\nu\rho\sigma} p_{1\rho} p_{2\sigma} .$$

This non-vanishing result implies that any definition of the determinant $\det \rlap{/}D$ breaks at least either axial current conservation or gauge invariance. Since gauge invariance is essential to the consistency of a gauge theory, we choose to break axial current conservation. Exchanging arguments, we obtain the value of

$$p_{1\mu}\Gamma_{\lambda\mu\nu}^{(3)}(k; p_1, p_2) = \frac{e^2}{6\pi^2} \epsilon_{\lambda\nu\rho\sigma} k_\rho p_{2\sigma} .$$

Instead, if we had used a gauge invariant regularization, the result for $\Gamma^{(3)}$ would have differed by a term $\delta\Gamma^{(3)}$ proportional to (23):

$$\delta\Gamma_{\lambda\mu\nu}^{(3)}(k; p_1, p_2) = K\epsilon_{\lambda\mu\nu\rho}(p_1 - p_2)_\rho .$$

The constant K then is determined by the condition of gauge invariance

$$p_{1\mu} \left[\Gamma_{\lambda\mu\nu}^{(3)}(k; p_1, p_2) + \delta\Gamma_{\lambda\mu\nu}^{(3)}(k; p_1, p_2) \right] = 0,$$

which yields

$$p_{1\mu}\delta\Gamma_{\lambda\mu\nu}^{(3)}(k; p_1, p_2) = -\frac{e^2}{6\pi^2} \epsilon_{\lambda\nu\rho\sigma} k_\rho p_{2\sigma} \quad \Rightarrow \quad K = e^2/(6\pi^2).$$

This gives an additional contribution to the divergence of the current

$$k_\lambda \delta\Gamma_{\lambda\mu\nu}^{(3)}(k; p_1, p_2) = \frac{e^2}{3\pi^2} \epsilon_{\mu\lambda\rho\sigma} p_{1\rho} p_{2\sigma} .$$

Therefore, in a QED-like gauge invariant field theory with massless fermions, the axial current is not conserved: this is called the chiral anomaly. For any gauge invariant regularization, one finds

$$k_\lambda \Gamma^{(3)}_{\lambda\mu\nu}(k; p_1, p_2) = \left(\frac{e^2}{2\pi^2} \equiv \frac{2\alpha}{\pi} \right) \epsilon_{\mu\nu\rho\sigma} p_{1\rho} p_{2\sigma} \,, \tag{27}$$

where α is the fine stucture constant. After Fourier transformation, (27) can be rewritten as an axial current non-conservation equation:

$$\langle \partial_\lambda J^5_\lambda(x) \rangle = -i\frac{\alpha}{4\pi} \epsilon_{\mu\nu\rho\sigma} F_{\mu\nu}(x) F_{\rho\sigma}(x) \,. \tag{28}$$

Since global chiral symmetry is not broken, the integral over the whole space of the anomalous term must vanish. This condition is indeed verified since the anomaly can immediately be written as a total derivative:

$$\epsilon_{\mu\nu\rho\sigma} F_{\mu\nu} F_{\rho\sigma} = 4\partial_\mu (\epsilon_{\mu\nu\rho\sigma} A_\nu \partial_\rho A_\sigma).$$

The space integral of the anomalous term depends only on the behaviour of the gauge field at the boundaries, and this property already indicates a connection between *topology and anomalies*.

Equation (28) also implies

$$\ln \det \left[e^{i\gamma_5 \theta(x)} \not{D} e^{i\gamma_5 \theta(x)} \right] = \ln \det \not{D} - i\frac{\alpha}{4\pi} \int \mathrm{d}^4 x \, \theta(x) \epsilon_{\mu\nu\rho\sigma} F_{\mu\nu}(x) F_{\rho\sigma}(x). \tag{29}$$

Remark. One might be surprised that in the calculation the divergence of the axial current does not vanish, though the regularization of the fermion propagator seems to be consistent with chiral symmetry. The reason is simple: if we add for example higher derivative terms to the action, the form of the axial current is modified and the additional contributions cancel the term we have found.

In the form we have organized the calculation, it generalizes without difficulty to general even dimensions $2n$. Note simply that the permutation $(\mathbf{p}_1, \mu) \leftrightarrow (\mathbf{p}_2, \nu)$ in (25) is replaced by a cyclic permutation. If gauge invariance is maintained, the anomaly in the divergence of the axial current $J^S_\lambda(x)$ in general is

$$\langle \partial_\lambda J^S_\lambda(x) \rangle = -2i\frac{e^n}{(4\pi)^n n!} \epsilon_{\mu_1\nu_1\ldots\mu_n\nu_n} F_{\mu_1\nu_1} \cdots F_{\mu_n\nu_n} \,, \tag{30}$$

where $\epsilon_{\mu_1\nu_1\ldots\mu_n\nu_n}$ is the completely antisymmetric tensor, and $J^S_\lambda \equiv J^{(2n+1)}_\lambda$ is the axial current.

Boson Regulator Fields. We have seen that we could also regularize by adding massive fermions and bosons with fermion spin, the unpaired boson affecting transformation properties under space-dependent chiral transformations. Denoting by ϕ the boson field and by M its mass, we perform in the regularized functional integral a change of variables of the form of a space-dependent chiral transformation acting in the same way on the fermion and boson field. The variation δS of the action at first order in θ is

$$\delta S = \int \mathrm{d}^4 x \left[\partial_\mu \theta(x) J_\mu^5(x) + 2iM\theta(x)\bar\phi(x)\gamma_5\phi(x) \right]$$

with

$$J_\mu^5(x) = i\bar\psi(x)\gamma_5\gamma_\mu\psi(x) + i\bar\phi(x)\gamma_5\gamma_\mu\phi(x).$$

Expanding in θ and identifying the coefficient of $\theta(x)$, we thus obtain the equation

$$\langle \partial_\mu J_\mu^5(x) \rangle = 2iM \langle \bar\phi(x)\gamma_5\phi(x) \rangle = -2iM \operatorname{tr} \gamma_5 \langle x| \slashed{D}^{-1} |x \rangle . \tag{31}$$

The divergence of the axial current comes here from the boson contribution. We know that in the large M limit it becomes quadratic in A. Expanding the r.h.s. in powers of A, keeping the quadratic term, we find after Fourier transformation

$$C^{(2)}(k) = -2iMe^2 \int \mathrm{d}^4 p_1 \, \mathrm{d}^4 p_2 \, A_\mu(p_1) A_\nu(p_2) \int \frac{\mathrm{d}^4 q}{(2\pi)^4}$$
$$\times \operatorname{tr} \left[\gamma_5 (\slashed{q} + \slashed{k} - iM)^{-1} \gamma_\mu (\slashed{q} - \slashed{p}_2 - iM)^{-1} \gamma_\nu (\slashed{q} - iM)^{-1} \right] . \tag{32}$$

The apparent divergence of this contribution is regularized by formally vanishing diagrams that we do not write, but which justify the following formal manipulations.

In the trace the formal divergences cancel and one obtains

$$C^{(2)}(k) \sim_{M\to\infty} 8M^2 e^2 \epsilon_{\mu\nu\rho\sigma} \int \mathrm{d}^4 p_1 \, \mathrm{d}^4 p_2 \, p_{1\rho} p_{2\sigma} A_\mu(p_1) A_\nu(p_2)$$
$$\times \frac{1}{(2\pi)^4} \int \frac{\mathrm{d}^4 q}{(q^2 + M^2)^3} .$$

The limit $M \to \infty$ corresponds to remove the regulator. The limit is finite because after rescaling of q the mass can be eliminated. One finds

$$C^{(2)}(k) \underset{M\to\infty}{\sim} \frac{e^2}{4\pi^2} \epsilon_{\mu\nu\rho\sigma} \int \mathrm{d}^4 p_1 \, \mathrm{d}^4 p_2 \, p_{1\rho} p_{2\sigma} A_\mu(p_1) A_\nu(p_2) ,$$

in agreement with (27).

Point-Splitting Regularization. Another calculation, based on regularization by point splitting, gives further insight into the mechanism that generates the anomaly. We thus consider the non-local operator

$$J_\mu^5(x, a) = i\bar\psi(x - a/2)\gamma_5\gamma_\mu\psi(x + a/2) \exp\left[ie \int_{x-a/2}^{x+a/2} A_\lambda(s)\mathrm{d}s_\lambda \right],$$

in the limit $|a| \to 0$. To avoid a breaking of rotation symmetry by the regularization, before taking the limit $|a| \to 0$ we will average over all orientations of the vector a. The multiplicative gauge factor (parallel transporter) ensures gauge invariance of the regularized operator (transformations (19)). The divergence of the operator for $|a| \to 0$ then becomes

$$\partial_\mu^x J_\mu^5(x, a) \sim -ea_\lambda \bar\psi(x - a/2)\gamma_5\gamma_\mu F_{\mu\lambda}(x)\psi(x + a/2)$$

$$\times \exp\left[ie \int_{x-a/2}^{x+a/2} A_\lambda(s)\mathrm{d}s_\lambda\right],$$

where the $\psi, \bar\psi$ field equations have been used. We now expand the expectation value of the equation in powers of A. The first term vanishes. The second term is quadratic in A and yields

$$\langle \partial_\mu^x J_\mu^5(x, a)\rangle \sim ie^2 a_\lambda F_{\mu\lambda}(x) \int \mathrm{d}^4 y\, A_\nu(y+x)\, \mathrm{tr}\, \gamma_5 \Delta_{\mathrm{F}}(y-a/2)\gamma_\nu \Delta_{\mathrm{F}}(-y-a/2)\gamma_\mu ,$$

where $\Delta_{\mathrm{F}}(y)$ is the fermion propagator:

$$\Delta_{\mathrm{F}}(y) = -\frac{i}{(2\pi)^4} \int \mathrm{d}^4 k\, \mathrm{e}^{iky}\frac{\slashed k}{k^2} = \frac{1}{2\pi^2}\frac{\slashed y}{y^4}.$$

We now take the trace. The propagator is singular for $|y| = O(|a|)$ and, therefore, we can expand $A_\nu(x + y)$ in powers of y. The first term vanishes for symmetry reasons ($y \mapsto -y$), and we obtain

$$\langle \partial_\mu^x J_\mu^5(x, a)\rangle \sim \frac{ie^2}{\pi^4}\epsilon_{\mu\nu\tau\sigma}a_\lambda F_{\mu\lambda}(x)\partial_\rho A_\nu(x) \int \mathrm{d}^4 y\frac{y_\rho y_\sigma a_\tau}{|y + a/2|^4 |y - a/2|^4}.$$

The integral over y gives a linear combination of $\delta_{\rho\sigma}$ and $a_\rho a_\sigma$ but the second term gives a vanishing contribution due to ϵ symbol. It follows that

$$\langle \partial_\mu^x J_\mu^5(x, a)\rangle \sim \frac{ie^2}{3\pi^4}\epsilon_{\mu\nu\tau\rho}a_\lambda a_\tau F_{\mu\lambda}(x)\partial_\rho A_\nu(x) \int \mathrm{d}^4 y\frac{y^2 - (y\cdot a)^2/a^2}{|y + a/2|^4 |y - a/2|^4}.$$

After integration, we then find

$$\langle \partial_\mu^x J_\mu^5(x, a)\rangle \sim \frac{ie^2}{4\pi^2}\epsilon_{\mu\nu\tau\rho}\frac{a_\lambda a_\tau}{a^2} F_{\mu\lambda}(x)F_{\rho\nu}(x).$$

Averaging over the a directions, we see that the divergence is finite for $|a| \to 0$ and, thus,

$$\lim_{|a|\to 0} \left\langle \partial_\mu^x \overline{J_\mu^5(x, a)}\right\rangle = \frac{ie^2}{16\pi^2}\epsilon_{\mu\nu\lambda\rho}F_{\mu\lambda}(x)F_{\rho\nu}(x),$$

in agreeement with the result (28).

On the lattice an averaging over a_μ is produced by summing over all lattice directions. Because the only expression quadratic in a_μ that has the symmetry of the lattice is a^2, the same result is found: the anomaly is lattice-independent.

A Direct Physical Application. In a phenomenological model of Strong Interaction physics, where a $SU(2) \times SU(2)$ chiral symmetry is softly broken by the pion mass, in the absence of anomalies the divergence of the neutral axial current is proportional to the π_0 field (corresponding to the neutral pion). A short formal calculation then indicates that the decay rate of π_0 into two photons should vanish at zero momentum. Instead, taking into account the axial anomaly (28), one obtains a non-vanishing contribution to the decay, in good agreement with experimental data.

Chiral Gauge Theory. A gauge theory is consistent only if the gauge field is coupled to a conserved current. An anomaly that affects the current destroys gauge invariance in the full quantum theory. Therefore, the theory with axial gauge symmetry, where the action in the fermion sector reads

$$S(\bar{\psi}, \psi; B) = - \int \mathrm{d}^4 x \, \bar{\psi}(x)(\slashed{\partial} + ig\gamma_5 \slashed{B})\psi(x),$$

is inconsistent. Indeed current conservation applies to the BBB vertex at one-loop order. Because now the three point vertex is symmetric the divergence is given by the expression (26), and thus does not vanish.

More generally, the anomaly prevents the construction of a theory that would have both an abelian gauge vector and axial symmetry, where the action in the fermion sector would read

$$S(\bar{\psi}, \psi; A, B) = - \int \mathrm{d}^4 x \, \bar{\psi}(x)(\slashed{\partial} + ie\slashed{A} + i\gamma_5 g\slashed{B})\psi(x).$$

A way to solve both problems is to cancel the anomaly by introducing another fermion of opposite chiral coupling. With more fermions other combinations of couplings are possible. Note, however, that a purely axial gauge theory with two fermions of opposite chiral charges can be rewritten as a vector theory by combining differently the chiral components of both fermions.

4.3   Two Dimensions

As an exercise and as a preliminary to the discussion of the $CP(N-1)$ models in Sect.5.2, we verify by explicit calculation the general expression (30) in the special example of dimension 2:

$$\langle \partial_\mu J_\mu^3 \rangle = -i \frac{e}{2\pi} \epsilon_{\mu\nu} F_{\mu\nu} . \tag{33}$$

The general form of the r.h.s. is again dictated by locality and power counting: the anomaly must have canonical dimension 2. The explicit calculation requires some care because massless fields may lead to IR divergences in two dimensions. One thus gives a mass m to fermions, which breaks chiral symmetry explicitly, and takes the massless limit at the end of the calculation. The calculation involves only one diagram:

$$\Gamma_{\mu\nu}^{(2)}(k,-k) = \frac{\delta}{\delta A_\nu(-k)} \langle J_\mu^3(k) \rangle \Big|_{A=0} = \frac{\delta}{\delta A_\nu(-k)} i \operatorname{tr} \left[\gamma_3 \gamma_\mu \slashed{D}^{-1}(k) \right] \Big|_{A=0}$$

$$= \frac{e}{(2\pi)^2} \operatorname{tr} \gamma_3 \gamma_\mu \int \mathrm{d}^2 q \frac{1}{i\slashed{q} + m} \gamma_\nu \frac{1}{i\slashed{q} + i\slashed{k} + m}.$$

Here the γ-matrices are simply the ordinary Pauli matrices. Then,

$$k_\mu \Gamma_{\mu\nu}^{(2)}(k,-k) = \frac{e}{(2\pi)^2} \operatorname{tr} \gamma_3 \slashed{k} \int \mathrm{d}^2 q \frac{1}{i\slashed{q} + m} \gamma_\nu \frac{1}{i\slashed{q} + i\slashed{k} + m}.$$

We use the method of the boson regulator field, which yields the two-dimensional analogue of (31). Here, it leads to the calculation of the difference between two diagrams (analogues of (32)) due to the explicit chiral symmetry breaking:

$$C_\mu(k) = 2m \frac{e}{(2\pi)^2} \operatorname{tr} \gamma_3 \int \mathrm{d}^2 q \frac{1}{i\slashed{q} + m} \gamma_\nu \frac{1}{i\slashed{q} + i\slashed{k} + m} - (m \mapsto M)$$

$$= 2m \frac{e}{(2\pi)^2} \operatorname{tr} \gamma_3 \int \mathrm{d}^2 q \frac{(m - i\slashed{q}) \gamma_\mu (m - i\slashed{q} - i\slashed{k})}{(q^2 + m^2)[(k+q)^2 + m^2]} - (m \mapsto M).$$

In the trace again the divergent terms cancel:

$$C_\mu(k) = 4em^2 \epsilon_{\mu\nu} k_\nu \frac{1}{(2\pi)^2} \int \frac{\mathrm{d}^2 q}{(q^2 + m^2)[(k+q)^2 + m^2]} - (m \mapsto M).$$

The two contributions are now separately convergent. When $m \to 0$, the m^2 factor dominates the logarithmic IR divergence and the contribution vanishes. In the second term, in the limit $M \to \infty$, one obtains

$$C_\mu(k)\big|_{m\to 0, M\to\infty} \sim -4eM^2 \epsilon_{\mu\nu} k_\nu \frac{1}{(2\pi)^2} \int \frac{\mathrm{d}^2 q}{(q^2 + M^2)^2} = -\frac{e}{\pi} \epsilon_{\mu\nu} k_\nu,$$

in agreement with (33).

4.4   Non-Abelian Vector Gauge Fields and Abelian Axial Current

We still consider an abelian axial current but now in the framework of a non-abelian gauge theory. The fermion fields transform non-trivially under a gauge group G and $\mathbf{A}_\mu$ is the corresponding gauge field. The action is

$$S(\bar{\psi}, \psi; A) = -\int \mathrm{d}^4 x\, \bar{\psi}(x) \slashed{D} \psi(x)$$

with the convention (9) and

$$\slashed{D} = \slashed{\partial} + \slashed{A}. \tag{34}$$

In a gauge transformation represented by a unitary matrix $\mathbf{g}(x)$, the gauge field $\mathbf{A}_\mu$ and the Dirac operator become

$$\mathbf{A}_\mu(x) \mapsto \mathbf{g}(x)\partial_\mu\mathbf{g}^{-1}(x) + \mathbf{g}(x)\mathbf{A}_\mu(x)\mathbf{g}^{-1}(x) \;\Rightarrow\; \slashed{D} \mapsto \mathbf{g}^{-1}(x)\slashed{D}\mathbf{g}(x) \,. \tag{35}$$

The axial current

$$J_\mu^5(x) = i\bar{\psi}(x)\gamma_5\gamma_\mu\psi(x)$$

is still gauge invariant. Therefore, no new calculation is needed; the result is completely determined by dimensional analysis, gauge invariance, and the preceding abelian calculation that yields the term of order $\mathbf{A}^2$:

$$\left\langle \partial_\lambda J_\lambda^5(x) \right\rangle = -\frac{i}{16\pi^2}\epsilon_{\mu\nu\rho\sigma} \operatorname{tr} \mathbf{F}_{\mu\nu}\mathbf{F}_{\rho\sigma} \,, \tag{36}$$

in which $\mathbf{F}_{\mu\nu}$ now is the corresponding curvature (10). Again this expression must be a total derivative. Indeed, one verifies that

$$\epsilon_{\mu\nu\rho\sigma}\operatorname{tr}\mathbf{F}_{\mu\nu}\mathbf{F}_{\rho\sigma} = 4\,\epsilon_{\mu\nu\rho\sigma}\partial_\mu\operatorname{tr}(\mathbf{A}_\nu\partial_\rho\mathbf{A}_\sigma + \frac{2}{3}\mathbf{A}_\nu\mathbf{A}_\rho\mathbf{A}_\sigma). \tag{37}$$

4.5   Anomaly and Eigenvalues of the Dirac Operator

We assume that the spectrum of $\slashed{D}$, the Dirac operator in a non-abelian gauge field (34), is discrete (putting temporarily the fermions in a box if necessary) and call d_n and $\varphi_n(x)$ the corresponding eigenvalues and eigenvectors:

$$\slashed{D}\varphi_n = d_n\varphi_n \,.$$

For a unitary or orthogonal group, the massless Dirac operator is anti-hermitian; therefore, the eigenvalues are imaginary and the eigenvectors orthogonal. In addition, we choose them with unit norm.

The eigenvalues are gauge invariant because, in a gauge transformation characterized by a unitary matrix $\mathbf{g}(x)$, the Dirac operator transforms like in (35), and thus simply

$$\varphi_n(x) \mapsto \mathbf{g}(x)\varphi_n(x).$$

The anticommutation $\slashed{D}\gamma_5 + \gamma_5\slashed{D} = 0$ implies

$$\slashed{D}\gamma_5\varphi_n = -d_n\gamma_5\varphi_n \,.$$

Therefore, either d_n is different from zero and $\gamma_5\varphi_n$ is an eigenvector of $\slashed{D}$ with eigenvalue $-d_n$, or d_n vanishes. The eigenspace corresponding to the eigenvalue 0 then is invariant under γ_5, which can be diagonalized: the eigenvectors of $\slashed{D}$ can be chosen eigenvectors of definite chirality, that is eigenvectors of γ_5 with eigenvalue ± 1:

$$\slashed{D}\varphi_n = 0 \,, \quad \gamma_5\varphi_n = \pm\varphi_n \,.$$

We call n_+ and n_- the dimensions of the eigenspace of positive and negative chirality, respectively.

We now consider the determinant of the operator $\not{D}+m$ regularized by mode truncation (mode regularization):

$$\det_N(\not{D}+m) = \prod_{n\leq N}(d_n+m),$$

keeping the N lowest eigenvalues of $\not{D}$ (in modulus), with $N-n_+-n_-$ even, in such a way that the corresponding subspace remains γ_5 invariant.

The regularization is gauge invariant because the eigenvalues of $\not{D}$ are gauge invariant.

Note that in the truncated space

$$\operatorname{tr}\gamma_5 = n_+ - n_-. \tag{38}$$

The trace of γ_5 equals $n_+ - n_-$, the *index* of the Dirac operator $\not{D}$. A non-vanishing index thus endangers axial current conservation.

In a chiral transformation (20) with constant θ, the determinant of $(\not{D}+m)$ becomes

$$\det_N(\not{D}+m) \mapsto \det_N\left(e^{i\theta\gamma_5}(\not{D}+m)e^{i\theta\gamma_5}\right).$$

We now consider the various eigenspaces.

If $d_n \neq 0$, the matrix γ_5 is represented by the Pauli matrix σ_1 in the sum of eigenspaces corresponding to the two eigenvalues $\pm d_n$ and $\not{D}+m$ by $d_n\sigma_3+m$. The determinant in the subspace then is

$$\det\left(e^{i\theta\sigma_1}(d_n\sigma_3+m)e^{i\theta\sigma_1}\right) = \det e^{2i\theta\sigma_1}\det(d_n\sigma_3+m) = m^2 - d_n^2,$$

because σ_1 is traceless.

In the eigenspace of dimension n_+ of vanishing eigenvalues d_n with eigenvectors with positive chirality, γ_5 is diagonal with eigenvalue 1 and, thus,

$$m^{n_+} \mapsto m^{n_+}e^{2i\theta n_+}.$$

Similarly, in the eigenspace of chirality -1 and dimension n_-,

$$m^{n_-} \mapsto m^{n_-}e^{-2i\theta n_-}.$$

We conclude

$$\det_N\left(e^{i\theta\gamma_5}(\not{D}+m)e^{i\theta\gamma_5}\right) = e^{2i\theta(n_+-n_-)}\det_N(\not{D}+m),$$

The ratio of the two determinants is independent of N. Taking the limit $N \to \infty$, one finds

$$\det\left[\left(e^{i\gamma_5\theta}(\not{D}+m)e^{i\gamma_5\theta}\right)(\not{D}+m)^{-1}\right] = e^{2i\theta(n_+-n_-)}. \tag{39}$$

Note that the l.h.s. of (39) is obviously 1 when $\theta = n\pi$, which implies that the coefficient of 2θ in the r.h.s. must indeed be an integer.

The variation of $\ln\det(\not{D}+m)$:

$$\ln\det\left[\left(e^{i\gamma_5\theta}(\not{D}+m)e^{i\gamma_5\theta}\right)(\not{D}+m)^{-1}\right] = 2i\theta\,(n_+ - n_-),$$

at first order in θ, is related to the variation of the action (18) (see (29)) and, thus, to the expectation value of the integral of the divergence of the axial current, $\int \mathrm{d}^4 x \left\langle \partial_\mu J_\mu^5(x) \right\rangle$ in four dimensions. In the limit $m = 0$, it is thus related to the space integral of the chiral anomaly (36).

We have thus found a local expression giving the index of the Dirac operator:

$$-\frac{1}{32\pi^2} \epsilon_{\mu\nu\rho\sigma} \int \mathrm{d}^4 x \, \mathrm{tr} \, \mathbf{F}_{\mu\nu} \mathbf{F}_{\rho\sigma} = n_+ - n_- \,. \tag{40}$$

Concerning this result several comments can be made:

(i) At first order in θ, in the absence of regularization, we have calculated ($\ln \det = \mathrm{tr} \ln$)

$$\ln \det \left[1 + i\theta \left(\gamma_5 + (\slashed{D} + m)\gamma_5(\slashed{D} + m)^{-1} \right) \right] \sim 2i\theta \, \mathrm{tr} \, \gamma_5 \,,$$

where the cyclic property of the trace has been used. Since the trace of the matrix γ_5 in the full space vanishes, one could expect, naively, a vanishing result. But trace here means trace in matrix space and in coordinate space and γ_5 really stands for $\gamma_5 \delta(x - y)$. The mode regularization gives a well-defined finite result for the ill-defined product $0 \times \delta^d(0)$.

(ii) The property that the integral (40) is quantized shows that the form of the anomaly is related to topological properties of the gauge field since the integral does not change when the gauge field is deformed continuously. The integral of the anomaly over the whole space, thus, depends only on the behaviour at large distances of the curvature tensor $\mathbf{F}_{\mu\nu}$ and the anomaly must be a total derivative as (37) confirms.

(iii) One might be surprised that $\det \slashed{D}$ is not invariant under global chiral transformations. However, we have just established that when the integral of the anomaly does not vanish, $\det \slashed{D}$ vanishes. This explains that, to give a meaning to the r.h.s. of (39), we have been forced to introduce a mass to find a non-trivial result. The determinant of $\slashed{D}$ in the subspace orthogonal to eigenvectors with vanishing eigenvalue, even in presence of a mass, is chiral invariant by parity doubling. But for $n_+ \neq n_-$, this is not the case for the determinant in the eigenspace of eigenvalue zero because the trace of γ_5 does not vanish in this eigenspace (38). In the limit $m \to 0$, the complete determinant vanishes but not the ratio of determinants for different values of θ because the powers of m cancel.

(iv) The discussion of the index of the Dirac operator is valid in any even dimension. Therefore, the topological character and the quantization of the space integral of the anomaly are general.

5   Instantons, Anomalies, and θ-Vacua

We now discuss the role of instantons in several examples where the classical potential has a periodic structure with an infinite set of degenerate minima. We exhibit their topological character, and in the presence of gauge fields relate them to anomalies and the index of the Dirac operator. Instantons imply that the eigenstates of the hamiltonian depend on an angle θ. In the quantum field theory the notion of θ-vacuum emerges.

5.1   The Periodic Cosine Potential

As a first example of the role of instantons when topology is involved, we consider a simple hamiltonian with a periodic potential

$$H = -\frac{g}{2}\left(\mathrm{d}/\mathrm{d}x\right)^2 + \frac{1}{2g}\sin^2 x. \tag{41}$$

The potential has an infinite number of degenerate minima for $x = n\pi$, $n \in \mathbb{Z}$. Each minimum is an equivalent starting point for a perturbative calculation of the eigenvalues of H. Periodicity implies that the perturbative expansions are identical to all orders in g, a property that seems to imply that the quantum hamiltonian has an infinite number of degenerate eigenstates. In reality, we know that the exact spectrum of the hamiltonian H is not degenerate, due to barrier penetration. Instead, it is continuous and has, at least for g small enough, a band structure.

The Structure of the Ground State. To characterize more precisely the structure of the spectrum of the hamiltonian (41), we introduce the operator T that generates an elementary translation of one period π:

$$T\psi(x) = \psi(x + \pi).$$

Since T commutes with the hamiltonian,

$$[T, H] = 0,$$

both operators can be diagonalized simultaneously. Because the eigenfunctions of H must be bounded at infinity, the eigenvalues of T are pure phases. Each eigenfunction of H thus is characterized by an angle θ (pseudo-momentum) associated with an eigenvalue of T:

$$T\left|\theta\right\rangle = \mathrm{e}^{i\theta}\left|\theta\right\rangle.$$

The corresponding eigenvalues $E_n(\theta)$ are periodic functions of θ and, for $g \to 0$, are close to the eigenvalues of the harmonic oscillator:

$$E_n(\theta) = n + 1/2 + O(g).$$

To all orders in powers of g, $E_n(\theta)$ is independent of θ and the spectrum of H is infinitely degenerate. Additional exponentially small contributions due to barrier penetration lift the degeneracy and introduce a θ dependence. To each value of n then corresponds a band when θ varies in $[0, 2\pi]$.

Path Integral Representation. The spectrum of H can be extracted from the calculation of the quantity

$$\mathcal{Z}_\ell(\beta) = \operatorname{tr} T^\ell e^{-\beta H} = \frac{1}{2\pi} \sum_{n=0}^{\infty} \int d\theta \, e^{i\ell\theta} e^{-\beta E_n(\theta)}.$$

Indeed,

$$\mathcal{Z}(\theta, \beta) \equiv \sum_\ell e^{i\ell\theta} \mathcal{Z}_\ell(\beta) = \sum_n e^{-\beta E_n(\theta)}, \tag{42}$$

where $\mathcal{Z}(\theta, \beta)$ is the partition function restricted to states with a fixed θ angle.

The path integral representation of $\mathcal{Z}_\ell(\beta)$ differs from the representation of the partition function $\mathcal{Z}_0(\beta)$ only by the boundary conditions. The operator T has the effect of translating the argument x in the matrix element $\langle x' | \operatorname{tr} e^{-\beta H} | x \rangle$ before taking the trace. It follows that

$$\mathcal{Z}_\ell(\beta) = \int [dx(t)] \exp\left[-S(x) \right], \tag{43}$$

$$S(x) = \frac{1}{2g} \int_{-\beta/2}^{\beta/2} \left[\dot{x}^2(t) + \sin^2\left(x(t) \right) \right] dt, \tag{44}$$

where one integrates over paths satisfying the boundary condition $x(\beta/2) = x(-\beta/2) + \ell\pi$. A careful study of the trace operation in the case of periodic potentials shows that $x(-\beta/2)$ varies over only one period (see Appendix A).

Therefore, from (42), we derive the path integral representation

$$\mathcal{Z}(\theta, \beta) = \sum_\ell \int_{x(\beta/2)=x(-\beta/2)+\ell\pi} [dx(t)] \exp\left[-S(x) + i\ell\theta \right]$$

$$= \int_{x(\beta/2)=x(-\beta/2) \ (\operatorname{mod} \pi)} [dx(t)] \exp\left[-S(x) + i\frac{\theta}{\pi} \int_{-\beta/2}^{\beta/2} dt \, \dot{x}(t) \right]. \tag{45}$$

Note that ℓ is a topological number since two trajectories with different values of ℓ cannot be related continuously. In the same way,

$$Q = \frac{1}{\pi} \int_{-\beta/2}^{\beta/2} dt \, \dot{x}(t)$$

is a topological charge; it depends on the trajectory only through the boundary conditions.

For β large and $g \to 0$, the path integral is dominated by the constant solutions $x_c(t) = 0 \mod \pi$ corresponding to the $\ell = 0$ sector. A non-trivial θ dependence can come only from instanton (non-constant finite action saddle points) contributions corresponding to quantum tunnelling. Note that, quite generally,

$$\int dt \left[\dot{x}(t) \pm \sin\left(x(t) \right) \right]^2 \geq 0 \implies S \geq \left| \cos\left(x(+\infty) \right) - \cos\left(x(-\infty) \right) \right| / g. \tag{46}$$

The action (44) is finite for $\beta \to \infty$ only if $x(\pm\infty) = 0 \bmod \pi$. The non-vanishing value of the r.h.s. of (46) is $2/g$. This minimum is reached for trajectories x_c that are solutions of

$$\dot{x}_c = \pm \sin x_c \;\Rightarrow\; x_c(t) = 2 \arctan \mathrm{e}^{\pm(t-t_0)},$$

and the corresponding classical action then is

$$\mathcal{S}(x_c) = 2/g\,.$$

The instanton solutions belong to the $\ell = \pm 1$ sector and connect two consecutive minima of the potential. They yield the leading contribution to barrier penetration for $g \to 0$. An explicit calculation yields

$$E_0(g) = E_{\text{pert.}}(g) - \frac{4}{\sqrt{\pi g}} \mathrm{e}^{-2/g} \cos\theta [1 + O(g)],$$

where $E_{\text{pert.}}(g)$ is the sum of the perturbative expansion in powers of g.

5.2 Instantons and Anomaly: CP(N-1) Models

We now consider a set of two-dimensional field theories, the $CP(N-1)$ models, where again instantons and topology play a role and the semi-classical vacuum has a similar periodic structure. The new feature is the relation between the topological charge and the two-dimensional chiral anomaly.

Here, we describe mainly the nature of the instanton solutions and refer the reader to the literature for a more detailed analysis. Note that the explicit calculation of instanton contributions in the small coupling limit in the $CP(N-1)$ models, as well as in the non-abelian gauge theories discussed in Sect. 5.3, remains to large extent an unsolved problem. Due to the scale invariance of the classical theory, instantons depend on a scale (or size) parameter. Instanton contributions then involve the running coupling constant at the instanton size. Both families of theories are UV asymptotically free. Therefore, the running coupling is small for small instantons and the semi-classical approximation is justified. However, in the absence of any IR cut-off, the running coupling becomes large for large instantons, and it is unclear whether a semi-classical approximation remains valid.

The CP(N-1) Manifolds. We consider a N-component complex vector φ of unit length:

$$\bar{\varphi} \cdot \varphi = 1\,.$$

This φ-space is also isomorphic to the quotient space $U(N)/U(N-1)$. In addition, two vectors φ and φ' are considered equivalent if

$$\varphi' \equiv \varphi \;\Leftrightarrow\; \varphi'_\alpha = \mathrm{e}^{i\Lambda}\varphi_\alpha\,. \tag{47}$$

This condition characterizes the symmetric space and complex Grassmannian manifold $U(N)/U(1)/U(N-1)$. It is isomorphic to the manifold $CP(N-1)$

(for $N-1$-dimensional Complex Projective), which is obtained from $\mathbb{C}^N$ by the equivalence relation

$$z_\alpha \equiv z'_\alpha \quad \text{if} \quad z'_\alpha = \lambda z_\alpha$$

where λ belongs to the Riemann sphere (compactified complex plane).

The CP(N-1) Models. A symmetric space admits a unique invariant metric and this leads to a unique action with two derivatives, up to a multiplicative factor. Here, one representation of the unique $U(N)$ symmetric classical action is

$$S(\varphi, A_\mu) = \frac{1}{g} \int \mathrm{d}^2 x \, \overline{D_\mu \varphi} \cdot D_\mu \varphi \,,$$

in which g is a coupling constant and D_μ the covariant derivative:

$$D_\mu = \partial_\mu + iA_\mu \,.$$

The field A_μ is a gauge field for the $U(1)$ transformations:

$$\varphi'(x) = e^{i\Lambda(x)} \varphi(x), \quad A'_\mu(x) = A_\mu(x) - \partial_\mu \Lambda(x). \tag{48}$$

The action is obviously $U(N)$ symmetric and the gauge symmetry ensures the equivalence (47).

Since the action contains no kinetic term for A_μ, the gauge field is not a dynamical but only an auxiliary field that can be integrated out. The action is quadratic in A and the gaussian integration results in replacing in the action A_μ by the solution of the A-field equation

$$A_\mu = i\bar{\varphi} \cdot \partial_\mu \varphi \,, \tag{49}$$

where (5.2) has been used. After this substitution, the field $\bar{\varphi} \cdot \partial_\mu \varphi$ acts as a composite gauge field.

For what follows, however, we find it more convenient to keep A_μ as an independent field.

Instantons. To prove the existence of locally stable non-trivial minima of the action, the following Bogomolnyi inequality can be used (note the analogy with (46)):

$$\int \mathrm{d}^2 x \, |D_\mu \varphi \mp i\epsilon_{\mu\nu} D_\nu \varphi|^2 \geq 0 \,,$$

($\epsilon_{\mu\nu}$ being the antisymmetric tensor, $\epsilon_{12} = 1$). After expansion, the inequality can be cast into the form

$$S(\varphi) \geq 2\pi |Q(\varphi)|/g$$

with

$$Q(\varphi) = -\frac{i}{2\pi} \epsilon_{\mu\nu} \int \mathrm{d}^2 x \, D_\mu \varphi \cdot \overline{D_\nu \varphi} = \frac{i}{2\pi} \int \mathrm{d}^2 x \, \epsilon_{\mu\nu} \bar{\varphi} \cdot D_\nu D_\mu \varphi \,. \tag{50}$$

Then,

$$i\epsilon_{\mu\nu}D_\nu D_\mu = \tfrac{1}{2}i\epsilon_{\mu\nu}[D_\nu, D_\mu] = \tfrac{1}{2}F_{\mu\nu}\,, \tag{51}$$

where $F_{\mu\nu}$ is the curvature:

$$F_{\mu\nu} = \partial_\mu A_\nu - \partial_\nu A_\mu\,.$$

Therefore, using (5.2),

$$Q(\varphi) = \frac{1}{4\pi}\int d^2x\,\epsilon_{\mu\nu}F_{\mu\nu}\,. \tag{52}$$

The integrand is proportional to the two-dimensional abelian chiral anomaly (33), and thus is a total divergence:

$$\tfrac{1}{2}\epsilon_{\mu\nu}F_{\mu\nu} = \partial_\mu\epsilon_{\mu\nu}A_\nu\,.$$

Substituting this form into (52) and integrating over a large disc of radius R, one obtains

$$Q(\varphi) = \frac{1}{2\pi}\lim_{R\to\infty}\oint_{|x|=R} dx_\mu\,A_\mu(x). \tag{53}$$

$Q(\varphi)$ thus depends only on the behaviour of the classical solution for $|x|$ large and is a topological charge. Finiteness of the action demands that at large distances $D_\mu\varphi$ vanishes and, therefore,

$$D_\mu\varphi = 0 \;\Rightarrow\; [D_\mu, D_\nu]\varphi = F_{\mu\nu}\varphi = 0\,.$$

Since $\varphi \neq 0$, this equation implies that $F_{\mu\nu}$ vanishes and, thus, that A_μ is a pure gauge (and φ a gauge transform of a constant vector):

$$A_\mu = \partial_\mu\Lambda(x) \;\Rightarrow\; Q(\varphi) = \frac{1}{2\pi}\lim_{R\to\infty}\oint_{|x|=R} dx_\mu\partial_\mu\Lambda(x)\,. \tag{54}$$

The topological charge measures the variation of the angle $\Lambda(x)$ on a large circle, which is a multiple of 2π because φ is regular. One is thus led to the consideration of the homotopy classes of mappings from $U(1)$, that is S_1 to S_1, which are characterized by an integer n, the winding number. This is equivalent to the statement that the homotopy group $\pi_1(S_1)$ is isomorphic to the additive group of integers $\mathbb{Z}$.

Then,

$$Q(\varphi) = n \;\Longrightarrow\; S(\varphi) \geq 2\pi|n|/g\,.$$

The equality $S(\varphi) = 2\pi|n|/g$ corresponds to a local minimum and implies that the classical solutions satisfy first order partial differential (self-duality) equations:

$$D_\mu\varphi = \pm i\epsilon_{\mu\nu}D_\nu\varphi\,. \tag{55}$$

For each sign, there is really only one equation, for instance $\mu = 1, \nu = 2$. It is simple to verify that both equations imply the φ-field equations, and combined with the constraint (5.2), the A-field equation (49). In complex coordinates $z = x_1 + ix_2$, $\bar{z} = x_1 - ix_2$, they can be written as

$$\partial_z \varphi_\alpha(z, \bar{z}) = -iA_z(z, \bar{z})\varphi_\alpha(z, \bar{z}),$$
$$\partial_{\bar{z}} \varphi_\alpha(z, \bar{z}) = -iA_{\bar{z}}(z, \bar{z})\varphi_\alpha(z, \bar{z}).$$

Exchanging the two equations just amounts to exchange φ and $\bar{\varphi}$. Therefore, we solve only the second equation which yields

$$\varphi_\alpha(z, \bar{z}) = \kappa(z, \bar{z})P_\alpha(z),$$

where $\kappa(z, \bar{z})$ is a particular solution of

$$\partial_{\bar{z}} \kappa(z, \bar{z}) = -iA_{\bar{z}}(z, \bar{z})\kappa(z, \bar{z}).$$

Vector solutions of (55) are proportional to holomorphic or anti-holomorphic (depending on the sign) vectors (this reflects the conformal invariance of the classical field theory). The function $\kappa(z, \bar{z})$, which gauge invariance allows to choose real (this corresponds to the $\partial_\mu A_\mu = 0$ gauge), then is constrained by the condition (5.2):

$$\kappa^2(z, \bar{z})\, P \cdot \bar{P} = 1\,.$$

The asymptotic conditions constrain the functions $P_\alpha(z)$ to be polynomials. Common roots to all P_α would correspond to non-integrable singularities for φ_α and, therefore, are excluded by the condition of finiteness of the action. Finally, if the polynomials have maximal degree n, asymptotically

$$P_\alpha(z) \sim c_\alpha z^n \;\Rightarrow\; \varphi_\alpha \sim \frac{c_\alpha}{\sqrt{\mathbf{c} \cdot \bar{\mathbf{c}}}} (z/\bar{z})^{n/2}\,.$$

When the phase of z varies by 2π, the phase of φ_α varies by $2n\pi$, showing that the corresponding winding number is n.

The Structure of the Semi-classical Vacuum. In contrast to our analysis of periodic potentials in quantum mechanics, here we have discussed the existence of instantons without reference to the structure of the classical vacuum. To find an interpretation of instantons in gauge theories, it is useful to express the results in the temporal gauge $A_2 = 0$. Then, the action is still invariant under space-dependent gauge transformations. The minima of the classical φ potential correspond to fields $\varphi(x_1)$, where x_1 is the space variable, gauge transforms of a constant vector:

$$\varphi(x_1) = e^{i\Lambda(x_1)}\mathbf{v}\,, \quad \bar{\mathbf{v}} \cdot \mathbf{v} = 1\,.$$

Moreover, if the vacuum state is invariant under space reflection, $\varphi(+\infty) = \varphi(-\infty)$ and, thus,

$$\Lambda(+\infty) - \Lambda(-\infty) = 2\nu\pi \quad \nu \in \mathbb{Z}\,.$$

Again ν is a topological number that classifies degenerate classical minima, and the semi-classical vacuum has a periodic structure. This analysis is consistent with Gauss's law, which implies only that states are invariant under infinitesimal gauge transformations and, thus, under gauge transformations of the class $\nu = 0$ that are continuously connected to the identity.

We now consider a large rectangle with extension R in the space direction and T in the euclidean time direction and by a smooth gauge transformation continue the instanton solution to the temporal gauge. Then, the variation of the pure gauge comes entirely from the sides at fixed time. For $R \to \infty$, one finds

$$\Lambda(+\infty, 0) - \Lambda(-\infty, 0) - [\Lambda(+\infty, T) - \Lambda(-\infty, T)] = 2n\pi .$$

Therefore, instantons interpolate between different classical minima. Like in the case of the cosine potential, to project onto a proper quantum eigenstate, the "θ-vacuum" corresponding to an angle θ, one adds, in analogy with the expression (45), a topological term to the classical action. Here,

$$S(\varphi) \mapsto S(\varphi) + i\frac{\theta}{4\pi} \int d^2x \, \epsilon_{\mu\nu} F_{\mu\nu}.$$

Remark. Replacing in the topological charge Q the gauge field by the explicit expression (49), one finds

$$Q(\varphi) = \frac{i}{2\pi} \int d^2x \, \epsilon_{\mu\nu} \partial_\mu \bar\varphi \cdot \partial_\nu \varphi = \frac{i}{2\pi} \int d\bar\varphi_\alpha \wedge d\varphi_\alpha ,$$

where the notation of exterior differential calculus has been used. We recognize the integral of a two-form, a symplectic form, and $4\pi Q$ is the area of a 2-surface embedded in $CP(N-1)$. A symplectic form is always closed. Here it is also exact, so that Q is the integral of a one-form (cf. (53)):

$$Q(\varphi) = \frac{i}{2\pi} \int \bar\varphi_\alpha d\varphi_\alpha = \frac{i}{4\pi} \int (\bar\varphi_\alpha d\varphi_\alpha - \varphi_\alpha d\bar\varphi_\alpha) .$$

The $O(3)$ Non-Linear σ-Model. The $CP(1)$ model is locally isomorphic to the $O(3)$ non-linear σ-model, with the identification

$$\phi^i(x) = \bar\varphi_\alpha(x)\sigma^i_{\alpha\beta}\varphi_\beta(x) ,$$

where σ^i are the three Pauli matrices.

Using, for example, an explicit representation of Pauli matrices, one indeed verifies

$$\phi^i(x)\phi^i(x) = 1 , \quad \partial_\mu\phi^i(x)\partial_\mu\phi^i(x) = 4\overline{D_\mu\varphi} \cdot D_\mu\varphi .$$

Therefore, the field theory can be expressed in terms of the field ϕ^i and takes the form of the non-linear σ-model. The fields ϕ are gauge invariant and the whole physical picture is a picture of confinement of the charged scalar "quarks" $\varphi_\alpha(x)$ and the propagation of neutral bound states corresponding to the fields ϕ^i.

Instantons in the ϕ description take the form of ϕ configurations with uniform limit for $|x| \to \infty$. Thus, they define a mapping from the compactified plane topologically equivalent to S_2 to the sphere S_2 (the ϕ^i configurations). Since $\pi_2(S_2) = \mathbb{Z}$, the φ and ϕ pictures are consistent.

In the example of $CP(1)$, a solution of winding number 1 is

$$\varphi_1 = \frac{1}{\sqrt{1 + z\bar{z}}}, \qquad \varphi_2 = \frac{z}{\sqrt{1 + z\bar{z}}}.$$

Translating the $CP(1)$ minimal solution into the $O(3)$ σ-model language, one finds

$$\phi_1 = \frac{z + \bar{z}}{1 + \bar{z}z}, \qquad \phi_2 = \frac{1}{i}\frac{z - \bar{z}}{1 + \bar{z}z}, \qquad \phi_3 = \frac{1 - \bar{z}z}{1 + \bar{z}z}.$$

This defines a stereographic mapping of the plane onto the sphere S_2, as one verifies by setting $z = \tan(\eta/2)e^{i\theta}$, $\eta \in [0, \pi]$.

In the $O(3)$ representation

$$Q = \frac{i}{2\pi}\int d\bar{\varphi}_\alpha \wedge d\varphi_\alpha = \frac{1}{8\pi}\epsilon_{ijk}\int \phi_i d\phi_j \wedge \phi_k \equiv \frac{1}{8\pi}\epsilon_{\mu\nu}\epsilon_{ijk}\int d^2x\, \phi_i \partial_\mu \phi_j \partial_\nu \phi_k.$$

The topological charge $4\pi Q$ has the interpretation of the area of the sphere S_2, multiply covered, and embedded in $\mathbb{R}^3$. Its value is a multiple of the area of S_2, which in this interpretation explains the quantization.

5.3 Instantons and Anomaly: Non-Abelian Gauge Theories

We now consider non-abelian gauge theories in four dimensions. Again, gauge field configurations can be found that contribute to the chiral anomaly and for which, therefore, the r.h.s. of (40) does not vanish. A specially interesting example is provided by instantons, that is finite action solutions of euclidean field equations.

To discuss this problem it is sufficient to consider pure gauge theories and the gauge group $SU(2)$, since a general theorem states that for a Lie group containing $SU(2)$ as a subgroup the instantons are those of the $SU(2)$ subgroup.

In the absence of matter fields it is convenient to use a $SO(3)$ notation. The gauge field $\mathbf{A}_\mu$ is a $SO(3)$ vector that is related to the element $\mathfrak{A}_\mu$ of the Lie algebra used previously as gauge field by

$$\mathfrak{A}_\mu = -\tfrac{1}{2}i\mathbf{A}_\mu \cdot \sigma,$$

where σ_i are the three Pauli matrices. The gauge action then reads

$$S(\mathbf{A}_\mu) = \frac{1}{4g^2}\int [\mathbf{F}_{\mu\nu}(x)]^2 d^4x,$$

(g is the gauge coupling constant) where the curvature

$$\mathbf{F}_{\mu\nu} = \partial_\mu \mathbf{A}_\nu - \partial_\nu \mathbf{A}_\mu + \mathbf{A}_\mu \times \mathbf{A}_\nu,$$

is also a $SO(3)$ vector.

The corresponding classical field equations are

$$\mathbf{D}_\nu \mathbf{F}_{\nu\mu} = \partial_\nu \mathbf{F}_{\nu\mu} + \mathbf{A}_\nu \times \mathbf{F}_{\nu\mu} = 0. \tag{56}$$

The existence and some properties of instantons in this theory follow from considerations analogous to those presented for the $CP(N-1)$ model.

We define the dual of the tensor $\mathbf{F}_{\mu\nu}$ by

$$\tilde{\mathbf{F}}_{\mu\nu} = \tfrac{1}{2}\epsilon_{\mu\nu\rho\sigma}\mathbf{F}_{\rho\sigma}.$$

Then, the Bogomolnyi inequality

$$\int d^4x \left[\mathbf{F}_{\mu\nu}(x) \pm \tilde{\mathbf{F}}_{\mu\nu}(x)\right]^2 \geq 0$$

implies

$$\mathcal{S}(\mathbf{A}_\mu) \geq 8\pi^2|Q(\mathbf{A}_\mu)|/g^2$$

with

$$Q(\mathbf{A}_\mu) = \frac{1}{32\pi^2}\int d^4x\,\mathbf{F}_{\mu\nu}\cdot\tilde{\mathbf{F}}_{\mu\nu}. \tag{57}$$

The expression $Q(\mathbf{A}_\mu)$ is proportional to the integral of the chiral anomaly (36), here written in $SO(3)$ notation.

We have already pointed out that the quantity $\mathbf{F}_{\mu\nu}\cdot\tilde{\mathbf{F}}_{\mu\nu}$ is a pure divergence (37):

$$\mathbf{F}_{\mu\nu}\cdot\tilde{\mathbf{F}}_{\mu\nu} = \partial_\mu V_\mu$$

with

$$V_\mu = -4\,\epsilon_{\mu\nu\rho\sigma}\,\mathrm{tr}\left(\mathfrak{A}_\nu\partial_\rho\mathfrak{A}_\sigma + \frac{2}{3}\mathfrak{A}_\nu\mathfrak{A}_\rho\mathfrak{A}_\sigma\right)$$
$$= 2\epsilon_{\mu\nu\rho\sigma}\left[\mathbf{A}_\nu\cdot\partial_\rho\mathbf{A}_\sigma + \tfrac{1}{3}\mathbf{A}_\nu\cdot(\mathbf{A}_\rho\times\mathbf{A}_\sigma)\right]. \tag{58}$$

The integral thus depends only on the behaviour of the gauge field at large distances and its values are quantized (40). Here again, as in the $CP(N-1)$ model, the bound involves a topological charge: $Q(\mathbf{A}_\mu)$.

Stokes theorem implies

$$\int_D d^4x\,\partial_\mu V_\mu = \int_{\partial D} d\Omega\,\hat{n}_\mu V_\mu,$$

where $d\Omega$ is the measure on the boundary ∂D of the four-volume D and $\hat{n}_\mu$ the unit vector normal to ∂D. We take for D a sphere of large radius R and find for the topological charge

$$Q(\mathbf{A}_\mu) = \frac{1}{32\pi^2}\int d^4x\,\mathrm{tr}\,\mathbf{F}_{\mu\nu}\cdot\tilde{\mathbf{F}}_{\mu\nu} = \frac{1}{32\pi^2}R^3\int_{r=R} d\Omega\,\hat{n}_\mu V_\mu, \tag{59}$$

The finiteness of the action implies that the classical solution must asymptotically become a pure gauge, that is, with our conventions,

$$\mathfrak{A}_\mu = -\tfrac{1}{2}i\mathbf{A}_\mu\cdot\sigma = \mathbf{g}(x)\partial_\mu\mathbf{g}^{-1}(x) + O\left(|x|^{-2}\right)\quad |x|\to\infty. \tag{60}$$

The element $\mathbf{g}$ of the $SU(2)$ group can be parametrized in terms of Pauli matrices:

$$\mathbf{g} = u_4 \mathbf{1} + i\mathbf{u} \cdot \sigma, \tag{61}$$

where the four-component real vector $(u_4, \mathbf{u})$ satisfies

$$u_4^2 + \mathbf{u}^2 = 1,$$

and thus belongs to the unit sphere S_3. Since $SU(2)$ is topologically equivalent to the sphere S_3, the pure gauge configurations on a sphere of large radius $|x| = R$ define a mapping from S_3 to S_3. Such mappings belong to different homotopy classes that are characterized by an integer called the *winding number*. Here, we identify the homotopy group $\pi_3(S_3)$, which again is isomorphic to the additive group of integers $\mathbb{Z}$.

The simplest one to one mapping corresponds to an element of the form

$$\mathbf{g}(x) = \frac{x_4 \mathbf{1} + i\mathbf{x} \cdot \sigma}{r}, \quad r = (x_4^2 + \mathbf{x}^2)^{1/2}$$

and thus

$$A_m^i \underset{r \to \infty}{\sim} 2\,(x_4 \delta_{im} + \epsilon_{imk} x_k)\, r^{-2}, \quad A_4^i = -2x_i r^{-2}.$$

Note that the transformation

$$\mathbf{g}(x) \mapsto \mathbf{U}_1 \mathbf{g}(x) \mathbf{U}_2^\dagger = \mathbf{g}(\mathbf{R}x),$$

where $\mathbf{U}_1$ and $\mathbf{U}_2$ are two constant $SU(2)$ matrices, induces a $SO(4)$ rotation of matrix $\mathbf{R}$ of the vector x_μ. Then,

$$\mathbf{U}_2 \partial_\mu \mathbf{g}^\dagger(x) \mathbf{U}_1^\dagger = R_{\mu\nu} \partial_\nu \mathbf{g}^\dagger(\mathbf{R}x), \quad \mathbf{U}_1 \mathbf{g}(x) \partial_\mu \mathbf{g}^\dagger(x) \mathbf{U}_1^\dagger = \mathbf{g}(\mathbf{R}x) R_{\mu\nu} \partial_\nu \mathbf{g}^\dagger(\mathbf{R}x)$$

and, therefore,

$$\mathbf{U}_1 \mathfrak{A}_\mu(x) \mathbf{U}_1^\dagger = R_{\mu\nu} \mathfrak{A}_\nu(\mathbf{R}x).$$

Introducing this relation into the definition (58) of V_μ, one verifies that the dependence on the matrix $\mathbf{U}_1$ cancels in the trace and, thus, V_μ transforms like a 4-vector. Since only one vector is available, and taking into account dimensional analysis, one concludes that

$$V_\mu \propto x_\mu / r^4.$$

For $r \to \infty$, $\mathbf{A}_\mu$ approaches a pure gauge (60) and, therefore, V_μ can be transformed into

$$V_\mu \underset{r \to \infty}{\sim} -\frac{1}{3} \epsilon_{\mu\nu\rho\sigma} \mathbf{A}_\nu \cdot (\mathbf{A}_\rho \times \mathbf{A}_\sigma).$$

It is sufficient to calculate V_1. We choose $\rho = 3, \sigma = 4$ and multiply by a factor six to take into account all other choices. Then,

$$V_1 \underset{r \to \infty}{\sim} 16\epsilon_{ijk}(x_4 \delta_{2i} + \epsilon_{i2l} x_l)(x_4 \delta_{3j} + \epsilon_{j3m} x_m) x_k / r^6 = 16x_1 / r^4$$

and, thus,

$$V_\mu \sim 16 x_\mu / r^4 = 16 \hat{n}_\mu / R^3.$$

The powers of R in (59) cancel and since $\int d\Omega = 2\pi^2$, the value of the topological charge is simply

$$Q(\mathbf{A}_\mu) = 1.$$

Comparing this result with (40), we see that we have indeed found the minimal action solution.

Without explicit calculation we know already, from the analysis of the index of the Dirac operator, that the topological charge is an integer:

$$Q(\mathbf{A}_\mu) = \frac{1}{32\pi^2} \int d^4x\, \mathbf{F}_{\mu\nu} \cdot \tilde{\mathbf{F}}_{\mu\nu} = n \in \mathbb{Z}.$$

As in the case of the $CP(N-1)$ model, this result has a geometric interpretation. In general, in the parametrization (61),

$$V_\mu \underset{r\to\infty}{\sim} \frac{8}{3} \epsilon_{\mu\nu\rho\sigma} \epsilon_{\alpha\beta\gamma\delta} u_\alpha \partial_\nu u_\beta \partial_\rho u_\gamma \partial_\sigma u_\delta .$$

A few algebraic manipulations starting from

$$\int_{S_3} R^3 d\Omega\, \hat{n}_\mu V_\mu = \frac{1}{6} \epsilon_{\mu\nu\rho\sigma} \int V_\mu du_\nu \wedge du_\rho \wedge du_\sigma ,$$

then yield

$$Q = \frac{1}{12\pi^2} \epsilon_{\mu\nu\rho\sigma} \int u_\mu du_\nu \wedge du_\rho \wedge du_\sigma , \tag{62}$$

where the notation of exterior differential calculus again has been used. The area Σ_p of the sphere S_{p-1} in the same notation can be written as

$$\Sigma_p = \frac{2\pi^{p/2}}{\Gamma(p/2)} = \frac{1}{(p-1)!} \epsilon_{\mu_1 \dots \mu_p} \int u_{\mu_1} du_{\mu_2} \wedge \dots \wedge du_{\mu_p} ,$$

when the vector u_μ describes the sphere S_{p-1} only once. In the r.h.s. of (62), one thus recognizes an expression proportional to the area of the sphere S_3. Because in general u_μ describes S_3 n times when x_μ describes S_3 only once, a factor n is generated.

The inequality (57) then implies

$$S(\mathbf{A}_\mu) \geq 8\pi^2 |n|/g^2 .$$

The equality, which corresponds to a local minimum of the action, is obtained for fields satisfying the *self-duality equations*

$$\mathbf{F}_{\mu\nu} = \pm \tilde{\mathbf{F}}_{\mu\nu} .$$

These equations, unlike the general classical field equations (56), are first order partial differential equations and, thus, easier to solve. The one-instanton solution, which depends on an arbitrary scale parameter λ, is

$$A^i_m = \frac{2}{r^2 + \lambda^2} (x_4 \delta_{im} + \epsilon_{imk} x_k), \quad m = 1, 2, 3, \quad A^i_4 = -\frac{2x_i}{r^2 + \lambda^2}. \tag{63}$$

The Semi-classical Vacuum. We now proceed in analogy with the analysis of the $CP(N-1)$ model. In the temporal gauge $\mathbf{A}_4 = 0$, the classical minima of the potential correspond to gauge field components $\mathbf{A}_i$, $i = 1, 2, 3$, which are pure gauge functions of the three space variables x_i:

$$\mathfrak{A}_m = -\tfrac{1}{2}i\mathbf{A}_m \cdot \sigma = \mathbf{g}(x_i)\partial_m\mathbf{g}^{-1}(x_i).$$

The structure of the classical minima is related to the homotopy classes of mappings of the group elements $\mathbf{g}$ into compactified $\mathbb{R}^3$ (because $\mathbf{g}(x)$ goes to a constant for $|x| \to \infty$), that is again of S_3 into S_3 and thus the semi-classical vacuum, as in the $CP(N-1)$ model, has a periodic structure. One verifies that the instanton solution (63), transported into the temporal gauge by a gauge transformation, connects minima with different winding numbers. Therefore, as in the case of the $CP(N-1)$ model, to project onto a θ-vacuum, one adds a term to the classical action of gauge theories:

$$\mathcal{S}_\theta(\mathbf{A}_\mu) = \mathcal{S}(\mathbf{A}_\mu) + \frac{i\theta}{32\pi^2}\int d^4x\, \mathbf{F}_{\mu\nu} \cdot \tilde{\mathbf{F}}_{\mu\nu},$$

and then integrates over all fields $\mathbf{A}_\mu$ without restriction. At least in the semi-classical approximation, the gauge theory thus depends on one additional parameter, the angle θ. For non-vanishing values of θ, the additional term violates CP conservation and is at the origin of the strong CP violation problem: Except if θ vanishes for some as yet unknown reason then, according to experimental data, it can only be unnaturally small.

5.4   Fermions in an Instanton Background

We now apply this analysis to QCD, the theory of strong interactions, where N_F Dirac fermions $\mathbf{Q}$, $\bar{\mathbf{Q}}$, the quark fields, are coupled to non-abelian gauge fields $\mathbf{A}_\mu$ corresponding to the $SU(3)$ colour group. We return here to standard $SU(3)$ notation with generators of the Lie Algebra and gauge fields being represented by anti-hermitian matrices. The action can then be written as

$$\mathcal{S}(\mathbf{A}_\mu, \bar{\mathbf{Q}}, \mathbf{Q}) = -\int d^4x \left[\frac{1}{4g^2}\,\mathrm{tr}\,\mathbf{F}_{\mu\nu}^2 + \sum_{f=1}^{N_f}\bar{\mathbf{Q}}_f\left(\slashed{D} + m_f\right)\mathbf{Q}_f \right].$$

The existence of abelian anomalies and instantons has several physical consequences. We mention here two of them.

The Strong CP Problem. According to the analysis of Sect. 4.5, only configurations with a non-vanishing index of the Dirac operator contribute to the θ-term. Then, the Dirac operator has at least one vanishing eigenvalue. If one fermion field is massless, the determinant resulting from the fermion integration thus vanishes, the instantons do not contribute to the functional integral and the strong CP violation problem is solved. However, such an hypothesis seems to be inconsistent with experimental data on quark masses. Another scheme is based on a scalar field, the *axion*, which unfortunately has remained, up to now, experimentally invisible.

The Solution of the $U(1)$ Problem. Experimentally it is observed that the masses of a number of pseudo-scalar mesons are smaller or even much smaller (in the case of pions) than the masses of the corresponding scalar mesons. This strongly suggests that pseudo-scalar mesons are almost Goldstone bosons associated with an approximate chiral symmetry realized in a phase of spontaneous symmetry breaking. (When a continuous (non gauge) symmetry is spontaneously broken, the spectrum of the theory exhibits massless scalar particles called Goldstone bosons.) This picture is confirmed by its many other phenomenological consequences.

In the Standard Model, this approximate symmetry is viewed as the consequence of the very small masses of the **u** and **d** quarks and the moderate value of the strange **s** quark mass.

Indeed, in a theory in which the quarks are massless, the action has a chiral $U(N_{\mathrm{F}}) \times U(N_{\mathrm{F}})$ symmetry, in which N_{F} is the number of flavours. The spontaneous breaking of chiral symmetry to its diagonal subgroup $U(N_{\mathrm{F}})$ leads to expect N_{F}^2 Goldstone bosons associated with all axial currents (corresponding to the generators of $U(N) \times U(N)$ that do not belong to the remaining $U(N)$ symmetry group). In the physically relevant theory, the masses of quarks are non-vanishing but small, and one expects this picture to survive approximately with, instead of Goldstone bosons, light pseudo-scalar mesons.

However, the experimental mass pattern is consistent only with a slightly broken $SU(2) \times SU(2)$ and more badly violated $SU(3) \times SU(3)$ symmetries.

From the preceding analysis, we know that the axial current corresponding to the $U(1)$ abelian subgroup has an anomaly. The WT identities, which imply the existence of Goldstone bosons, correspond to constant group transformations and, thus, involve only the space integral of the divergence of the current. Since the anomaly is a total derivative, one might have expected the integral to vanish. However, non-abelian gauge theories have configurations that give non-vanishing values of the form (40) to the space integral of the anomaly (36). For small couplings, these configurations are in the neighbourhood of instanton solutions (as discussed in Sect. 5.3). This indicates (though no satisfactory calculation of the instanton contribution has been performed yet) that for small, but non-vanishing, quark masses the $U(1)$ axial current is far from being conserved and, therefore, no corresponding light almost Goldstone boson is generated.

Instanton contributions to the anomaly thus resolve a long standing experimental puzzle.

Note that the usual derivation of WT identities involves only global chiral transformations and, therefore, there is no need to introduce axial currents. In the case of massive quarks, chiral symmetry is explicitly broken by soft mass terms and WT identities involve insertions of the operators

$$\mathcal{M}_f = m_f \int \mathrm{d}^4x \, \bar{\mathbf{Q}}_f(x)\gamma_5\mathbf{Q}_f(x),$$

which are the variations of the mass terms in an infinitesimal chiral transformation. If the contributions of $\mathcal{M}_f$ vanish when $m_f \to 0$, as one would normally expect, then a situation of approximate chiral symmetry is realized (in a symmetric or spontaneously broken phase). However, if one integrates over fermions first, at fixed gauge fields, one finds (disconnected) contributions proportional to

$$\langle \mathcal{M}_f \rangle = m_f \operatorname{tr} \gamma_5 \left(\slashed{D} + m_f \right)^{-1}.$$

We have shown in Sect. 4.5) that, for topologically non-trivial gauge field configurations, $\slashed{D}$ has zero eigenmodes, which for $m_f \to 0$ give the leading contributions

$$\langle \mathcal{M}_f \rangle = m_f \sum_n \int \mathrm{d}^4 x \, \varphi_n^*(x) \gamma_5 \varphi_n(x) \frac{1}{m_f} + O(m_f)$$
$$= (n_+ - n_-) + O(m_f).$$

These contributions do not vanish for $m_f \to 0$ and are responsible, after integration over gauge fields, of a violation of chiral symmetry.

6   Non-Abelian Anomaly

We first consider the problem of conservation of a general axial current in a non-abelian vector gauge theory and, then, the issue of obstruction to gauge invariance in chiral gauge theories.

6.1   General Axial Current

We now discuss the problem of the conservation of a general axial current in the example of an action with N massless Dirac fermions in the background of non-abelian vector gauge fields. The corresponding action can be written as

$$\mathcal{S}(\psi, \bar{\psi}; A) = -\int \mathrm{d}^4 x \, \bar{\psi}_i(x) \slashed{D} \psi_i(x).$$

In the absence of gauge fields, the action $\mathcal{S}(\psi, \bar{\psi}; 0)$ has a $U(N) \times U(N)$ symmetry corresponding to the transformations

$$\psi' = \left[\tfrac{1}{2}(1 + \gamma_5) \mathbf{U}_+ + \tfrac{1}{2}(1 - \gamma_5) \mathbf{U}_- \right] \psi \,,$$
$$\bar{\psi}' = \bar{\psi} \left[\tfrac{1}{2}(1 + \gamma_5) \mathbf{U}_-^\dagger + \tfrac{1}{2}(1 - \gamma_5) \mathbf{U}_+^\dagger \right], \tag{64}$$

where $\mathbf{U}_\pm$ are $N \times N$ unitary matrices. We denote by $\mathbf{t}^\alpha$ the anti-hermitian generators of $U(N)$:

$$\mathbf{U} = \mathbf{1} + \theta_\alpha \mathbf{t}^\alpha + O(\theta^2).$$

Vector currents correspond to the diagonal $U(N)$ subgroup of $U(N) \times U(N)$, that is to transformations such that $\mathbf{U}_+ = \mathbf{U}_-$ as one verifies from (64). We couple gauge fields A_μ^α to all vector currents and define

$$\mathbf{A}_\mu = \mathbf{t}^\alpha A_\mu^\alpha.$$

We define axial currents in terms of the infinitesimal space-dependent chiral transformations

$$\mathbf{U}_\pm = \mathbf{1} \pm \theta_\alpha(x)\mathbf{t}^\alpha + O(\theta^2) \;\Rightarrow\; \delta\psi = \theta_\alpha(x)\gamma_5\mathbf{t}^\alpha\psi, \quad \delta\bar\psi = \theta_\alpha(x)\bar\psi\gamma_5\mathbf{t}^\alpha.$$

The variation of the action then reads

$$\delta\mathcal{S} = \int \mathrm{d}^4x \left\{ J_\mu^{5\alpha}(x)\partial_\mu\theta_\alpha(x) + \theta_\alpha(x)\bar\psi(x)\gamma_5\gamma_\mu[\mathbf{A}_\mu, \mathbf{t}^\alpha]\psi(x) \right\},$$

where $J_\mu^{5\alpha}(x)$ is the axial current:

$$J_\mu^{5\alpha}(x) = \bar\psi\gamma_5\gamma_\mu\mathbf{t}^\alpha\psi.$$

Since the gauge group has a non-trivial intersection with the chiral group, the commutator $[\mathbf{A}_\mu, \mathbf{t}^\alpha]$ no longer vanishes. Instead,

$$[\mathbf{A}_\mu, \mathbf{t}^\alpha] = A_\mu^\beta f_{\beta\alpha\gamma}\mathbf{t}^\gamma,$$

where the $f_{\beta\alpha\gamma}$ are the totally antisymmetric structure constants of the Lie algebra of $U(N)$. Thus,

$$\delta\mathcal{S} = \int \mathrm{d}^4x \, \theta_\alpha(x) \left\{ -\partial_\mu J_\mu^{5\alpha}(x) + f_{\beta\alpha\gamma}A_\mu^\beta(x)J_\mu^{5\gamma}(x) \right\}.$$

The classical current conservation equation is replaced by the gauge covariant conservation equation

$$\mathbf{D}_\mu J_\mu^{5\alpha} = 0,$$

where we have defined the covariant divergence of the current by

$$\left(\mathbf{D}_\mu J_\mu^5\right)_\alpha \equiv \partial_\mu J_\mu^{5\alpha} + f_{\alpha\beta\gamma}A_\mu^\beta J_\mu^{5\gamma}.$$

In the contribution to the anomaly, the terms quadratic in the gauge fields are modified, compared to the expression (36), only by the appearance of a new geometric factor. Then the complete form of the anomaly is dictated by gauge covariance. One finds

$$\mathbf{D}_\lambda J_\lambda^{5\alpha}(x) = -\frac{1}{16\pi^2}\epsilon_{\mu\nu\rho\sigma} \operatorname{tr}\mathbf{t}^\alpha\mathbf{F}_{\mu\nu}\mathbf{F}_{\rho\sigma}.$$

This is the result for the most general chiral and gauge transformations. If we restrict both groups in such a way that the gauge group has an empty intersection with the chiral group, the anomaly becomes proportional to $\operatorname{tr}\mathbf{t}^\alpha$, where $\mathbf{t}^\alpha$ are the generators of the chiral group $G \times G$ and is, therefore, different from zero only for the abelian factors of G.

6.2 Obstruction to Gauge Invariance

We now consider left-handed (or right-handed) fermions coupled to a non-abelian gauge field. The action takes the form

$$\mathcal{S}(\bar{\psi}, \psi; A) = -\int d^4x \, \bar{\psi}(x)\tfrac{1}{2}\left(1 + \gamma_5\right)\slashed{D}\,\psi(x)$$

(the discussion with $\tfrac{1}{2}(1 - \gamma_5)$ is similar).

The gauge theory is consistent only if the partition function

$$\mathcal{Z}(\mathbf{A}_\mu) = \int [d\psi d\bar{\psi}]\exp\left[-\mathcal{S}(\psi, \bar{\psi}; A)\right]$$

is gauge invariant.

We introduce the generators $\mathbf{t}^\alpha$ of the gauge group in the fermion representation and define the corresponding current by

$$J_\mu^\alpha(x) = \bar{\psi}\tfrac{1}{2}\left(1 + \gamma_5\right)\gamma_\mu \mathbf{t}^\alpha \psi.$$

Again, the invariance of $\mathcal{Z}(\mathbf{A}_\mu)$ under an infinitesimal gauge transformation implies for the current $\mathbf{J}_\mu = J_\mu^\alpha \mathbf{t}^\alpha$ the covariant conservation equation

$$\langle \mathbf{D}_\mu \mathbf{J}_\mu \rangle = 0$$

with

$$\mathbf{D}_\mu = \partial_\mu + [\mathbf{A}_\mu, \bullet].$$

The calculation of the quadratic contribution to the anomaly is simple: the first regularization adopted for the calculation in Sect. 4.2 is also suited to the present situation since the current-gauge field three-point function is symmetric in the external arguments. The group structure is reflected by a simple geometric factor. The global factor can be taken from the abelian calculation. It differs from result (26) by a factor $1/2$ that comes from the projector $\tfrac{1}{2}(1 + \gamma_5)$. The general form of the term of degree 3 in the gauge field can also easily be found while the calculation of the global factor is somewhat tedious. We show in Sect. 6.3 that it can be obtained from consistency conditions. The complete expression then reads

$$(\mathbf{D}_\mu \mathbf{J}_\mu(x))^\alpha = -\frac{1}{24\pi^2}\partial_\mu \epsilon_{\mu\nu\rho\sigma}\, \mathrm{tr}\left[\mathbf{t}^\alpha\left(\mathbf{A}_\nu \partial_\rho \mathbf{A}_\sigma + \tfrac{1}{2}\mathbf{A}_\nu \mathbf{A}_\rho \mathbf{A}_\sigma\right)\right]. \qquad (65)$$

If the projector $\tfrac{1}{2}(1 + \gamma_5)$ is replaced by $\tfrac{1}{2}(1 - \gamma_5)$, the sign of the anomaly changes.

Unless the anomaly vanishes identically, there is an obstruction to the construction of the gauge theory. The first term is proportional to

$$d_{\alpha\beta\gamma} = \tfrac{1}{2}\,\mathrm{tr}\left[\mathbf{t}^\alpha\left(\mathbf{t}^\beta \mathbf{t}^\gamma + \mathbf{t}^\gamma \mathbf{t}^\beta\right)\right].$$

The second term involves the product of four generators, but taking into account the antisymmetry of the ϵ tensor, one product of two consecutive can be replaced by a commutator. Therefore, the term is also proportional to $d_{\alpha\beta\gamma}$.

For a unitary representation the generators $\mathbf{t}^\alpha$ are, with our conventions, antihermitian. Therefore, the coefficients $d_{\alpha\beta\gamma}$ are purely imaginary:

$$d^*_{\alpha\beta\gamma} = \tfrac{1}{2} \operatorname{tr} \left[\mathbf{t}^\alpha \left(\mathbf{t}^\beta \mathbf{t}^\gamma + \mathbf{t}^\gamma \mathbf{t}^\beta\right)\right]^\dagger = -d_{\alpha\beta\gamma}\,.$$

These coefficients vanish for all representations that are real: the $\mathbf{t}^\alpha$ antisymmetric, or pseudo-real, that is $\mathbf{t}^\alpha = -S\,{}^T\mathbf{t}^\alpha S^{-1}$. It follows that the only non-abelian groups that can lead to anomalies in four dimensions are $SU(N)$ for $N \geq 3$, $SO(6)$, and E_6.

6.3 Wess–Zumino Consistency Conditions

In Sect. 6.2, we have calculated the part of the anomaly that is quadratic in the gauge field and asserted that the remaining non-quadratic contributions could be obtained from geometric arguments. The anomaly is the variation of a functional under an infinitesimal gauge transformation. This implies compatibility conditions, which here are constraints on the general form of the anomaly, the Wess–Zumino consistency conditions. One convenient method to derive these constraints is based on BRS transformations: one expresses that BRS transformations are nilpotent.

In a BRS transformation, the variation of the gauge field $\mathbf{A}_\mu$ takes the form

$$\delta_{\mathrm{BRS}}\mathbf{A}_\mu(x) = \mathbf{D}_\mu\mathbf{C}(x)\bar\varepsilon\,, \tag{66}$$

where $\mathbf{C}$ is a fermion spinless "ghost" field and $\bar\varepsilon$ an anticommuting constant. The corresponding variation of $\ln \mathcal{Z}(\mathbf{A}_\mu)$ is

$$\delta_{\mathrm{BRS}} \ln \mathcal{Z}(\mathbf{A}_\mu) = - \int \mathrm{d}^4x \, \langle\mathbf{J}_\mu(x)\rangle \, \mathbf{D}_\mu\mathbf{C}(x)\bar\varepsilon\,. \tag{67}$$

The anomaly equation has the general form

$$\langle\mathbf{D}_\mu\mathbf{J}_\mu(x)\rangle = \mathcal{A}\left(\mathbf{A}_\mu; x\right).$$

In terms of $\mathcal{A}$, the equation (67), after an integration by parts, can be rewritten as

$$\delta_{\mathrm{BRS}} \ln \mathcal{Z}(\mathbf{A}_\mu) = \int \mathrm{d}^4x \, \mathcal{A}\left(\mathbf{A}_\mu; x\right) \mathbf{C}(x)\bar\varepsilon\,.$$

Since the r.h.s. is a BRS variation, it satisfies a non-trivial constraint obtained by expressing that the square of the BRS operator δ_{BRS} vanishes (it has the property of a cohomology operator):

$$\delta^2_{\mathrm{BRS}} = 0$$

and called the Wess–Zumino consistency conditions.

To calculate the BRS variation of $\mathcal{A}\mathbf{C}$, we need also the BRS transformation of the fermion ghost $\mathbf{C}(x)$:

$$\delta_{\mathrm{BRS}}\mathbf{C}(x) = \bar\varepsilon\,\mathbf{C}^2(x)\,. \tag{68}$$

The condition that $\mathcal{A}C$ is BRS invariant,

$$\delta_{\mathrm{BRS}} \int \mathrm{d}^4 x\, \mathcal{A}\left(\mathbf{A}_\mu; x\right) \mathbf{C}(x) = 0\,,$$

yields a constraint on the possible form of anomalies that determines the term cubic in $\mathbf{A}$ in the r.h.s. of (65) completely. One can verify that

$$\delta_{\mathrm{BRS}}\, \epsilon_{\mu\nu\rho\sigma} \int \mathrm{d}^4 x\, \mathrm{tr}\left[\mathbf{C}(x)\partial_\mu \left(\mathbf{A}_\nu \partial_\rho \mathbf{A}_\sigma + \tfrac{1}{2}\mathbf{A}_\nu \mathbf{A}_\rho \mathbf{A}_\sigma\right)\right] = 0\,.$$

Explicitly, after integration by parts, the equation takes the form

$$\epsilon_{\mu\nu\rho\sigma}\, \mathrm{tr} \int \mathrm{d}^4 x\, \left\{\partial_\mu \mathbf{C}^2(x)\mathbf{A}_\nu \partial_\rho \mathbf{A}_\sigma + \partial_\mu \mathbf{C}\mathbf{D}_\nu \mathbf{C}\partial_\rho \mathbf{A}_\sigma + \partial_\mu \mathbf{C}\mathbf{A}_\nu \partial_\rho \mathbf{D}_\sigma \mathbf{C} \right.$$
$$\left. + \tfrac{1}{2}\partial_\mu \mathbf{C}^2(x)\mathbf{A}_\nu \mathbf{A}_\rho \mathbf{A}_\sigma + \tfrac{1}{2}\partial_\mu \mathbf{C}\left(\mathbf{D}_\nu \mathbf{C}\mathbf{A}_\rho \mathbf{A}_\sigma + \mathbf{A}_\nu \mathbf{D}_\rho \mathbf{C}\mathbf{A}_\sigma + \mathbf{A}_\nu \mathbf{A}_\rho \mathbf{D}_\sigma \mathbf{C}\right)\right\} = 0\,.$$

The terms linear in A, after integrating by parts the first term and using the antisymmetry of the ϵ symbol, cancels automatically:

$$\epsilon_{\mu\nu\rho\sigma}\, \mathrm{tr} \int \mathrm{d}^4 x\, \left(\partial_\mu \mathbf{C}\partial_\nu \mathbf{C}\partial_\rho \mathbf{A}_\sigma + \partial_\mu \mathbf{C}\mathbf{A}_\nu \partial_\rho \partial_\sigma \mathbf{C}\right) = 0\,.$$

In the same way, the cubic terms cancel (the anticommuting properties of $\mathbf{C}$ have to be used):

$$\epsilon_{\mu\nu\rho\sigma}\, \mathrm{tr} \int \mathrm{d}^4 x\, \left\{\left(\partial_\mu \mathbf{C}\mathbf{C} + \mathbf{C}\partial_\mu \mathbf{C}\right)\mathbf{A}_\nu \mathbf{A}_\rho \mathbf{A}_\sigma + \partial_\mu \mathbf{C}\left([\mathbf{A}_\nu, \mathbf{C}]\mathbf{C}\mathbf{A}_\rho \mathbf{A}_\sigma \right.\right.$$
$$\left.\left. + \mathbf{A}_\nu[\mathbf{A}_\rho, \mathbf{C}]\mathbf{A}_\sigma + \mathbf{A}_\nu \mathbf{A}_\rho[\mathbf{A}_\sigma, \mathbf{C}]\right)\right\} = 0\,.$$

It is only the quadratic terms that give a relation between the quadratic and cubic terms in the anomaly, both contributions being proportional to

$$\epsilon_{\mu\nu\rho\sigma}\, \mathrm{tr} \int \mathrm{d}^4 x\, \partial_\mu \mathbf{C}\partial_\nu \mathbf{C}\mathbf{A}_\rho \mathbf{A}_\sigma\,.$$

7   Lattice Fermions: Ginsparg–Wilson Relation

Notation. We now return to the problem of lattice fermions discussed in Sect. 3.4. For convenience we set the lattice spacing $a = 1$ and use for the fields the notation $\psi(x) \equiv \psi_x$.

Ginsparg–Wilson Relation. It had been noted, many years ago, that a potential way to avoid the doubling problem while still retaining chiral properties in the continuum limit was to look for lattice Dirac operators $\mathbf{D}$ that, instead of anticommuting with γ_5, would satisfy the relation

$$\mathbf{D}^{-1}\gamma_5 + \gamma_5 \mathbf{D}^{-1} = \gamma_5 \mathbf{1} \tag{69}$$

where **1** stands for the identity both for lattice sites and in the algebra of γ-matrices. More explicitly,

$$(\mathbf{D}^{-1})_{xy}\gamma_5 + \gamma_5(\mathbf{D}^{-1})_{xy} = \gamma_5\delta_{xy}\,.$$

More generally, the r.h.s. can be replaced by any local positive operator on the lattice: locality of a lattice operator is defined by a decrease of its matrix elements that is at least exponential when the points x, y are separated. The anti-commutator being local, it is expected that it does not affect correlation functions at large distance and that chiral properties are recovered in the continuum limit. Note that when **D** is the Dirac operator in a gauge background, the condition (69) is gauge invariant.

However, lattice Dirac operators solutions to the Ginsparg–Wilson relation (69) have only recently been discovered because the demands that both **D** and the anticommutator $\{\mathbf{D}^{-1}, \gamma_5\}$ should be local seemed difficult to satisfy, specially in the most interesting case of gauge theories.

Note that while relation (69) implies some generalized form of chirality on the lattice, as we now show, it does not guarantee the absence of doublers, as examples illustrate. But the important point is that in this class solutions can be found without doublers.

7.1   Chiral Symmetry and Index

We first discuss the main properties of a Dirac operator satisfying relation (69) and then exhibit a generalized form of chiral transformations on the lattice.

Using the relation, quite generally true for any euclidean Dirac operator satisfying hermiticity and reflection symmetry (see textbooks on symmetries of euclidean fermions),

$$\mathbf{D}^\dagger = \gamma_5\mathbf{D}\gamma_5\,, \tag{70}$$

one can rewrite relation (69), after multiplication by γ_5, as

$$\mathbf{D}^{-1} + (\mathbf{D}^{-1})^\dagger = \mathbf{1}$$

and, therefore,

$$\mathbf{D} + \mathbf{D}^\dagger = \mathbf{D}\mathbf{D}^\dagger = \mathbf{D}^\dagger\mathbf{D}\,. \tag{71}$$

This implies that the lattice operator **D** has an index and, in addition, that

$$\mathbf{S} = \mathbf{1} - \mathbf{D} \tag{72}$$

is unitary:

$$\mathbf{S}\mathbf{S}^\dagger = \mathbf{1}\,.$$

The eigenvalues of **S** lie on the unit circle. The eigenvalue one corresponds to the pole of the Dirac propagator.

Note also the relations

$$\gamma_5\mathbf{S} = \mathbf{S}^\dagger\gamma_5\,, \quad (\gamma_5\mathbf{S})^2 = \mathbf{1}\,. \tag{73}$$

The matrix $\gamma_5 \mathbf{S}$ is hermitian and $\frac{1}{2}(1 \pm \gamma_5 \mathbf{S})$ are two orthogonal projectors. If $\mathbf{D}$ is a Dirac operator in a gauge background, these projectors depend on the gauge field.

It is then possible to construct lattice actions that have a chiral symmetry that corresponds to local but non point-like transformations. In the abelian example,

$$\psi'_x = \sum_y \left(e^{i\theta\gamma_5 \mathbf{S}}\right)_{xy} \psi_y \,, \quad \bar{\psi}'_x = \bar{\psi}_x e^{i\theta\gamma_5} \,. \tag{74}$$

(The reader is reminded that in the formalism of functional integrals, ψ and $\bar{\psi}$ are independent integration variables and, thus, can be transformed independently.) Indeed, the invariance of the lattice action $\mathcal{S}(\bar{\psi}, \psi)$,

$$\mathcal{S}(\bar{\psi}, \psi) = \sum_{x,y} \bar{\psi}_x \mathbf{D}_{xy} \psi_y = \mathcal{S}(\bar{\psi}', \psi'),$$

is implied by

$$e^{i\theta\gamma_5} \mathbf{D} e^{i\theta\gamma_5 \mathbf{S}} = \mathbf{D} \iff \mathbf{D} e^{i\theta\gamma_5 \mathbf{S}} = e^{-i\theta\gamma_5} \mathbf{D} \,.$$

Using the second relation in (73), we expand the exponentials and reduce the equation to

$$\mathbf{D}\gamma_5 \mathbf{S} = -\gamma_5 \mathbf{D} \,, \tag{75}$$

which is another form of relation (69).

However, the transformations (74), no longer leave the integration measure of the fermion fields,

$$\prod_x \mathrm{d}\psi_x \mathrm{d}\bar{\psi}_x \,,$$

automatically invariant. The jacobian of the change of variables $\psi \mapsto \psi'$ is

$$J = \det e^{i\theta\gamma_5} e^{i\theta\gamma_5 \mathbf{S}} = \det e^{i\theta\gamma_5(2-\mathbf{D})} = 1 + i\theta \operatorname{tr} \gamma_5(2 - \mathbf{D}) + O(\theta^2), \tag{76}$$

where trace means trace in the space of γ matrices and in the lattice indices. This leaves open the possibility of generating the expected anomalies, when the Dirac operator of the free theory is replaced by the covariant operator in the background of a gauge field, as we now show.

Eigenvalues of the Dirac Operator in a Gauge Background. We briefly discuss the index of a lattice Dirac operator $\mathbf{D}$ satisfying relation (69), in a gauge background. We assume that its spectrum is discrete (this is certainly true on a finite lattice where $\mathbf{D}$ is a matrix). The operator $\mathbf{D}$ is related by (72) to a unitary operator $\mathbf{S}$ whose eigenvalues have modulus one. Therefore, if we denote by $|n\rangle$ its nth eigenvector,

$$\mathbf{D}|n\rangle = (1 - \mathbf{S})|n\rangle = (1 - e^{i\theta_n})|n\rangle \Rightarrow \mathbf{D}^\dagger |n\rangle = (1 - e^{-i\theta_n})|n\rangle \,.$$

Then, using (70), we infer

$$\mathbf{D}\gamma_5 |n\rangle = (1 - e^{-i\theta_n})\gamma_5 |n\rangle \,.$$

The discussion that follows then is analogous to the discussion of Sect. 4.5 to which we refer for details. We note that when the eigenvalues are not real, $\theta_n \neq 0$ (mod π), $\gamma_5 |n\rangle$ is an eigenvector different from $|n\rangle$ because the eigenvalues are different. Instead, in the two subspaces corresponding to the eigenvalues 0 and 2, we can choose eigenvectors with definite chirality

$$\gamma_5 |n\rangle = \pm |n\rangle .$$

We call below $n_\pm$ the number of eigenvalues 0, and $\nu_\pm$ the number of eigenvalues 2 with chirality ± 1.

Note that on a finite lattice δ_{xy} is a finite matrix and, thus,

$$\text{tr} \, \gamma_5 \delta_{xy} = 0 .$$

Therefore,

$$\text{tr} \, \gamma_5 (2 - \mathbf{D}) = - \text{tr} \, \gamma_5 \mathbf{D} ,$$

which implies

$$\sum_n \langle n | \gamma_5 (2 - \mathbf{D}) | n \rangle = - \sum_n \langle n | \gamma_5 \mathbf{D} | n \rangle .$$

In the equation all complex eigenvalues cancel because the vectors $|n\rangle$ and $\gamma_5 |n\rangle$ are orthogonal. The sum reduces to the subspace of real eigenvalues, where the eigenvectors have definite chirality. On the l.h.s. only the eigenvalue 0 contributes, and on the r.h.s. only the eigenvalue 2. We find

$$n_+ - n_- = -(\nu_+ - \nu_-).$$

This equation tells us that the difference between the number of states of different chirality in the zero eigenvalue sector is cancelled by the difference in the sector of eigenvalue two (which corresponds to very massive states).

Remark. It is interesting to note the relation between the spectrum of $\mathbf{D}$ and the spectrum of $\gamma_5 \mathbf{D}$, which from relation (70) is a hermitian matrix,

$$\gamma_5 \mathbf{D} = \mathbf{D}^\dagger \gamma_5 = (\gamma_5 \mathbf{D})^\dagger ,$$

and, thus, diagonalizable with real eigenvalues. It is simple to verify the following two equations, of which the second one is obtained by changing θ into $\theta + 2\pi$,

$$\gamma_5 \mathbf{D} (1 - i e^{i\theta_n/2} \gamma_5) |n\rangle = 2 \sin(\theta_n/2)(1 - i e^{i\theta_n/2} \gamma_5) |n\rangle ,$$
$$\gamma_5 \mathbf{D} (1 + i e^{i\theta_n/2} \gamma_5) |n\rangle = -2 \sin(\theta_n/2)(1 + i e^{i\theta_n/2} \gamma_5) |n\rangle .$$

These equations imply that the eigenvalues $\pm 2 \sin(\theta_n/2)$ of $\gamma_5 \mathbf{D}$ are paired except for $\theta_n = 0$ (mod π) where $|n\rangle$ and $\gamma_5 |n\rangle$ are proportional. For $\theta_n = 0$, $\gamma_5 \mathbf{D}$ has also eigenvalue 0. For $\theta_n = \pi$, $\gamma_5 \mathbf{D}$ has eigenvalue ± 2 depending on the chirality of $|n\rangle$.

In the same way,

$$\gamma_5 (2 - \mathbf{D})(1 + e^{i\theta_n/2} \gamma_5) |n\rangle = 2 \cos(\theta_n/2)(1 + e^{i\theta_n/2} \gamma_5) |n\rangle ,$$
$$\gamma_5 (2 - \mathbf{D})(1 - e^{i\theta_n/2} \gamma_5) |n\rangle = -2 \cos(\theta_n/2)(1 - e^{i\theta_n/2} \gamma_5) |n\rangle .$$

Jacobian and Lattice Anomaly. The variation of the jacobian (76) can now be evaluated. Opposite eigenvalues of $\gamma_5(2 - \mathbf{D})$ cancel. The eigenvalues for $\theta_n = \pi$ give factors one. Only $\theta_n = 0$ gives a non-trivial contribution:

$$J = \det e^{i\theta\gamma_5(2-\mathbf{D})} = e^{2i\theta(n_+ - n_-)}.$$

The quantity $\operatorname{tr}\gamma_5(2 - \mathbf{D})$, coefficient of the term of order θ, is a sum of terms that are local, gauge invariant, pseudoscalar, and topological as the continuum anomaly (36) since

$$\operatorname{tr}\gamma_5(2 - \mathbf{D}) = \sum_n \langle n| \gamma_5(2 - \mathbf{D}) |n\rangle = 2(n_+ - n_-).$$

Non-Abelian Generalization. We now consider the non-abelian chiral transformations

$$\psi_U = \left[\tfrac{1}{2}(1 + \gamma_5 \mathbf{S})\mathbf{U}_+ + \tfrac{1}{2}(1 - \gamma_5 \mathbf{S})\mathbf{U}_-\right] \psi,$$
$$\bar\psi_U = \bar\psi \left[\tfrac{1}{2}(1 + \gamma_5)\mathbf{U}_-^\dagger + \tfrac{1}{2}(1 - \gamma_5)\mathbf{U}_+^\dagger\right], \tag{77}$$

where $\mathbf{U}_\pm$ are matrices belonging to some unitary group G. Near the identity

$$\mathbf{U} = 1 + \mathbf{\Theta} + O(\mathbf{\Theta}^2),$$

where $\mathbf{\Theta}$ is an element of the Lie algebra.

We note that this amounts to define differently chiral components of $\bar\psi$ and ψ, for ψ the definition being even gauge field dependent.

We assume that G is a vector symmetry of the fermion action, and thus the Dirac operator commutes with all elements of the Lie algebra:

$$[\mathbf{D}, \mathbf{\Theta}] = 0.$$

Then, again, the relation (69) in the form (75) implies the invariance of the fermion action:

$$\bar\psi_U \mathbf{D} \psi_U = \bar\psi \mathbf{D} \psi.$$

The jacobian of an infinitesimal chiral transformation $\mathbf{\Theta} = \mathbf{\Theta}_+ = -\mathbf{\Theta}_-$ is

$$J = 1 + \operatorname{tr}\gamma_5 \mathbf{\Theta}(2 - \mathbf{D}) + O(\mathbf{\Theta}^2).$$

Wess–Zumino Consistency Conditions. To determine anomalies in the case of gauge fields coupling differently to fermion chiral components, one can on the lattice also play with the property that BRS transformations are nilpotent. They take the form

$$\delta\mathbf{U}_{xy} = \bar\varepsilon \left(\mathbf{C}_x \mathbf{U}_{xy} - \mathbf{U}_{xy}\mathbf{C}_y\right),$$
$$\delta\mathbf{C}_x = \bar\varepsilon \mathbf{C}_x^2,$$

instead of (66), (68). Moreover, the matrix elements $\mathbf{D}_{xy}$ of the gauge covariant Dirac operator transform like $\mathbf{U}_{xy}$.

7.2   Explicit Construction: Overlap Fermions

An explicit solution of the Ginsparg–Wilson relation without doublers can be derived from operators $\mathbf{D_W}$ that share the properties of the Wilson–Dirac operator of (17), that is which avoid doublers at the price of breaking chiral symmetry explicitly. Setting

$$\mathbf{A} = 1 - \mathbf{D_W}/M\,, \tag{78}$$

where $M > 0$ is a mass parameter that must chosen, in particular, such that $\mathbf{A}$ has no zero eigenvalue, one takes

$$\mathbf{S} = \mathbf{A}\left(\mathbf{A}^\dagger \mathbf{A}\right)^{-1/2} \;\Rightarrow\; \mathbf{D} = 1 - \mathbf{A}\left(\mathbf{A}^\dagger \mathbf{A}\right)^{-1/2}. \tag{79}$$

The matrix $\mathbf{A}$ is such that

$$\mathbf{A}^\dagger = \gamma_5 \mathbf{A}\gamma_5 \;\Rightarrow\; \mathbf{B} = \gamma_5 \mathbf{A} = \mathbf{B}^\dagger.$$

The hermitian matrix $\mathbf{B}$ has real eigenvalues. Moreover,

$$\mathbf{B}^\dagger \mathbf{B} = \mathbf{B}^2 = \mathbf{A}^\dagger \mathbf{A} \;\Rightarrow\; \left(\mathbf{A}^\dagger \mathbf{A}\right)^{1/2} = |\mathbf{B}|.$$

We conclude

$$\gamma_5 \mathbf{S} = \operatorname{sgn}\mathbf{B}\,,$$

where $\operatorname{sgn}\mathbf{B}$ is the matrix with the same eigenvectors as $\mathbf{B}$, but all eigenvalues replaced by their sign. In particular this shows that $(\gamma_5 \mathbf{S})^2 = 1$.

With this ansatz $\mathbf{D}$ has a zero eigenmode when $\mathbf{A}\left(\mathbf{A}^\dagger \mathbf{A}\right)^{-1/2}$ has the eigenvalue one. This can happen when $\mathbf{A}$ and $\mathbf{A}^\dagger$ have the same eigenvector with a *positive* eigenvalue.

This is the idea of overlap fermions, the name overlap refering only to the way this Dirac operator was initially introduced.

Free Fermions. We now verify the absence of doublers for vanishing gauge fields. The Fourier representation of a Wilson–Dirac operator has the general form

$$D_W(p) = \alpha(p) + i\gamma_\mu \beta_\mu(p), \tag{80}$$

where $\alpha(p)$ and $\beta_\mu(p)$ are real, periodic, smooth functions. In the continuum limit, one must recover the usual massless Dirac operator, which implies

$$\beta_\mu(p) \underset{|p|\to 0}{\sim} p_\mu\,, \quad \alpha(p) \geq 0\,, \quad \alpha(p) \underset{|p|\to 0}{=} O(p^2),$$

and $\alpha(p) > 0$ for all values of p_μ such that $\beta_\mu(p) = 0$ for $|p| \neq 0$ (i.e. all values that correspond to doublers). Equation (17) in the limit $m = 0$ provides an explicit example.

Doublers appear if the determinant of the overlap operator $\mathbf{D}$ (78, 79) vanishes for $|\mathbf{p}| \neq 0$. In the example of the operator (80), a short calculation shows that this happens when

$$\left[\sqrt{\left(M - \alpha(p)\right)^2 + \beta_\mu^2(p)} - M + \alpha(p)\right]^2 + \beta_\mu^2(p) = 0\,.$$

This implies $\beta_\mu(p) = 0$, an equation that necessarily admits doubler solutions, and

$$|M - \alpha(p)| = M - \alpha(p).$$

The solutions to this equation depend on the value of $\alpha(p)$ with respect to M for the doubler modes, that is for the values of p such that $\beta_\mu(p) = 0$. If $\alpha(p) \leq M$ the equation is automatically satisfied and the corresponding doubler survives. As mentioned in the introduction to this section, the relation (69) alone does not guarantee the absence of doublers. Instead, if $\alpha(p) > M$, the equation implies $\alpha(p) = M$, which is impossible. Therefore, by rescaling $\alpha(p)$, if necessary, we can keep the wanted $p_\mu = 0$ mode while eliminating all doublers. The modes associated to doublers for $\alpha(p) \leq M$ then, instead, correspond to the eigenvalue 2 for $\mathbf{D}$, and the doubling problem is solved, at least in a free theory.

In presence of a gauge field, the argument can be generalized provided the plaquette terms in the lattice action are constrained to remain sufficiently close to one.

Remark. Let us stress that, if it seems that the doubling problem has been solved from the formal point of view, from the numerical point of view the calculation of the operator $(\mathbf{A}^\dagger \mathbf{A})^{-1/2}$ in a gauge background represents a major challenge.

8   Supersymmetric Quantum Mechanics and Domain Wall Fermions

Because the construction of lattice fermions without doublers we have just described is somewhat artificial, one may wonder whether there is a context in which they would appear more naturally. Therefore, we now briefly outline how a similar lattice Dirac operator can be generated by embedding first four-dimensional space in a larger five-dimensional space. This is the method of domain wall fermions.

Because the general idea behind domain wall fermions has emerged first in another context, as a preparation, we first recall a few properties of the spectrum of the hamiltonian in supersymmetric quantum mechanics, a topic also related to the index of the Dirac operator (Sect. 4.5), and very directly to stochastic dynamics in the form of Langevin or Fokker–Planck equations.

8.1   Supersymmetric Quantum Mechanics

We now construct a quantum theory that exhibits the simplest form of supersymmetry where space–time reduces to time only. We know that this reduces fields to paths and, correspondingly, quantum field theory to simple quantum mechanics.

We first introduce a first order differential operator $\mathfrak{D}$ acting on functions of one real variable, which is a 2×2 matrix (σ_i still are the Pauli matrices):

$$\mathfrak{D} \equiv \sigma_1 \mathrm{d}_x - i\sigma_2 A(x) \tag{81}$$

$(d_x \equiv d/dx)$. The function $A(x)$ is real and, thus, the operator $\mathfrak{D}$ is anti-hermitian.

The operator $\mathfrak{D}$ shares several properties with the Dirac operator of Sect. 4.5. In particular, it satisfies

$$\sigma_3 \mathfrak{D} + \mathfrak{D}\sigma_3 = 0,$$

and, thus, has an index (σ_3 playing the role of γ_5). We introduce the operator

$$D = d_x + A(x) \;\Rightarrow\; D^\dagger = -d_x + A(x),$$

and

$$Q = D \begin{pmatrix} 0 & 0 \\ 1 & 0 \end{pmatrix} \Rightarrow Q^\dagger = D^\dagger \begin{pmatrix} 0 & 1 \\ 0 & 0 \end{pmatrix}.$$

Then,

$$\mathfrak{D} = Q - Q^\dagger,$$
$$Q^2 = (Q^\dagger)^2 = 0. \tag{82}$$

We consider now the positive semi-definite hamiltonian, anticommutator of Q and $Q^\dagger$,

$$H = QQ^\dagger + Q^\dagger Q = -\mathfrak{D}^2 = \begin{pmatrix} D^\dagger D & 0 \\ 0 & DD^\dagger \end{pmatrix}.$$

The relations (82) imply that

$$[H, Q] = [H, Q^\dagger] = 0.$$

The operators $Q, Q^\dagger$ are the generators of the simplest form of a supersymmetric algebra and the hamiltonian H is supersymmetric.

The eigenvectors of H have the form $\psi_+(x)(1,0)$ and $\psi_-(x)(0,1)$ and satisfy, respectively,

$$D^\dagger D \,|\psi_+\rangle = \varepsilon_+ \,|\psi_+\rangle, \quad \text{and} \quad DD^\dagger \,|\psi_-\rangle = \varepsilon_- \,|\psi_-\rangle, \quad \varepsilon_\pm \geq 0, \tag{83}$$

where

$$D^\dagger D = -d_x^2 + A^2(x) - A'(x), \quad DD^\dagger = -d_x^2 + A^2(x) + A'(x).$$

Moreover, if x belongs to a bounded interval or $A(x) \to \infty$ for $|x| \to \infty$, then the spectrum of H is discrete.

Multiplying the first equation in (83) by D, we conclude that if $D\,|\psi_+\rangle \neq 0$ and, thus, ε_+ does not vanish, it is an eigenvector of $DD^\dagger$ with eigenvalue ε_+, and conversely. Therefore, except for a possible ground state with vanishing eigenvalue, the spectrum of H is doubly degenerate.

This observation is consistent with the analysis of Sect. 4.5 applied to the operator $\mathfrak{D}$. We know from that section that either eigenvectors are paired $|\psi\rangle, \sigma_3|\psi\rangle$ with opposite eigenvalues $\pm i\sqrt{\varepsilon}$, or they correspond to the eigenvalue zero and can be chosen with definite chirality

$$\mathfrak{D}\,|\psi\rangle = 0, \quad \sigma_3\,|\psi\rangle = \pm|\psi\rangle.$$

It is convenient to now introduce the function $S(x)$:

$$S'(x) = A(x),$$

and for simplicity discuss only the situation of operators on the entire real line.

We assume that

$$S(x)/|x| \underset{x \to \pm\infty}{\geq} \ell > 0.$$

Then the function $S(x)$ is such that $e^{-S(x)}$ is a normalizable wave function: $\int dx\, e^{-2S(x)} < \infty$.

In the stochastic interpretation where $D^\dagger D$ has the interpretation of a Fokker–Planck hamiltonian generating the time evolution of some probability distribution, $e^{-2S(x)}$ is the equilibrium distribution.

When $e^{-S(x)}$ is normalizable, we know one eigenvector with vanishing eigenvalue and chirality $+1$, which corresponds to the isolated ground state of $D^\dagger D$

$$\mathcal{D}\,|\psi_+, 0\rangle = 0 \iff D\,|\psi_+\rangle = 0, \quad \sigma_3\,|\psi_+, 0\rangle = |\psi_+, 0\rangle$$

with

$$\psi_+(x) = e^{-S(x)}.$$

On the other hand, the formal solution of $D^\dagger|\psi_-\rangle = 0$,

$$\psi_-(x) = e^{S(x)},$$

is not normalizable and, therefore, no eigenvector with negative chirality is found.

We conclude that the operator $\mathcal{D}$ has only one eigenvector with zero eigenvalue corresponding to positive chirality: the index of $\mathcal{D}$ is one. Note that expressions for the index of the Dirac operator in a general background have been derived. In the present example, they yield

$$\text{Index} = \tfrac{1}{2}\left[\text{sgn}\,A(+\infty) - \text{sgn}\,A(-\infty)\right]$$

in agreement with the explicit calculation.

The Resolvent. For later purpose it is useful to exhibit some properties of the resolvent

$$\mathcal{G} = (\mathcal{D} - k)^{-1},$$

for real values of the parameter k. Parametrizing $\mathcal{G}$ as a 2×2 matrix:

$$\mathcal{G} = \begin{pmatrix} G_{11} & G_{12} \\ G_{21} & G_{22} \end{pmatrix},$$

one obtains

$$G_{11} = -k\left(D^\dagger D + k^2\right)^{-1}$$
$$G_{21} = -D\left(D^\dagger D + k^2\right)^{-1}$$
$$G_{12} = D^\dagger\left(DD^\dagger + k^2\right)^{-1}$$
$$G_{22} = -k\left(DD^\dagger + k^2\right)^{-1}.$$

For k^2 real one verifies $G_{21} = -G_{12}^\dagger$.

A number of properties then follow directly from the analysis presented in Appendix B.

When $k \to 0$ only G_{11} has a pole, $G_{11} = O(1/k)$, G_{22} vanishes as k and $G_{12}(x, y) = -G_{21}(y, x)$ have finite limits:

$$\mathfrak{G}(x, y) \underset{k \to 0}{\sim} \begin{pmatrix} -\frac{1}{k}\psi_+(x)\psi_+(y)/\|\psi_+\|^2 & -G_{21}(y, x) \\ G_{12}(x, y) & 0 \end{pmatrix} \sim -\frac{1}{2k}\frac{\psi_+(x)\psi_+(y)}{\|\psi_+\|^2}(1+\sigma_3).$$

Another limit of interest is the limit $y \to x$. The non-diagonal elements are discontinuous but the limit of interest for domain wall fermions is the average of the two limits

$$\overline{\mathfrak{G}(x, x)} = \tfrac{1}{2}(1 + \sigma_3)G_{11}(x, x) + \tfrac{1}{2}(1 - \sigma_3)G_{22}(x, x) + i\sigma_2\overline{G_{12}(x, x)}.$$

When the function $A(x)$ is odd, $A(-x) = -A(x)$, in the limit $x = 0$ the matrix $\mathfrak{G}(x, x)$ reduces to

$$\overline{\mathfrak{G}(0, 0)} = \tfrac{1}{2}(1 + \sigma_3)G_{11}(0, 0) + \tfrac{1}{2}(1 - \sigma_3)G_{22}(0, 0).$$

Examples.

(i) In the example of the function $S(x) = \tfrac{1}{2}x^2$, the two components of the hamiltonian H become

$$DD^\dagger = -d_x^2 + x^2 + 1, \quad D^\dagger D = -d_x^2 + x^2 - 1.$$

We recognize two shifted harmonic oscillators and the spectrum of $\mathfrak{D}$ contains one eigenvalue zero, and a spectrum of opposite eigenvalues $\pm i\sqrt{2n}$, $n \geq 1$.

(ii) Another example useful for later purpose is $S(x) = |x|$. Then $A(x) = \mathrm{sgn}(x)$ and $A'(x) = 2\delta(x)$. The two components of the hamiltonian H become

$$DD^\dagger = -d_x^2 + 1 + 2\delta(x), \quad D^\dagger D = -d_x^2 + 1 - 2\delta(x). \tag{84}$$

Here one finds one isolated eigenvalue zero, and a continuous spectrum $\varepsilon \geq 1$.

(iii) A less singular but similar example that can be solved analytically corresponds to $A(x) = \mu \tanh(x)$, where μ is for instance a positive constant. It leads to the potentials

$$V(x) = A^2(x) \pm A'(x) = \mu^2 - \frac{\mu(\mu \mp 1)}{\cosh^2(x)}.$$

The two operators have a continuous spectrum starting at μ^2 and a discrete spectrum

$$\mu^2 - (\mu - n)^2, \quad n \in \mathbb{N} \leq \mu, \qquad \mu^2 - (\mu - n - 1)^2, \quad n \in \mathbb{N} \leq \mu - 1.$$

8.2   Field Theory in Two Dimensions

A natural realization in quantum field theory of such a situation corresponds to a two-dimensional model of a Dirac fermion in the background of a static soliton (finite energy solution of the field equations).

We consider the action $S(\bar{\psi}, \psi, \varphi)$, $\psi, \bar{\psi}$ being Dirac fermions, and φ a scalar boson:

$$S(\bar{\psi}, \psi, \varphi) = \int \mathrm{d}x \, \mathrm{d}t \left[-\bar{\psi} \left(\not{\partial} + m + M\varphi \right) \psi + \tfrac{1}{2} \left(\partial_\mu \varphi \right)^2 + V(\varphi) \right].$$

We assume that $V(\varphi)$ has degenerate minima, like $(\varphi^2 - 1)^2$ or $\cos \varphi$, and field equations thus admit soliton solutions $\varphi(x)$, static solitons being the instantons of the one-dimensional quantum φ model.

Let us now study the spectrum of the corresponding Dirac operator

$$\mathcal{D} = \sigma_1 \partial_x + \sigma_2 \partial_t + m + M\varphi(x).$$

We assume for definiteness that $\varphi(x)$ goes from -1 for $x = -\infty$ to $+1$ for $x = +\infty$, a typical example being

$$\varphi(x) = \tanh(x).$$

Since time translation symmetry remains, we can introduce the (euclidean) time Fourier components and study

$$\mathcal{D} = \sigma_1 \mathrm{d}_x + i\omega\sigma_2 + m + M\varphi(x).$$

The zero eigenmodes of $\mathcal{D}$ are also the solutions of the eigenvalue equation

$$\mathfrak{D} \, |\psi\rangle = \omega \, |\psi\rangle, \quad \mathfrak{D} = \omega + i\sigma_2 \mathcal{D} = \sigma_3 \mathrm{d}_x + i\sigma_2 \big(m + M\varphi(x) \big),$$

which differs from (81) by an exchange between the matrices σ_3 and σ_1. The possible zero eigenmodes of $\mathfrak{D}$ ($\omega = 0$) thus satisfy

$$\sigma_1 \, |\psi\rangle = \epsilon \, |\psi\rangle, \quad \epsilon = \pm 1$$

and, therefore, are proportional to $\psi_\epsilon(x)$, which is a solution of

$$\epsilon \psi'_\epsilon + \big(m + M\varphi(x) \big) \psi_\epsilon = 0.$$

This equation has a normalizable solution only if $|m| < |M|$ and $\epsilon = +1$. Then we find one fermion zero-mode.

A soliton solution breaks space translation symmetry and thus generates a zero-mode (similar to Goldstone modes). Straightforward perturbation expansion around a soliton then would lead to IR divergences. Instead, the correct method is to remove the zero-mode by taking the position of the soliton as a collective coordinate. The integration over the position of the soliton then restores translation symmetry.

The implications of the fermion zero-mode require further analysis. It is found that it is associated with a double degeneracy of the soliton state, which carries $1/2$ fermion number.

8.3   Domain Wall Fermions

Continuum Formulation. One now considers four-dimensional space (but the strategy applies to all even dimensional spaces) as a surface embedded in five-dimensional space. We denote by x_μ the usual four coordinates, and by t the coordinate in the fifth dimension. Physical space corresponds to $t = 0$. We then study the five-dimensional Dirac operator $\mathcal{D}$ in the background of a classical scalar field $\varphi(t)$ that depends only on t. The fermion action reads

$$\mathcal{S}(\bar{\psi}, \psi) = - \int \mathrm{d}t \, \mathrm{d}^4x \, \bar{\psi}(t, x) \mathcal{D}\psi(t, x)$$

with

$$\mathcal{D} = \slashed{\partial} + \gamma_5 \mathrm{d}_t + M\varphi(Mt),$$

where the parameter M is a mass large with respect to the masses of all physical particles.

Since translation symmetry in four-space is not broken, we introduce the corresponding Fourier representation, and $\mathcal{D}$ then reads

$$\mathcal{D} = ip_\mu \gamma_\mu + \gamma_5 \mathrm{d}_t + M\varphi(Mt).$$

To find the mass spectrum corresponding to $\mathcal{D}$, it is convenient to write it as

$$\mathcal{D} = \gamma_{\mathbf{p}} \left[i|\mathbf{p}| + \gamma_{\mathbf{p}} \gamma_5 \mathrm{d}_t + \gamma_{\mathbf{p}} M\varphi(Mt) \right],$$

where $\gamma_{\mathbf{p}} = p_\mu \gamma_\mu / |\mathbf{p}|$ and thus $\gamma_{\mathbf{p}}^2 = 1$. The eigenvectors with vanishing eigenvalue of $\mathcal{D}$ are also those of the operator

$$\mathfrak{D} = i\gamma_{\mathbf{p}} \mathcal{D} + |\mathbf{p}| = i\gamma_{\mathbf{p}} \gamma_5 \mathrm{d}_t + i\gamma_{\mathbf{p}} M\varphi(Mt),$$

with eigenvalue $|\mathbf{p}|$.

We then note that $i\gamma_{\mathbf{p}}\gamma_5$, $\gamma_{\mathbf{p}}$, and $-\gamma_5$ are hermitian matrices that form a representation of the algebra of Pauli matrices. The operator $\mathfrak{D}$ can then be compared with the operator (81), and $M\varphi(Mt)$ corresponds to $A(x)$. Under the same conditions, $\mathfrak{D}$ has an eigenvector with an isolated vanishing eigenvalue corresponding to an eigenvector with positive chirality. All other eigenvalues, for dimensional reasons are proportional to M and thus correspond to fermions of large masses. Moreover, the eigenfunction with eigenvalue zero decays on a scale $t = O(1/M)$. Therefore, for M large one is left with a fermion that has a single chiral component, confined on the $t = 0$ surface.

One can imagine for the function $\varphi(t)$ some physical interpretation: φ may be an additional scalar field and $\varphi(t)$ may be a solution of the corresponding field equations that connects two minima $\varphi = \pm 1$ of the φ potential. In the limit of very sharp transition, one is led to the hamiltonian (84). Note that such an interpretation is possible only for even dimensions $d \geq 4$; in dimension 2, zero-modes related to breaking of translation symmetry due to the presence of the wall, would lead to IR divergences. These potential divergences thus forbid a static wall, a property analogous to the one encountered in the quantization of solitons in Sect. 8.2.

More precise results follow from the study of Sect. 8.1. We have noticed that $\mathfrak{G}(t_1, t_2; p)$, the inverse of the Dirac operator in Fourier representation, has a short distance singularity for $t_2 \to t_1$ in the form of a discontinuity. Here, this is an artifact of treating the fifth dimension differently from the four others. In real space for the function $\mathfrak{G}(t_1, t_2; x_1 - x_2)$ with separate points on the surface, $x_1 \neq x_2$, the limit $t_1 = t_2$ corresponds to points in five dimensions that do not coincide and this singularity is absent. A short analysis shows that this amounts in Fourier representation to take the average of the limiting values (a property that can easily be verified for the free propagator). Then, if $\varphi(t)$ is an odd function, for $t_1 = t_2 = 0$ one finds

$$\mathcal{D}^{-1}(p) = \frac{i}{2\not{p}} \left[d_1(p^2)(1 + \gamma_5) + (1 - \gamma_5)p^2 d_2(p^2) \right],$$

where d_1, d_2 are regular functions of p^2. Therefore, $\mathcal{D}^{-1}$ anticommutes with γ_5 and chiral symmetry is realized in the usual way. However, if $\varphi(t)$ is of more general type, one finds

$$\mathcal{D}^{-1} = \frac{i}{2\not{p}} \left[d_1(p^2)(1 + \gamma_5) + (1 - \gamma_5)p^2 d_2(p^2) \right] + d_3(p^2),$$

where d_3 is regular. As a consequence,

$$\gamma_5 \mathcal{D}^{-1} + \mathcal{D}^{-1}\gamma_5 = 2d_3(p^2)\gamma_5,$$

which is a form of Ginsparg–Wilson's relation because the r.h.s. is local.

Domain Wall Fermions: Lattice. We now replace four-dimensional continuum space by a lattice but keep the fifth dimension continuous. We replace the Dirac operator by the Wilson–Dirac operator (80) to avoid doublers. In Fourier representation, we find

$$\mathcal{D} = \alpha(\mathbf{p}) + i\beta_\mu(\mathbf{p})\gamma_\mu + \gamma_5 d_t + M\varphi(Mt).$$

This has the effect of replacing p_μ by $\beta_\mu(\mathbf{p})$ and shifting $M\varphi(Mt) \mapsto M\varphi(Mt) + \alpha(\mathbf{p})$. To ensure the absence of doublers, we require that for the values for which $\beta_\mu(\mathbf{p}) = 0$ and $\mathbf{p} \neq 0$ none of the solutions to the zero eigenvalue equation is normalizable. This is realized if $\varphi(t)$ is bounded for $|t| \to \infty$, for instance,

$$|\varphi(t)| \leq 1$$

and $M < |\alpha(\mathbf{p})|$.

The inverse Dirac operator on the surface $t = 0$ takes the general form

$$\mathcal{D}^{-1} = i\not{\beta} \left[\delta_1(p^2)(1 + \gamma_5) + (1 - \gamma_5)\delta_2(p^2) \right] + \delta_3(p^2),$$

where δ_1 is the only function that has a pole for $p = 0$, and where δ_2, δ_3 are regular. The function δ_3 does not vanish even if $\varphi(t)$ is odd because the addition of $\alpha(p^2)$ breaks the symmetry. We then always find Ginsparg–Wilson's relation

$$\gamma_5 \mathcal{D}^{-1} + \mathcal{D}^{-1}\gamma_5 = 2\delta_3(p^2)\gamma_5$$

More explicit expressions can be obtained in the limit $\varphi(t) = \mathrm{sgn}(t)$ (a situation analogous to (84)), using the analysis of the Appendix B.

Of course, computer simulations of domain walls require also discretizing the fifth dimension.

Acknowledgments

Useful discussions with T.W. Chiu, P. Hasenfratz and H. Neuberger are gratefully acknowledged. The author thanks also B. Feuerbacher, K. Schwenzer and F. Steffen for a very careful reading of the manuscript.

Appendix A. Trace Formula for Periodic Potentials

We consider a hamiltonian H corresponding to a real periodic potential $V(x)$ with period X:

$$V(x + X) = V(x).$$

Eigenfunctions $\psi_\theta(x)$ are then also eigenfunctions of the translation operator T:

$$T\psi_\theta(x) \equiv \psi_\theta(x + X) = \mathrm{e}^{i\theta}\psi_\theta(x). \tag{85}$$

We first restrict space to a box of size NX with periodic boundary conditions. This implies a quantization of the angle θ

$$\mathrm{e}^{iN\theta} = 1 \Rightarrow \theta = \theta_p \equiv 2\pi p/N, \quad 0 \leq p < N.$$

We call $\psi_{p,n}$ the normalized eigenfunctions of H corresponding to the band n and the pseudo-momentum θ_p,

$$\int_0^{NX} \mathrm{d}x\, \psi^*_{p,m}(x)\psi_{q,n}(x) = \delta_{mn}\delta_{pq},$$

and $E_n(\theta_p)$ the corresponding eigenvalues. Reality implies

$$E_n(\theta) = E_n(-\theta).$$

This leads to a decomposition of the identity operator in $[0, NX]$

$$\delta(x - y) = \sum_{p,n} \psi_{p,n}(x)\psi^*_{p,n}(y).$$

We now consider an operator O that commutes with T:

$$[T, O] = 0 \;\Rightarrow\; \langle x| O |y\rangle = \langle x + X| O |y + X\rangle .$$

Then,

$$\langle q, n| O |p, m\rangle = \int_0^{NX} dx \, dy \, \psi_{q,n}^*(x) \, \langle x| O |y\rangle \, \psi_{p,m}(y) = \delta_{pq} O_{mn}(\theta_p).$$

Its trace can be written as

$$\operatorname{tr} O = \int_0^{NX} dx \, \langle x| O |x\rangle = N \int_0^{X} dx \, \langle x| O |x\rangle = \sum_{p,n} O_{nn}(\theta_p).$$

We then take the infinite box limit $N \to \infty$. Then,

$$\frac{1}{N} \sum_p \;\to\; \frac{1}{2\pi} \int_0^{2\pi} d\theta$$

and, thus, we find

$$\int_0^X dx \, \langle x| O |x\rangle = \sum_n \frac{1}{2\pi} \int_0^{2\pi} O_{nn}(\theta) d\theta . \tag{86}$$

We now apply this general result to the operator

$$O = T^\ell e^{-\beta H} .$$

Then,

$$\int_0^X \langle x| T^\ell e^{-\beta H} |x\rangle \, dx = \frac{1}{2\pi} \sum_n \int_0^{2\pi} e^{i\ell\theta - \beta E_n(\theta)} d\theta ,$$

which using the definition of T can be rewritten as

$$\int_0^X \langle x + \ell X| e^{-\beta H} |x\rangle \, dx = \frac{1}{2\pi} \sum_n \int_0^{2\pi} e^{i\ell\theta - \beta E_n(\theta)} d\theta .$$

In the path integral formulation, this leads to a representation of the form

$$\int_{x(\beta/2)=x(-\beta/2)+\ell X} [dx(t)] \exp\left[-\mathcal{S}(x)\right] = \frac{1}{2\pi} \sum_n \int_0^{2\pi} e^{i\ell\theta - \beta E_n(\theta)} d\theta ,$$

where $x(-\beta/2)$ varies only in $[0, X]$, justifying the representation (43).

Appendix B.
Resolvent of the Hamiltonian in Supersymmetric QM

The resolvent $G(z) = (H + z)^{-1}$ of the hermitian operator

$$H = -\mathrm{d}_x^2 + V(x),$$

where $-z$ is outside the spectrum of H, satisfies the differential equation:

$$\left(-\mathrm{d}_x^2 + V(x) + z\right) G(z; x, y) = \delta(x - y). \tag{87}$$

We recall how $G(z; x, y)$ can be expressed in terms of two independent solutions of the homogeneous equation

$$\left(-\mathrm{d}_x^2 + V(x) + z\right) \varphi_{1,2}(x) = 0. \tag{88}$$

If one partially normalizes by choosing the value of the wronskian

$$W(\varphi_1, \varphi_2) \equiv \varphi_1'(x)\varphi_2(x) - \varphi_1(x)\varphi_2'(x) = 1$$

and, moreover, imposes the boundary conditions

$$\varphi_1(x) \to 0 \text{ for } x \to -\infty, \quad \varphi_2(x) \to 0 \text{ for } x \to +\infty,$$

then one verifies that $G(z; x, y)$ is given by

$$G(z; x, y) = \varphi_1(y)\varphi_2(x)\,\theta(x - y) + \varphi_1(x)\varphi_2(y)\,\theta(y - x). \tag{89}$$

After some algebra, one verifies that the diagonal elements $G(z; x; x)$ satisfy a third order linear differential equation.

If the potential is an even function, $V(-x) = V(x)$,

$$\varphi_2(x) \propto \varphi_1(-x).$$

Application. We now apply this result to the operator

$$H = \mathbf{D}\mathbf{D}^\dagger \quad \text{with} \quad z = k^2.$$

The functions φ_i then satisfy

$$\left(\mathbf{D}\mathbf{D}^\dagger + k^2\right) \varphi_i(x) \equiv \left[-\mathrm{d}_x^2 + A^2(x) + A'(x) + k^2\right] \varphi_i(x) = 0,$$

and (89) yields the resolvent $G_-(k^2; x, y)$, related to the matrix elements (84) by

$$G_{22}(k^2; x, y) = -kG_-(k^2; x, y).$$

The corresponding solutions for the operator $\mathbf{D}^\dagger\mathbf{D} + k^2$ follow since

$$\mathbf{D}^\dagger\left(\mathbf{D}\mathbf{D}^\dagger + k^2\right) \varphi_i = 0 = \left(\mathbf{D}^\dagger\mathbf{D} + k^2\right)\mathbf{D}^\dagger\varphi_i = 0.$$

The wronskian of the two functions

$$\chi_i(x) = \mathbf{D}^\dagger \varphi_i(x),$$

needed for normalization purpose, is simply

$$W(\chi_1, \chi_2) \equiv \chi_1'(x)\chi_2(x) - \chi_1(x)\chi_2'(x) = -k^2.$$

Thus, the corresponding resolvent G_+ (in (84) $G_{11} = -kG_+$) reads

$$G_+(k^2; x, y) = -\frac{1}{k^2} \left[\chi_1(y)\chi_2(x)\,\theta(x-y) + \chi_1(x)\chi_2(y)\,\theta(y-x) \right].$$

The limits $x = y$ are

$$G_-(k^2; x, x) = \varphi_1(x)\varphi_2(x), \quad G_+(k^2; x, x) = -\frac{1}{k^2}\chi_1(x)\chi_2(x).$$

If the potential is even, here this implies that $A(x)$ is odd, $G_\pm(k^2; x, x)$ are even functions.

We also need $\mathbf{D}^\dagger G_-(k^2; x, y)$:

$$\mathbf{D}^\dagger G_-(k^2; x, y) = \varphi_1(y)\mathbf{D}^\dagger\varphi_2(x)\theta(x-y) + \varphi_2(y)\mathbf{D}^\dagger\varphi_1(x)\theta(y-x).$$

We note that $\mathbf{D}^\dagger G_-(k^2; x, y)$ is not continuous at $x = y$:

$$\lim_{y \to x_+} \mathbf{D}^\dagger G_-(k^2; x, y) = \varphi_2(x)\mathbf{D}^\dagger\varphi_1(x), \quad \lim_{y \to x_-} \mathbf{D}^\dagger G_-(k^2; x, y) = \varphi_1(x)\mathbf{D}^\dagger\varphi_2(x)$$

and, therefore, from the wronskian,

$$\lim_{y \to x_-} \mathbf{D}^\dagger G_-(k^2; x, y) - \lim_{y \to x_+} \mathbf{D}^\dagger G_-(k^2; x, y) = 1.$$

The half sum is given by

$$\overline{\mathbf{D}^\dagger G_-(k^2; x, x)} = \tfrac{1}{2} \lim_{y \to x_-} \mathbf{D}^\dagger G_-(k^2; x, y) + \tfrac{1}{2} \lim_{y \to x_+} \mathbf{D}^\dagger G_-(k^2; x, y)$$
$$= \tfrac{1}{2}\mathbf{D}^\dagger\varphi_1(x)\varphi_2(x) + \tfrac{1}{2}\mathbf{D}^\dagger\varphi_2(x)\varphi_1(x)$$
$$= \tfrac{1}{2}\left(\varphi_1\varphi_2\right)'(x) + A(x)\varphi_1(x)\varphi_2(x).$$

This function is odd when $A(x)$ is odd.

In the limit $k \to 0$, one finds

$$\varphi_1(x) = N\mathrm{e}^{S(x)} \int_{-\infty}^x \mathrm{d}u\, \mathrm{e}^{-2S(u)}, \quad \varphi_2(x) = N\mathrm{e}^{S(x)} \int_x^\infty \mathrm{d}u\, \mathrm{e}^{-2S(u)}$$

with

$$N^2 \int_{-\infty}^{+\infty} \mathrm{d}u\, \mathrm{e}^{-2S(u)} = 1.$$

Moreover,

$$\mathbf{D}^\dagger\varphi_1(x) = -N\mathrm{e}^{-S(x)}, \quad \mathbf{D}^\dagger\varphi_2(x) = N\mathrm{e}^{-S(x)}.$$

Therefore, as expected

$$G_+(k^2; x, y) \underset{k \to 0}{\sim} \frac{1}{k^2} N^2 e^{-S(x)-S(y)}.$$

Finally,

$$\mathbf{D}^\dagger G_-(0; x, y) = N^2 \left[\theta(x - y) e^{-S(x)+S(y)} \int_{-\infty}^y \mathrm{d}u \, e^{-2S(u)} + (x \leftrightarrow y) \right]$$

and, therefore,

$$\overline{\mathbf{D}^\dagger G_-(0; x, x)} = \tfrac{1}{2} N^2 \int_{-\infty}^\infty \mathrm{d}t \, \mathrm{sgn}(x - t) e^{-2S(t)}.$$

References

1. The first part of these lectures is an expansion of several sections of J. Zinn-Justin, *Quantum Field Theory and Critical Phenomena*, Clarendon Press (Oxford 1989, fourth ed. 2002), to which the reader is referred for background in particular about euclidean field theory and general gauge theories, and references. For an early reference on Momentum cut-off regularization see W. Pauli and F. Villars, *Rev. Mod. Phys.* 21 (1949) 434.

2. Renormalizability of gauge theories has been proven using momentum regularization in B.W. Lee and J. Zinn-Justin, *Phys. Rev.* D5 (1972) 3121, 3137, 3155; D7 (1973) 1049.

3. The proof has been generalized using BRS symmetry and the master equation in J. Zinn-Justin in *Trends in Elementary Particle Physics*, ed. by H. Rollnik and K. Dietz, Lect. Notes Phys. **37** (Springer-Verlag, Berlin heidelberg 1975); in *Proc. of the 12th School of Theoretical Physics, Karpacz 1975*, Acta Universitatis Wratislaviensis 368.

4. A short summary can be found in J. Zinn-Justin *Mod. Phys. Lett.* A19 (1999) 1227.

5. Dimensional regularization has been introduced by: J. Ashmore, *Lett. Nuovo Cimento* 4 (1972) 289; G. 't Hooft and M. Veltman, *Nucl. Phys.* B44 (1972) 189; C.G. Bollini and J.J. Giambiagi, *Phys. Lett.* 40B (1972) 566, *Nuovo Cimento* 12B (1972) 20.

6. See also E.R. Speer, *J. Math. Phys.* 15 (1974) 1; M.C. Bergère and F. David, *J. Math. Phys.* 20 (1979 1244.

7. Its use in problems with chiral anomalies has been proposed in D.A. Akyeampong and R. Delbourgo, *Nuovo Cimento* 17A (1973) 578.

8. For an early review see G. Leibbrandt, *Rev. Mod. Phys.* 47 (1975) 849.

9. For dimensional regularization and other schemes, see also E.R. Speer in *Renormalization Theory*, Erice 1975, G. Velo and A.S. Wightman eds. (D. Reidel, Dordrecht, Holland 1976).

10. The consistency of the lattice regularization is rigorously established (except for theories with chiral fermions) in T. Reisz, *Commun. Math. Phys.* 117 (1988) 79, 639.

11. The generality of the doubling phenomenon for lattice fermions has been proven by H.B. Nielsen and M. Ninomiya, *Nucl. Phys.* B185 (1981) 20.

12. Wilson's solution to the fermion doubling problem is described in K.G. Wilson in *New Phenomena in Subnuclear Physics*, Erice 1975, A. Zichichi ed. (Plenum, New York 1977).

13. Staggered fermions have been proposed in T. Banks, L. Susskind and J. Kogut, *Phys. Rev.* D13 (1976) 1043.

14. The problem of chiral anomalies is discussed in J.S. Bell and R. Jackiw, *Nuovo Cimento* A60 (1969) 47; S.L. Adler, *Phys. Rev.* 177 (1969) 2426; W.A. Bardeen, *Phys. Rev.* 184 (1969) 1848; D.J. Gross and R. Jackiw, *Phys. Rev.* D6 (1972) 477; H. Georgi and S.L. Glashow, *Phys. Rev.* D6 (1972) 429; C. Bouchiat, J. Iliopoulos and Ph. Meyer, *Phys. Lett.* 38B (1972) 519.

15. See also the lectures S.L. Adler, in *Lectures on Elementary Particles and Quantum Field Theory*, S. Deser *et al* eds. (MIT Press, Cambridge 1970); M. E. Peskin, in *Recent Advances in Field Theory and Statistical Mechanics*, Les Houches 1982, R. Stora and J.-B. Zuber eds. (North-Holland, Amsterdam 1984); L. Alvarez-Gaumé, in *Fundamental problems of gauge theory*, Erice 1985 G. Velo and A.S. Wightman eds. (Plenum Press, New-York 1986).

16. The index of the Dirac operator in a gauge background is related to Atiyah–Singer's theorem M. Atiyah, R. Bott and V. Patodi, *Invent. Math.* 19 (1973) 279.

17. It is at the basis of the analysis relating anomalies to the regularization of the fermion measure K. Fujikawa, *Phys. Rev.* D21 (1980) 2848; D22 (1980) 1499(E).

18. The same strategy has been applied to the conformal anomaly K. Fujikawa, *Phys. Rev. Lett.* 44 (1980) 1733.

19. For non-perturbative global gauge anomalies see E. Witten, *Phys. Lett.* B117 (1982) 324; *Nucl. Phys.* B223 (1983) 422; S. Elitzur, V.P. Nair, *Nucl. Phys.* B243 (1984) 205.

20. The gravitational anomaly is discussed in L. Alvarez-Gaumé and E. Witten, *Nucl. Phys.* B234 (1984) 269.

21. See also the volumes S.B. Treiman, R. Jackiw, B. Zumino and E. Witten, *Current Algebra and Anomalies* (World Scientific, Singapore 1985) and references therein; R.A. Bertlman, *Anomalies in Quantum Field Theory*, Oxford Univ. Press, Oxford 1996.

22. Instanton contributions to the cosine potential have been calculated with increasing accuracy in E. Brézin, G. Parisi and J. Zinn-Justin, *Phys. Rev.* D16 (1977) 408; E.B. Bogomolny, *Phys. Lett.* 91B (1980) 431; J. Zinn-Justin, *Nucl. Phys.* B192 (1981) 125; B218 (1983) 333; *J. Math. Phys.* 22 (1981) 511; 25 (1984) 549.

23. Classical references on instantons in the $CP(N-1)$ models include A. Jevicki *Nucl. Phys.* B127 (1977) 125; D. Förster, *Nucl. Phys.* B130 (1977) 38; M. Lüscher, *Phys. Lett.* 78B (1978) 465; A. D'Adda, P. Di Vecchia and M. Lüscher, *Nucl. Phys.* B146 (1978) 63; B152 (1979) 125; H. Eichenherr, *Nucl. Phys.* B146 (1978) 215; V.L. Golo and A. Perelomov, *Phys. Lett.* 79B (1978) 112; A.M. Perelemov, *Phys. Rep.* 146 (1987) 135.

24. For instantons in gauge theories see A.A. Belavin, A.M. Polyakov, A.S. Schwartz and Yu S. Tyupkin, *Phys. Lett.* 59B (1975) 85; G. 't Hooft, *Phys. Rev. Lett.* 37 (1976) 8; *Phys. Rev.* D14 (1976) 3432 (Erratum *Phys. Rev.* D18 (1978) 2199); R. Jackiw and C. Rebbi, *Phys. Rev. Lett.* 37 (1976) 172; C.G. Callan, R.F. Dashen and D.J. Gross, *Phys. Lett.* 63B (1976) 334; A.A. Belavin and A.M. Polyakov, *Nucl. Phys.* B123 (1977) 429; F.R. Ore, *Phys. Rev.* D16 (1977) 2577; S. Chadha, P. Di Vecchia, A. D'Adda and F. Nicodemi, *Phys. Lett.* 72B (1977) 103; T. Yoneya, *Phys. Lett.* 71B (1977) 407; I.V. Frolov and A.S. Schwarz, *Phys. Lett.* 80B (1979) 406; E. Corrigan, P. Goddard and S. Templeton, *Nucl. Phys.* B151 (1979) 93.

25. For a solution of the $U(1)$ problem based on anomalies and instantons see G. 't Hooft, *Phys. Rep.* 142 (1986) 357.

26. The strong CP violation is discussed in R.D. Peccei, Helen R. Quinn, *Phys. Rev. D16* (1977) 1791; S.Weinberg, *Phys. Rev. Lett.* 40 (1978) 223, (Also in Mohapatra, R.N. (ed.), Lai, C.H. (ed.): Gauge Theories Of Fundamental Interactions, 396-399).

27. The Bogomolnyi bound is discussed in E.B. Bogomolnyi, *Sov. J. Nucl. Phys.* 24 (1976) 449; M.K. Prasad and C.M. Sommerfeld, *Phys. Rev. Lett.* 35 (1975) 760.

28. For more details see Coleman lectures in S. Coleman, *Aspects of symmetry*, Cambridge Univ. Press (Cambridge 1985).

29. BRS symmetry has been introduced in C. Becchi, A. Rouet and R. Stora, *Comm. Math. Phys.* 42 (1975) 127; *Ann. Phys.* (NY) 98 (1976) 287.

30. It has been used to determine the non-abelian anomaly by J. Wess and B. Zumino, *Phys. Lett.* 37B (1971) 95.

31. The overlap Dirac operator for chiral fermions is constructed explicitly in H. Neuberger, *Phys. Lett.* B417 (1998) 141 [hep-lat/9707022], *ibidem* B427 (1998) 353 [hep-lat/9801031].

32. The index theorem in lattice gauge theory is discussed in P. Hasenfratz, V. Laliena, F. Niedermayer, *Phys. Lett.* B427 (1998) 125 [hep-lat/9801021].

33. A modified exact chiral symmetry on the lattice was exhibited in M. Lüscher, *Phys. Lett.* B428 (1998) 342 [hep-lat/9802011], [hep-lat/9811032].

34. The overlap Dirac operator was found to provide solutions to the Ginsparg–Wilson relation P.H. Ginsparg and K.G. Wilson, *Phys. Rev.* D25 (1982) 2649.

35. See also D.H. Adams, *Phys. Rev. Lett.* 86 (2001) 200 [hep-lat/9910036], *Nucl. Phys.* B589 (2000) 633 [hep-lat/0004015]; K. Fujikawa, M. Ishibashi, H. Suzuki, [hep-lat/0203016].

36. In the latter paper the problem of CP violation is discussed.

37. Supersymmetric quantum mechanics is studied in E. Witten, *Nucl. Phys.* B188 (1981) 513.

38. General determinations of the index of the Dirac operator can be found in C. Callias, *Comm. Math. Phys.* 62 (1978) 213.

39. The fermion zero-mode in a soliton background in two dimensions is investigated in R. Jackiw and C. Rebbi, *Phys. Rev.* D13 (1976) 3398.

40. Special properties of fermions in presence of domain walls were noticed in C.G. Callan and J.A. Harvey, *Nucl. Phys.* B250 (1985) 427.

41. Domain wall fermions on the lattice were discussed in D.B. Kaplan, *Phys. Lett.* B288 (1992) 342.

42. See also (some deal with the delicate problem of the continuum limit when the fifth dimension is first discretized) M. Golterman, K. Jansen and D.B. Kaplan, *Phys.Lett.* B301 (1993) 219 [hep-lat/9209003); Y. Shamir, *Nucl. Phys.* B406 (1993) 90 [hep-lat/9303005] *ibidem* B417 (1994) 167 [hep-lat/9310006]; V. Furman, Y. Shamir, *Nucl. Phys.* B439 (1995) 54 [hep-lat/9405004]. Y. Kikukawa, T. Noguchi, [hep-lat/9902022].

43. For more discussions and references Proceedings of the workshop "Chiral 99", *Chinese Journal of Physics*, 38 (2000) 521-743; K. Fujikawa, *Int. J. Mod. Phys.* A16 (2001) 331; T.-W. Chiu, *Phys. Rev.* D58(1998) 074511 [hep-lat/9804016], *Nucl. Phys.* B588 (2000) 400 [hep-lat/0005005]; M. Lüscher, Lectures given at International School of Subnuclear Physics *Theory and Experiment Heading for New Physics*, Erice 2000, [hep-th/0102028] and references therein; P. Hasenfratz, Proceedings of "Lattice 2001", *Nucl. Phys. Proc. Suppl.* 106 (2002) 159 [hep-lat/0111023].

44. In particular the $U(1)$ problem has been discussed analytically and studied numerically on the lattice. For a recent reference see for instance L. Giusti, G.C. Rossi, M. Testa, G. Veneziano, [hep-lat/0108009].

45. Early simulations have used domain wall fermions. For a review see P.M. Vranas, *Nucl. Phys. Proc. Suppl.* 94 (2001) 177 [hep-lat/0011066]

46. A few examples are S. Chandrasekharan *et al, Phys. Rev. Lett.* 82 (1999) 2463 [hep-lat/9807018]; T. Blum *et al.,* [hep-lat/0007038]; T. Blum, RBC Collaboration, *Nucl. Phys. Proc. Suppl.* 106 (2002) 317 [hep-lat/0110185]; J-I. Noaki *et al* CP-PACS Collaboration, [hep-lat/0108013].

47. More recently overlap fermion simulations have been reported R.G. Edwards, U.M. Heller, J. Kiskis, R. Narayanan, *Phys. Rev.* D61 (2000) 074504 [hep-lat/9910041]; P. Hernández, K. Jansen, L. Lellouch, *Nucl. Phys. Proc. Suppl.* 83 (2000) 633 [hep-lat/9909026]; S.J. Dong, F.X. Lee, K.F. Liu, J.B. Zhang, *Phys. Rev. Lett.* 85 (2000) 5051 [hep-lat/0006004]; T. DeGrand, *Phys. Rev.* D63 (2001) 034503 [hep-lat/0007046]; R.V. Gavai, S. Gupta, R. Lacaze, [hep-lat/0107022]; L. Giusti, C. Hoelbling, C. Rebbi, [hep-lat/0110184]; *ibidem* [hep-lat/0108007].

Supersymmetric Solitons and Topology

M. Shifman

William I. Fine Theoretical Physics Institute, School of Physics and Astronomy, University of Minnesota, 116 Church Street SE, Minneapolis MN 55455, USA

Abstract. This lecture is devoted to solitons in supersymmetric theories. The emphasis is put on special features of supersymmetric solitons such as "BPS-ness". I explain why only zero modes are important in the quantization of the BPS solitons. Hybrid models (Landau–Ginzburg models on curved target spaces) are discussed in some detail. Topology of the target space plays a crucial role in the classification of the BPS solitons in these models. The phenomenon of multiplet shortening is considered. I present various topological indices (analogs of Witten's index) which count the number of solitons in various models.

1 Introduction

The term "soliton" was introduced in the 1960's, but the scientific research of solitons had started much earlier, in the nineteenth century when a Scottish engineer, John Scott-Russell, observed a large solitary wave in a canal near Edinburgh.

For the purpose of my lecture I will adopt a narrow interpretation of solitons. Let us assume that a field theory under consideration possesses a few (more than one) degenerate vacuum states. Then these vacua represent distinct phases of the theory. A field configuration smoothly interpolating between the distinct phases which is topologically stable will be referred to as soliton.[1] This definition is over-restrictive – for instance, it does not include vortices, which present a famous example of topologically stable solitons. I would be happy to discuss supersymmetric vortices and flux tubes. However, because of time limitations, I have to abandon this idea limiting myself to supersymmetric kinks and domain walls.

In non-supersymmetric field theories the vacuum degeneracy usually requires spontaneous breaking of some global symmetry – either discrete or continuous. In supersymmetric field theories (if supersymmetry – SUSY – is unbroken) all vacua *must* have a vanishing energy density and are thus degenerate.

This is the first reason why SUSY theories are so special as far as topological solitons are concerned. Another (more exciting) reason explaining the enormous interest in topological solitons in supersymmetric theories is the existence of a special class of solitons, which are called "critical" or "Bogomol'nyi–Prasad–Sommerfield saturated" (BPS for short).

A birds' eye view on the development of supersymmetry beginning from its inception in 1971 [1] is presented in Appendix B. A seminal paper which

---

[1] More exactly, we will call it "topological soliton".

M. Shifman, Supersymmetric Solitons and Topology, Lect. Notes Phys. **659**, 237–284 (2005)
http://www.springerlink.com/

opened for investigation the currently flourishing topic of BPS saturated solitons is that of Witten and Olive [2] where the authors noted that in many instances (supporting topological solitons) topological charges coincide with the so-called central charges [3] of superalgebras. This allows one to formulate the Bogomol'nyi–Prasad–Sommerfield construction [4] in algebraic terms and to extend the original classical formulation to the quantum level, making it exact. All these statements will be explained in detail below.

In high energy physics theorists traditionally deal with a variety of distinct solitons in various space-time dimensions D. Some of the most popular ones are:

(i) kinks in D = 1+1 (being elevated to D = 1+3 they represent domain walls);

(ii) vortices in D = 1+2 (being elevated to D = 1+3 they represent strings or flux tubes);

(iii) magnetic monopoles in D = 1+3.

In the three cases above the topologically stable solutions are known from the 1930's, '50's, and '70's, respectively. Then it was shown that all these solitons can be embedded in supersymmetric theories. To this end one adds an appropriate fermion sector, and if necessary, expands the boson sector. In this lecture we will limit ourselves to critical (or BPS-saturated) kinks and domain walls. Noncritical solitons are typically abundant, but we will not touch this theme at all.

The presence of fermions leads to a variety of novel physical phenomena which are inherent to BPS-saturated solitons. These phenomena are one of the prime subjects of my lecture.

Before I will be able to explain why supersymmetric solitons are special and interesting, I will have to review briefly well-known facts about solitons in bosonic theories and provide a general introduction to supersymmetry in appropriate models. I will start with the simplest model – one (real) scalar field in two dimensions plus the minimal set of superpartners.

2   D = 1+1; $\mathcal{N} = 1$

In this part we will consider the simplest supersymmetric model in D = 1+1 dimensions that admits solitons. The Lagrangian of this model is

$$\mathcal{L} = \frac{1}{2}\left\{\partial_\mu\phi\,\partial^\mu\phi + \bar\psi\, i\partial\!\!\!/\,\psi - \left(\frac{\partial W}{\partial\phi}\right)^2 - \frac{\partial^2 W}{\partial\phi^2}\,\bar\psi\psi\right\}, \tag{2.1}$$

where ϕ is a real scalar field and ψ is a Majorana spinor,

$$\psi = \begin{pmatrix}\psi_1 \\ \psi_2\end{pmatrix} \tag{2.2}$$

with $\psi_{1,2}$ real. Needless to say that the gamma matrices must be chosen in the Majorana representation. A convenient choice is

$$\gamma^0 = \sigma_2\,, \quad \gamma^1 = i\sigma_3\,, \tag{2.3}$$

where $\sigma_{2,3}$ are the Pauli matrices. For future reference we will introduce a "γ_5" matrix, $\gamma^5 = \gamma^0\gamma^1 = -\sigma_1$. Moreover,

$$\bar{\psi} = \psi\gamma^0 .$$

The *superpotential function* $\mathcal{W}(\phi)$ is, in principle, arbitrary. The model (2.1) with any $\mathcal{W}(\phi)$ is supersymmetric, provided that $\mathcal{W}' \equiv \partial\mathcal{W}/\partial\phi$ vanishes at some value of ϕ. The points ϕ_i where

$$\frac{\partial\mathcal{W}}{\partial\phi} = 0$$

are called critical. As can be seen from (2.1), the scalar potential is related to the superpotential as $U(\phi) = (1/2)(\partial\mathcal{W}/\partial\phi)^2$. Thus, the critical points correspond to a vanishing energy density,

$$U(\phi_i) = \frac{1}{2}\left(\frac{\partial\mathcal{W}}{\partial\phi}\right)^2_{\phi=\phi_i} = 0 . \tag{2.4}$$

The critical points accordingly are the classical minima of the potential energy – the classical vacua. For our purposes, the soliton studies, we require the existence of at least two distinct critical points in the problem under consideration. The kink will interpolate between distinct vacua.

Two popular choices of the superpotential function are:

$$\mathcal{W}(\phi) = \frac{m^2}{4\lambda}\phi - \frac{\lambda}{3}\phi^3 , \tag{2.5}$$

and

$$\mathcal{W}(\phi) = mv^2 \sin\frac{\phi}{v} . \tag{2.6}$$

Here m, λ, and v are real (positive) parameters. The first model is referred to as superpolynomial (SPM), the second as super–sine–Gordon (SSG). The classical vacua in the SPM are at $\phi = \pm m(2\lambda)^{-1} \equiv \phi^{\pm}_*$. I will assume that $\lambda/m \ll 1$ to ensure the applicability of the quasiclassical treatment. This is the weak coupling regime for the SPM. A kink solution interpolates between $\phi^-_* = -m/2\lambda$ at $z \to -\infty$ and $\phi^+_* = m/2\lambda$ at $z \to \infty$, an anti-kink solution between $\phi^+_* = m/2\lambda$ at $z \to -\infty$ and $\phi^-_* = -m/2\lambda$ at $z \to \infty$. The classical kink solution has the form

$$\phi_0 = \frac{m}{2\lambda} \tanh\frac{mz}{2} . \tag{2.7}$$

The weak coupling regime in the SSG case is attained at $v \gg 1$. In the super–sine–Gordon model there are infinitely many vacua; they lie at

$$\phi^k_* = v\left(\frac{\pi}{2} + k\pi\right) , \tag{2.8}$$

where k is an integer, either positive or negative. Correspondingly, there exist solitons connecting any pair of vacua. In this case we will limit ourselves to consideration of the "elementary" solitons, which connect adjacent vacua, e.g. $\phi_*^{0,-1} = \pm\pi v/2$,

$$\phi_0 = v \arcsin\left[\tanh(mz)\right] . \tag{2.9}$$

In D = 1+1 the real scalar field represents one degree of freedom (bosonic), and so does the two-component Majorana spinor (fermionic). Thus, the number of bosonic and fermionic degrees of freedom is identical, which is a necessary condition for supersymmetry. One can show in many different ways that the Lagrangian (2.1) does actually possess supersymmetry. For instance, let us consider the *supercurrent*,

$$J^\mu = (\slashed{\partial}\phi)\gamma^\mu\psi + i\frac{\partial W}{\partial\phi}\gamma^\mu\psi . \tag{2.10}$$

This object is linear in the fermion field; therefore, it is obviously fermionic. On the other hand, it is conserved. Indeed,

$$\partial_\mu J^\mu = (\partial^2\phi)\psi + (\slashed{\partial}\phi)(\slashed{\partial}\psi) + i\frac{\partial^2 W}{\partial\phi^2}(\slashed{\partial}\phi)\psi + i\frac{\partial W}{\partial\phi}\slashed{\partial}\psi . \tag{2.11}$$

The first, second, and third terms can be expressed by virtue of the equations of motion, which immediately results in various cancelations. After these cancelations only one term is left in the divergence of the supercurrent,

$$\partial_\mu J^\mu = -\frac{1}{2}\frac{\partial^3 W}{\partial\phi^3}(\bar\psi\psi)\,\psi . \tag{2.12}$$

If one takes into account (i) the fact that the spinor ψ is real and two-component, and (ii) the Grassmannian nature of $\psi_{1,2}$, one immediately concludes that the right-hand side in (2.12) vanishes.

The supercurrent conservation implies the existence of *two* conserved charges,[2]

$$Q_\alpha = \int dz\, J_\alpha^0 = \int dz\,\left\{\left(\slashed{\partial}\phi + i\frac{\partial W}{\partial\phi}\right)(\gamma^0\psi)\right\}_\alpha , \qquad \alpha = 1,2. \tag{2.13}$$

These *supercharges* form a doublet with respect to the Lorentz group in D = 1+1. They generate supertransformations of the fields, for instance,

$$[Q_\alpha,\phi] = -i\psi_\alpha , \qquad \{Q_\alpha,\bar\psi_\beta\} = (\slashed{\partial})_{\alpha\beta}\phi + i\frac{\partial W}{\partial\phi}\delta_{\alpha\beta} , \tag{2.14}$$

and so on. In deriving (2.14) I used the canonical commutation relations

$$\left[\phi(t,z),\dot\phi(t,z')\right] = i\delta(z-z') , \qquad \{\psi_\alpha(t,z),\bar\psi_\beta(t,z')\} = (\gamma^0)_{\alpha\beta}\delta(z-z'). \tag{2.15}$$

Note that by acting with Q on the bosonic field we get a fermionic one and *vice versa*. This fact demonstrates, once again, that the supercharges are symmetry generators of fermionic nature.

---

[2] Two-dimensional theories with two conserved supercharges are referred to as $\mathcal{N} = 1$.

Given the expression for the supercharges (2.13) and the canonical commutation relations (2.15) it is not difficult to find the superalgebra,

$$\{Q_\alpha, \bar{Q}_\beta\} = 2 \, (\gamma^\mu)_{\alpha\beta} \, P_\mu + 2i \, (\gamma^5)_{\alpha\beta} \, \mathcal{Z} \, . \tag{2.16}$$

Here P_μ is the operator of the total energy and momentum,

$$P^\mu = \int dz T^{\mu 0} \, , \tag{2.17}$$

where $T^{\mu\nu}$ is the energy-momentum tensor,

$$T^{\mu\nu} = \partial^\mu \phi \, \partial^\nu \phi + \frac{1}{2} \bar{\psi} \gamma^\mu \, i\partial^\nu \psi - \frac{1}{2} \, g^{\mu\nu} \left[\partial_\gamma \phi \, \partial^\gamma \phi - (\mathcal{W}')^2 \right] \, , \tag{2.18}$$

and $\mathcal{Z}$ is the central charge,

$$\mathcal{Z} = \int dz \, \partial_z \mathcal{W}(\phi) = \mathcal{W}[\phi(z = \infty)] - \mathcal{W}[\phi(z = -\infty)] \, . \tag{2.19}$$

The local form of the superalgebra (2.16) is

$$\left\{ J_\alpha^\mu, \bar{Q}_\beta \right\} = 2 \, (\gamma_\nu)_{\alpha\beta} \, T^{\mu\nu} + 2i \, (\gamma^5)_{\alpha\beta} \, \zeta^\mu \, , \tag{2.20}$$

where ζ^μ is the conserved topological current,

$$\zeta^\mu = \epsilon^{\mu\nu} \partial_\nu \mathcal{W} \, . \tag{2.21}$$

Symmetrization (antisymmetrization) over the bosonic (fermionic) operators in the products is implied in the above expressions.

I pause here to make a few comments. Equation (2.16) can be viewed as a general definition of supersymmetry. Without the second term on the right-hand side, i.e. in the form $\{Q_\alpha, \bar{Q}_\beta\} = 2 \, (\gamma^\mu)_{\alpha\beta} \, P_\mu$, it was obtained by two of the founding fathers of supersymmetry, Golfand and Likhtman, in 1971 [1]. The $\mathcal{Z}$ term in (2.16) is referred to as the central extension. At a naive level of consideration one might be tempted to say that this term vanishes since it is the integral of a full derivative. Actually, it does *not* vanish in problems in which one deals with topological solitons. We will see this shortly. The occurrence of the central charge $\mathcal{Z}$ is in one-to-one correspondence with the topological charges – this fact was noted by Witten and Olive [2]. Even before the work of Witten and Olive, the possibility of central extensions of the defining superalgebra was observed, within a purely algebraic consideration, by Haag, Lopuszanski, and Sohnius [3]. The theories with centrally extended superalgebras are special: they admit critical solitons. Since the central charge is the integral of the full derivative, it is independent of details of the soliton solution and is determined only by the boundary conditions. To ensure that $\mathcal{Z} \neq 0$ the field ϕ must tend to distinct limits at $z \to \pm\infty$.

2.1   Critical (BPS) Kinks

A kink in D = 1+1 is a particle. Any given soliton solution obviously breaks translational invariance. Since $\{Q, \bar{Q}\} \propto P$, typically both supercharges are broken on the soliton solutions,

$$Q_\alpha|\text{sol}\rangle \neq 0, \qquad \alpha = 1, 2. \tag{2.22}$$

However, for certain special kinks, one can preserve 1/2 of supersymmetry, i.e.

$$Q_1|\text{sol}\rangle \neq 0 \quad \text{and} \quad Q_2|\text{sol}\rangle = 0, \tag{2.23}$$

or *vice versa*. Such kinks are called critical, or *BPS-saturated*.[3]

The critical kink must satisfy a first order differential equation – this fact, as well as the particular form of the equation, follows from the inspection of (2.13) or the second equation in (2.14). Indeed, for static fields $\phi = \phi(z)$, the supercharges Q_α are proportional to

$$Q_\alpha \propto \begin{pmatrix} \partial_z\phi + \mathcal{W}' & 0 \\ 0 & -\partial_z\phi + \mathcal{W}' \end{pmatrix}. \tag{2.24}$$

One of the supercharges vanishes provided that

$$\frac{\partial\phi(z)}{\partial z} = \pm \frac{\partial\mathcal{W}(\phi)}{\partial\phi}, \tag{2.25}$$

or, for short,

$$\partial_z\phi = \pm\mathcal{W}'. \tag{2.26}$$

The plus and minus signs correspond to kink and anti-kink, respectively. Generically, the equations that express the conditions for the vanishing of certain supercharges are called *BPS equations*.

The first order BPS equation (2.26) implies that the kink automatically satisfies the general second order equation of motion. Indeed, let us differentiate both sides of (2.26) with respect to z. Then one gets

$$\partial_z^2\phi = \pm\partial_z\mathcal{W}' = \pm\mathcal{W}'' \partial_z\phi$$

$$= \mathcal{W}''\mathcal{W}' = \frac{\partial U}{\partial\phi}. \tag{2.27}$$

---

[3] More exactly, in the case at hand we deal with 1/2 BPS-saturated kinks. As I have already mentioned, BPS stands for Bogomol'nyi, Prasad, and Sommerfield [4]. In fact, these authors considered solitons in a non-supersymmetric setting. They found, however, that under certain conditions they can be described by first order differential equations, rather than second order equations of motion. Moreover, under these conditions the soliton mass was shown to be proportional to the topological charge. We understand now that the limiting models considered in [4] are bosonic sectors of supersymmetric models.

The latter presents the equation of motion for static (time independent) field configurations. This is a general feature of supersymmetric theories: compliance with the BPS equations entails compliance with the equations of motion.

The inverse statement is generally speaking wrong – not all solitons which are static solutions of the second order equations of motion satisfy the BPS equations. However, in the model at hand, with a *single* scalar field, the inverse statement *is* true. In this model any static solution of the equation of motion satisfies the BPS equation. This is due to the fact that there exists an "integral of motion." Indeed, let us reinterpret z as a "time," for a short while. Then the equation $\partial_z^2 \phi - U' = 0$ can be reinterpreted as $\ddot{\phi} - U' = 0$, i.e. the one-dimensional motion of a particle of mass 1 in the potential $-U(\phi)$. The conserved "energy" is $(1/2)\,\dot{\phi}^2 - U$. At $-\infty$ both the "kinetic" and "potential" terms tend to zero. This boundary condition emerges because the kink solution interpolates between two critical points, the vacua of the model, while supersymmetry ensures that $U(\phi_*) = 0$. Thus, on the kink configuration $(1/2)\,\dot{\phi}^2 = U$ implying that $\dot{\phi} = \pm \mathcal{W}'$.

We have already learned that the BPS saturation in the supersymmetric setting means the preservation of a part of supersymmetry. Now, let us ask ourselves why this feature is so precious.

To answer this question let us have a closer look at the superalgebra (2.16). In the kink rest frame it reduces to

$$(Q_1)^2 = M + \mathcal{Z}, \qquad (Q_2)^2 = M - \mathcal{Z}$$

$$\{Q_1, Q_2\} = 0, \tag{2.28}$$

where M is the kink mass. Since Q_2 vanishes on the critical kink, we see that

$$M = \mathcal{Z}. \tag{2.29}$$

Thus, the kink mass is equal to the central charge, a nondynamical quantity which is determined only by the boundary conditions on the field ϕ (more exactly, by the values of the superpotential in the vacua between which the kink under consideration interpolates).

2.2   The Kink Mass (Classical)

The classical expression for the central charge is given in (2.19). (Anticipating a turn of events I hasten to add that a quantum anomaly will modify this classical expression; see Sect. 2.6.) Now we will discuss the critical kink mass.

In the SPM

$$\phi_*^\pm = \pm \frac{m}{2\lambda}, \qquad \mathcal{W}_0^\pm \equiv \mathcal{W}[\phi_*^\pm] = \pm \frac{m^3}{12\lambda^2} \tag{2.30}$$

and, hence,

$$M_{\mathrm{SPM}} = \frac{m^3}{6\lambda^2}. \tag{2.31}$$

In the SSG model

$$\phi_*^\pm = \pm v \, \frac{\pi}{2} \,, \qquad \mathcal{W}_0^\pm \equiv \mathcal{W}[\phi_*^\pm] = \pm m v^2 \,. \qquad (2.32)$$

Therefore,

$$M_{\text{SSG}} = 2 m v^2 \,. \qquad (2.33)$$

Applicability of the quasiclassical approximation demands $m/\lambda \gg 1$ and $v \gg 1$, respectively.

2.3   Interpretation of the BPS Equations. Morse Theory

In the model described above we deal with a single scalar field. Since the BPS equation is of first order, it can always be integrated in quadratures. Examples of the solution for two popular choices of the superpotential are given in (2.7) and (2.9).

The one-field model is the simplest but certainly not the only model with interesting applications. The generic multi-field $\mathcal{N} = 1$ SUSY model of the Landau–Ginzburg type has a Lagrangian of the form

$$\mathcal{L} = \frac{1}{2} \left\{ \partial_\mu \phi^a \, \partial^\mu \phi^a + i \bar\psi^a \gamma^\mu \partial_\mu \psi^a - \frac{\partial \mathcal{W}}{\partial \phi^a} \frac{\partial \mathcal{W}}{\partial \phi^a} - \frac{\partial^2 \mathcal{W}}{\partial \phi^a \partial \phi^b} \, \bar\psi^a \psi^b \right\} \,, \qquad (2.34)$$

where the superpotential $\mathcal{W}$ now depends on n variables, $\mathcal{W} = \mathcal{W}(\phi^a)$; in what follows a, b will be referred to as "flavor" indices, $a, b = 1, ..., n$. The sum over both a and b is implied in (2.34). The vacua (critical points) of the generic model are determined by a set of equations

$$\frac{\partial \mathcal{W}}{\partial \phi^a} = 0 \,, \qquad a = 1, ..., n \,. \qquad (2.35)$$

If one views $\mathcal{W}(\phi^a)$ as a "mountain profile," the critical points are the extremal points of this profile – minima, maxima, and saddle points. At the critical points the potential energy

$$U(\phi^a) = \frac{1}{2} \left(\frac{\partial \mathcal{W}}{\partial \phi^a} \right)^2 \qquad (2.36)$$

is minimal – $U(\phi_*^a)$ vanishes. The kink solution is a trajectory $\phi^a(z)$ interpolating between a selected pair of critical points.

The BPS equations take the form

$$\frac{\partial \phi^a}{\partial z} = \pm \frac{\partial \mathcal{W}}{\partial \phi^a} \,, \qquad a = 1, ..., n \,. \qquad (2.37)$$

For $n > 1$ not all solutions of the equations of motion are the solutions of the BPS equations, generally speaking. In this case the critical kinks represent a subclass of all possible kinks. Needless to say, as a general rule the set of equations (2.37) cannot be analytically integrated.

A mechanical analogy exists which allows one to use the rich intuition one has with mechanical motion in order to answer the question whether or not a solution interpolating between two given critical points exist. Indeed, let us again interpret z as a "time." Then (2.37) can be read as follows: the velocity vector is equal to the force (the gradient of the superpotential profile). This is the equation describing the flow of a very viscous fluid, such as honey. One places a droplet of honey at a given extremum of the profile $\mathcal{W}$ and then one asks oneself whether or not this droplet will flow into another given extremum of this profile. If there is no obstruction in the form of an abyss or an intermediate extremum, the answer is yes. Otherwise it is no.

Mathematicians developed an advanced theory regarding gradient flows. It is called Morse theory. Here I will not go into further details referring the interested reader to Milnor's well-known textbook [5].

2.4   Quantization. Zero Modes: Bosonic and Fermionic

So far we were discussing classical kink solutions. Now we will proceed to quantization, which will be carried out in the quasiclassical approximation (i.e. at weak coupling).

The quasiclassical quantization procedure is quite straightforward. With the classical solution denoted by ϕ_0, one represents the field ϕ as a sum of the classical solution plus small deviations,

$$\phi = \phi_0 + \chi\,. \tag{2.38}$$

One then expands χ, and the fermion field ψ, in modes of appropriately chosen differential operators, in such a way as to diagonalize the Hamiltonian. The coefficients in the mode expansion are the canonical coordinates to be quantized. The zero modes in the mode expansion – they are associated with the collective coordinates of the kink – must be treated separately. As we will see, for critical solitons all nonzero modes cancel (this is a manifestation of the Bose–Fermi cancelation inherent to supersymmetric theories). In this sense, the quantization of supersymmetric solitons is simpler than the one of their non-supersymmetric brethren. We have to deal exclusively with the zero modes. The cancelation of the nonzero modes will be discussed in the next section.

To properly define the mode expansion we have to discretize the spectrum, i.e. introduce an infrared regularization. To this end we place the system in a large spatial box, i.e., we impose the boundary conditions at $z = \pm L/2$, where L is a large auxiliary size (at the very end, $L \to \infty$). The conditions we choose are

$$[\partial_z \phi - \mathcal{W}'(\phi)]_{z=\pm L/2} = 0\,, \qquad \psi_1|_{z=\pm L/2} = 0\,,$$

$$[\partial_z - \mathcal{W}''(\phi)]\,\psi_2|_{z=\pm L/2} = 0\,, \tag{2.39}$$

where $\psi_{1,2}$ denote the components of the spinor ψ_α. The first line is nothing but a supergeneralization of the BPS equation for the classical kink solution. The second line is a consequence of the Dirac equation of motion: if ψ satisfies the Dirac equation, there are essentially no boundary conditions for ψ_2. Therefore, the second line is not an independent boundary condition – it follows from the first line. These boundary conditions fully determine the eigenvalues and the eigenfunctions of the appropriate differential operators of the second order; see (2.40) below.

The above choice of the boundary conditions is definitely not unique, but it is particularly convenient because it is compatible with the residual supersymmetry in the presence of the BPS soliton. The boundary conditions (2.39) are consistent with the classical solutions, both for the spatially constant vacuum configurations and for the kink. In particular, the soliton solution ϕ_0 given in (2.7) (for the SPM) or (2.9) (for the SSG model) satisfies $\partial_z\phi - \mathcal{W}' = 0$ everywhere. Note that the conditions (2.39) are not periodic.

Now, for the mode expansion we will use the second order Hermitean differential operators L_2 and $\tilde{L}_2$,

$$L_2 = P^\dagger P \,, \qquad \tilde{L}_2 = PP^\dagger \,, \tag{2.40}$$

where

$$P = \partial_z - \mathcal{W}''|_{\phi=\phi_0(z)} \,, \qquad P^\dagger = -\partial_z - \mathcal{W}''|_{\phi=\phi_0(z)} \,. \tag{2.41}$$

The operator L_2 defines the modes of $\chi \equiv \phi - \phi_0$, and those of the fermion field ψ_2, while $\tilde{L}_2$ does this job for ψ_1. The boundary conditions for $\psi_{1,2}$ are given in (2.39), for χ they follow from the expansion of the first condition in (2.39),

$$[\partial_z - \mathcal{W}''(\phi_0(z))]\,\chi|_{z=\pm L/2} = 0\,. \tag{2.42}$$

It would be natural at this point if you would ask me why it is the differential operators L_2 and $\tilde{L}_2$ that are chosen for the mode expansion. In principle, any Hermitean operator has an orthonormal set of eigenfunctions. The choice above is singled out because it ensures diagonalization. Indeed, the quadratic form following from the Lagrangian (2.1) for small deviations from the classical kink solution is

$$S^{(2)} \to \frac{1}{2} \int d^2x \left\{ -\chi L_2 \chi - i\psi_1 P\psi_2 + i\psi_2 P^\dagger \psi_1 \right\}\,. \tag{2.43}$$

where I neglected time derivatives and used the fact that $d\phi_0/dz = \mathcal{W}'(\phi_0)$ for the kink under consideration. If diagonalization is not yet transparent, wait for an explanatory comment in the next section.

It is easy to verify that there is only one zero mode $\chi_0(z)$ for the operator L_2. It has the form

$$\chi_0 \propto \frac{d\phi_0}{dz} \propto \mathcal{W}'|_{\phi=\phi_0(z)} \propto \begin{cases} \dfrac{1}{\cosh^2(mz/2)} & \text{(SPM)}\,. \\[2mm] \dfrac{1}{\cosh(mz)} & \text{(SSG)}\,. \end{cases} \tag{2.44}$$

It is quite obvious that this zero mode is due to translations. The corresponding collective coordinate z_0 can be introduced through the substitution $z \longrightarrow z - z_0$

in the classical kink solution. Then

$$\chi_0 \propto \frac{\partial \phi_0(z - z_0)}{\partial z_0}. \tag{2.45}$$

The existence of the zero mode for the fermion component ψ_2, which is proportional to the same function $\partial \phi_0 / \partial z_0$ as the zero mode in χ, (in fact, this is the zero mode in P), is due to supersymmetry. The translational bosonic zero mode entails a fermionic one usually referred to as "supersymmetric (or supertranslational) mode."

The operator $\tilde{L}_2$ has no zero modes at all.

The translational and supertranslational zero modes discussed above imply that the kink[4] is described by two collective coordinates: its center z_0 and a fermionic "center" η, which is a Grassmann parameter,

$$\phi = \phi_0(z - z_0) + \text{nonzero modes}, \quad \psi_2 = \eta \chi_0 + \text{nonzero modes}, \tag{2.46}$$

where χ_0 is the normalized mode obtained from (2.44) by normalization. The nonzero modes in (2.46) are those of the operator L_2. As for the component ψ_1 of the fermion field, we decompose ψ_1 in modes of the operator $\tilde{L}_2$; thus, ψ_1 is given by the sum over nonzero modes of this operator ($\tilde{L}_2$ has no zero modes).

Now, we are ready to derive a Lagrangian describing the moduli dynamics. To this end we substitute (2.46) in the original Lagrangian (2.1) ignoring the nonzero modes and assuming that time dependence enters only through (an adiabatically slow) time dependence of the moduli, z_0 and η,

$$\mathcal{L}_{\text{QM}} = -M + \frac{1}{2} \dot{z}_0^2 \int dz \left(\frac{d\phi_0(z)}{dz} \right)^2 + \frac{i}{2} \eta \dot{\eta} \int dz \, (\chi_0(z))^2$$

$$= -M + \frac{M}{2} \dot{z}_0^2 + \frac{i}{2} \eta \dot{\eta}, \tag{2.47}$$

where M is the kink mass and the subscript QM emphasizes the fact that the original field theory is now reduced to quantum mechanics of the kink moduli. The bosonic part of this Lagrangian is quite evident: it corresponds to a free non-relativistic motion of a particle with mass M.

A priori one might expect the fermionic part of $\mathcal{L}_{\text{QM}}$ to give rise to a Fermi–Bose doubling. While generally speaking this is the case, in the simplest example at hand there is no doubling, and the "fermion center" modulus does not manifest itself.

Indeed, the (quasiclassical) quantization of the system amounts to imposing the commutation (anticommutation) relations

$$[p, z_0] = -i, \quad \eta^2 = \frac{1}{2}, \tag{2.48}$$

where $p = M \dot{z}_0$ is the canonical momentum conjugated to z_0. It means that in the quantum dynamics of the soliton moduli z_0 and η, the operators p and η can

---

[4] Remember, in two dimensions the kink is a particle!

be realized as

$$p = M\dot{z}_0 = -i\frac{d}{dz_0}, \qquad \eta = \frac{1}{\sqrt{2}}. \tag{2.49}$$

(It is clear that we could have chosen $\eta = -1/\sqrt{2}$ as well. The two choices are physically equivalent.)

Thus, η reduces to a constant; the Hamiltonian of the system is

$$H_{\mathrm{QM}} = M - \frac{1}{2M}\frac{d^2}{dz_0^2}. \tag{2.50}$$

The wave function on which this Hamiltonian acts is *single-component*.

One can obtain the same Hamiltonian by calculating supercharges. Substituting the mode expansion in the supercharges (2.13) we arrive at

$$Q_1 = 2\sqrt{\mathcal{Z}}\,\eta + \dots, \qquad Q_2 = \sqrt{\mathcal{Z}}\,\dot{z}_0\,\eta + \dots, \tag{2.51}$$

and $Q_2^2 = H_{\mathrm{QM}} - M$. (Here the ellipses stand for the omitted nonzero modes.) The supercharges depend only on the canonical momentum p,

$$Q_1 = \sqrt{2\mathcal{Z}}, \qquad Q_2 = \frac{p}{\sqrt{2\mathcal{Z}}}. \tag{2.52}$$

In the rest frame in which we perform our consideration $\{Q_1, Q_2\} = 0$; the only value of p consistent with it is $p = 0$. Thus, for the kink at rest, $Q_1 = \sqrt{2\mathcal{Z}}$ and $Q_2 = 0$, which is in full agreement with the general construction. The representation (2.52) can be used at nonzero p as well. It reproduces the superalgebra (2.16) in the non-relativistic limit, with p having the meaning of the total spatial momentum P_1.

The conclusion that there is no Fermi–Bose doubling for the supersymmetric kink rests on the fact that there is only *one* (real) fermion zero mode in the kink background, and, consequently, a single fermionic modulus. This is totally counterintuitive and is, in fact, a manifestation of an *anomaly*. We will discuss this issue in more detail later (Sect. 2.7).

2.5   Cancelation of Nonzero Modes

Above we have omitted the nonzero modes altogether. Now I want to show that for the kink in the ground state the impact of the bosonic nonzero modes is canceled by that of the fermionic nonzero modes.

For each given nonzero eigenvalue, there is one bosonic eigenfunction (in the operator L_2), the same eigenfunction in ψ_2, and one eigenfunction in ψ_1 (of the operator $\tilde{L}_2$) with the same eigenvalue. The operators L_2 and $\tilde{L}_2$ have the same spectrum (except for the zero modes) and their eigenfunctions are related.

Indeed, let χ_n be a (normalized) eigenfunction of L_2,

$$L_2\chi_n(z) = \omega_n^2\chi_n(z). \tag{2.53}$$

Introduce

$$\tilde{\chi}_n(z) = \frac{1}{\omega_n} P \chi_n(z).$$
(2.54)

Then, $\tilde{\chi}_n(z)$ is a (normalized) eigenfunction of $\tilde{L}_2$ with the same eigenvalue,

$$\tilde{L}_2 \tilde{\chi}_n(z) = PP^\dagger \frac{1}{\omega_n} P \chi_n(z) = \frac{1}{\omega_n} P \omega_n^2 \chi_n(z) = \omega_n^2 \tilde{\chi}_n(z).$$
(2.55)

In turn,

$$\chi_n(z) = \frac{1}{\omega_n} P^\dagger \tilde{\chi}_n(z).$$
(2.56)

The quantization of the nonzero modes is quite standard. Let us denote the Hamiltonian *density* by $\mathcal{H}$,

$$H = \int dz\, \mathcal{H}.$$

Then in the quadratic in the quantum fields χ approximation the Hamiltonian density takes the following form:

$$\mathcal{H} - \partial_z \mathcal{W} = \frac{1}{2} \{\dot{\chi}^2 + [(\partial_z - \mathcal{W}'')\chi]^2$$

$$+ i\, \psi_2(\partial_z + \mathcal{W}'')\psi_1 + i\psi_1(\partial_z - \mathcal{W}'')\psi_2\},$$
(2.57)

where $\mathcal{W}''$ is evaluated at $\phi = \phi_0$. We recall that the prime denotes differentiation over ϕ,

$$\mathcal{W}'' = \frac{\partial^2 \mathcal{W}}{\partial \phi^2}.$$

The expansion in eigenmodes has the form,

$$\chi(x) = \sum_{n \neq 0} b_n(t)\, \chi_n(z), \quad \psi_2(x) = \sum_{n \neq 0} \eta_n(t)\, \chi_n(z),$$

$$\psi_1(x) = \sum_{n \neq 0} \xi_n(t)\, \tilde{\chi}_n(z).$$
(2.58)

Note that the summation does not include the zero mode $\chi_0(z)$. This mode is not present in ψ_1 at all. As for the expansions of χ and ψ_2, the inclusion of the zero mode would correspond to a shift in the collective coordinates z_0 and η. Their quantization has been already considered in the previous section. Here we set $z_0 = 0$.

The coefficients b_n, η_n and ξ_n are time-dependent operators. Their equal time commutation relations are determined by the canonical commutators (2.15),

$$[b_m, \dot{b}_n] = i\delta_{mn}, \quad \{\eta_m, \eta_n\} = \delta_{mn}, \quad \{\xi_m, \xi_n\} = \delta_{mn}.$$
(2.59)

Thus, the mode decomposition reduces the dynamics of the system under consideration to quantum mechanics of an infinite set of supersymmetric harmonic oscillators (in higher orders the oscillators become anharmonic). The *ground state* of the quantum kink corresponds to setting each oscillator in the set to the ground state.

Constructing the creation and annihilation operators in the standard way, we find the following nonvanishing expectations values of the bilinears built from the operators b_n, η_n, and ξ_n in the ground state:

$$\langle \dot{b}_n^2 \rangle_{\text{sol}} = \frac{\omega_n}{2}, \qquad \langle b_n^2 \rangle_{\text{sol}} = \frac{1}{2\omega_n}, \qquad \langle \eta_n \xi_n \rangle_{\text{sol}} = \frac{i}{2}. \tag{2.60}$$

The expectation values of other bilinears obviously vanish. Combining (2.57), (2.58), and (2.60) we get

$$\langle \text{sol} \, | \mathcal{H}(z) - \partial_z \mathcal{W} | \, \text{sol} \rangle$$

$$= \frac{1}{2} \sum_{n \neq 0} \left\{ \frac{\omega_n}{2} \chi_n^2 + \frac{1}{2\omega_n} [(\partial_z - \mathcal{W}'')\chi_n]^2 - \frac{\omega_n}{2} \chi_n^2 \right.$$

$$\left. - \frac{1}{2\omega_n} [(\partial_z - \mathcal{W}'')\chi_n]^2 \right\} \equiv 0. \tag{2.61}$$

In other words, for the critical kink (in the ground state) the Hamiltonian density is *locally* equal to $\partial_z W$ – this statement is valid at the level of quantum corrections!

The four terms in the braces in (2.61) are in one-to-one correspondence with those in (2.57). Note that in proving the vanishing of the right-hand side we did not perform integrations by parts. The vanishing of the right-hand side of (2.57) demonstrates explicitly the residual supersymmetry – i.e. the conservation of Q_2 and the fact that $M = \mathcal{Z}$. Equation (2.61) must be considered as a local version of BPS saturation (i.e. conservation of a residual supersymmetry).

The multiplet shortening guarantees that the equality $M = \mathcal{Z}$ is not corrected in higher orders. For critical solitons, quantum corrections cancel altogether; $M = \mathcal{Z}$ is exact.

What lessons can one draw from the considerations of this section? In the case of the polynomial model the target space is noncompact, while the one in the sine–Gordon case can be viewed as a compact target manifold S^1. In these both cases we get one and the same result: a short (one-dimensional) soliton multiplet defying the fermion parity (further details will be given in Sect. 2.7).

2.6   Anomaly I

We have explicitly demonstrated that the equality between the kink mass M and the central charge $\mathcal{Z}$ survives at the quantum level. The classical expression for the central charge is given in (2.19). If one takes proper care of ultraviolet regularization one can show [6] that quantum corrections do modify (2.19). Here we will present a simple argument demonstrating the emergence of an anomalous term in the central charge. We also discuss its physical meaning.

To begin with, let us consider $\gamma^\mu J_\mu$ where J_μ is the supercurrent defined in (2.10). This quantity is related to the superconformal properties of the model under consideration. At the classical level

$$(\gamma^\mu J_\mu)_{\text{class}} = 2i\,W'\psi\,. \tag{2.62}$$

Note that the first term in the supercurrent (2.10) gives no contribution in (2.62) due to the fact that in two dimensions $\gamma_\mu\gamma^\nu\gamma^\mu = 0$.

The local form of the superalgebra is given in (2.20). Multiplying (2.20) by γ_μ from the left, we get the supertransformation of $\gamma_\mu J^\mu$,

$$\frac{1}{2}\left\{\gamma^\mu J_\mu, \bar{Q}\right\} = T^\mu_\mu + i\gamma_\mu\gamma^5\,\zeta^\mu\,, \qquad \gamma^5 = \gamma^0\gamma^1 = -\sigma_1\,. \tag{2.63}$$

This equation establishes a supersymmetric relation between $\gamma^\mu J_\mu$, T^μ_μ, and ζ^μ and, as was mentioned above, remains valid with quantum corrections included. But the expressions for these operators can (and will) be changed. Classically the trace of the energy-momentum tensor is

$$(T^\mu_\mu)_{\text{class}} = (W')^2 + \frac{1}{2}W''\,\bar{\psi}\psi\,, \tag{2.64}$$

as follows from (2.18). The zero component of the topological current ζ^μ in the second term in (2.63) classically coincides with the density of the central charge, $\partial_z W$, see (2.21). It is seen that the trace of the energy-momentum tensor and the density of the central charge appear in this relation together.

It is well-known that in renormalizable theories with ultraviolet logarithmic divergences, both the trace of the energy-momentum tensor and $\gamma^\mu J_\mu$ have anomalies. We will use this fact, in conjunction with (2.63), to establish the general form of the anomaly in the density of the central charge.

To get an idea of the anomaly, it is convenient to use dimensional regularization. If we assume that the number of dimensions is $D = 2 - \varepsilon$ rather than $D = 2$, the first term in (2.10) does generate a nonvanishing contribution to $\gamma^\mu J_\mu$, proportional to $(D-2)(\partial_\nu\phi)\,\gamma^\nu\psi$. At the quantum level this operator gets an ultraviolet logarithm (i.e. $(D-2)^{-1}$ in dimensional regularization), so that $D-2$ cancels, and we are left with an anomalous term in $\gamma^\mu J_\mu$.

To do the one-loop calculation, we apply here (as well as in some other instances below) the background field technique: we substitute the field ϕ by its background and quantum parts, ϕ and χ, respectively,

$$\phi \longrightarrow \phi + \chi\,. \tag{2.65}$$

Specifically, for the anomalous term in $\gamma^\mu J_\mu$,

$$(\gamma^\mu J_\mu)_{\text{anom}} = (D-2)(\partial_\nu\phi)\,\gamma^\nu\psi = -(D-2)\,\chi\gamma^\nu\partial_\nu\psi$$

$$= i(D-2)\,\chi\,W''(\phi+\chi)\,\psi\,, \tag{2.66}$$

where an integration by parts has been carried out, and a total derivative term is omitted (on dimensional grounds it vanishes in the limit $D = 2$). We also used the equation of motion for the ψ field. The quantum field χ then forms a loop and we get for the anomaly,

$$
(\gamma^\mu J_\mu)_{\text{anom}} = i\,(D-2)\,\langle 0|\chi^2|0\rangle\,\mathcal{W}'''(\phi)\,\psi
$$

$$
= -(D-2)\int \frac{d^D p}{(2\pi)^D}\,\frac{1}{p^2-m^2}\,\mathcal{W}'''(\phi)\,\psi
$$

$$
= \frac{i}{2\pi}\mathcal{W}'''(\phi)\,\psi\,. \tag{2.67}
$$

The supertransformation of the anomalous term in $\gamma^\mu J_\mu$ is

$$
\frac{1}{2}\left\{(\gamma^\mu J_\mu)_{\text{anom}}\,,\bar{Q}\right\} = \left(\frac{1}{8\pi}\mathcal{W}''''\bar{\psi}\psi + \frac{1}{4\pi}\mathcal{W}'''\mathcal{W}'\right)
$$

$$
+ i\gamma_\mu\gamma^5\epsilon^{\mu\nu}\partial_\nu\left(\frac{1}{4\pi}\mathcal{W}''\right)\,. \tag{2.68}
$$

The first term on the right-hand side is the anomaly in the trace of the energy-momentum tensor, the second term represents the anomaly in the topological current. The *corrected current* has the form

$$
\zeta^\mu = \epsilon^{\mu\nu}\partial_\nu\left(\mathcal{W} + \frac{1}{4\pi}\mathcal{W}''\right)\,. \tag{2.69}
$$

Consequently, at the quantum level, after the inclusion of the anomaly, the central charge becomes

$$
\mathcal{Z} = \left(\mathcal{W} + \frac{1}{4\pi}\mathcal{W}''\right)_{z=+\infty} - \left(\mathcal{W} + \frac{1}{4\pi}\mathcal{W}''\right)_{z=-\infty}\,. \tag{2.70}
$$

2.7 Anomaly II (Shortening Supermultiplet Down to One State)

In the model under consideration, see (2.1), the fermion field is real which implies that the fermion number is not defined. What is defined, however, is the fermion parity G. Following a general tradition, G is sometimes denoted as $(-1)^F$, in spite of the fact that in the case at hand the fermion number F does not exist. The tradition originates, of course, in models with complex fermions, where the fermion number F does exist, but we will not dwell on this topic.

The action of G reduces to changing the sign for the fermion operators leaving the boson operators intact, for instance,

$$
G\,Q_\alpha\,G^{-1} = -Q_\alpha\,, \qquad G\,P_\mu\,G^{-1} = P_\mu\,. \tag{2.71}
$$

The fermion parity G realizes Z_2 symmetry associated with changing the sign of the fermion fields. This symmetry is obvious at the classical level (and, in fact, in any finite order of perturbation theory). This symmetry is very intuitive – this is the Z_2 symmetry which distinguishes fermion states from the boson states in the model at hand, with the Majorana fermions.

Here I will try to demonstrate (without delving too deep into technicalities) that in the soliton sector the very classification of states as either bosonic or fermionic is broken. The disappearance of the fermion parity in the BPS soliton sector is a global anomaly [7].

Let us consider the algebra (2.28) in the special case $M^2 = Z^2$. Assuming Z to be positive, we consider the BPS soliton, $M = Z$, for which the supercharge Q_2 is trivial, $Q_2 = 0$. Thus, we are left with a single supercharge Q_1 realized nontrivially. The algebra reduces to a single relation

$$(Q_1)^2 = 2\, Z\,. \tag{2.72}$$

The irreducible representations of this algebra are one-dimensional. There are two such representations,

$$Q_1 = \pm \sqrt{2Z}\,, \tag{2.73}$$

i.e., two types of solitons,

$$Q_1 |\,\text{sol}_+ \,\rangle = \sqrt{2Z}\,|\,\text{sol}_+ \,\rangle\,, \qquad Q_1 |\,\text{sol}_- \,\rangle = -\sqrt{2Z}\,|\,\text{sol}_- \,\rangle\,. \tag{2.74}$$

It is clear that these two representations are unitary non-equivalent.

The one-dimensional irreducible representation of supersymmetry implies multiplet shortening: the short BPS supermultiplet contains only one state while non-BPS supermultiplets contain two. The possibility of such supershort one-dimensional multiplets was discarded in the literature for years. It is for a reason: while the fermion parity $(-1)^F$ is granted in any local field theory based on fermionic and bosonic fields, it is *not defined* in the one-dimensional irreducible representation. Indeed, if it were defined, it would be -1 for Q_1, which is incompatible with any of the equations (2.74). The only way to recover $(-1)^F$ is to have a reducible representation containing both $|\,\text{sol}_+\,\rangle$ and $|\,\text{sol}_-\,\rangle$. Then,

$$Q_1 = \sigma_3 \sqrt{2Z}\,, \qquad (-1)^F = \sigma_1\,, \tag{2.75}$$

where $\sigma_{1,2,3}$ stand for the Pauli matrices.

Does this mean that the one-state supermultiplet is not a possibility in the local field theory? As I argued above, in the simplest two-dimensional supersymmetric model (2.1) the BPS solitons do exist and do realize such supershort multiplets defying $(-1)^F$. These BPS solitons are neither bosons nor fermions. Further details can be found in [7], in which a dedicated research of this particular global anomaly is presented. The important point is that short multiplets of BPS states are protected against becoming non-BPS under small perturbations. Although the overall sign of Q_1 on the irreducible representation is not observable, the relative sign is. For instance, there are two types of reducible representations of dimension two: one is $\{+, -\}$ (see (2.75)), and the other $\{+, +\}$ (equivalent to $\{-, -\}$). In the first case, two states can pair up and leave the BPS bound as soon as appropriate perturbations are introduced. In the second case, the BPS relation $M = Z$ is "bullet-proof."

To reiterate, the discrete Z_2 symmetry $G = (-1)^F$ discussed above is nothing but the change of sign of all fermion fields, $\psi \to -\psi$. This symmetry is seemingly present in any theory with fermions. How on earth can this symmetry be lost in the soliton sector?

Technically the loss of $G = (-1)^F$ is due to the fact that there is only *one* (real) fermion zero mode on the soliton in the model at hand. Normally, the fermion degrees of freedom enter in holomorphic pairs, $\{\bar\psi, \psi\}$. In our case of a single fermion zero mode we have "one half" of such a pair. The second fermion zero mode, which would produce the missing half, turns out to be delocalized. More exactly, it is not localized on the soliton, but, rather, on the boundary of the "large box" one introduces for quantization (see Sect. 2.5). For physical measurements made far away from the auxiliary box boundary, the fermion parity G is lost, and the supermultiplet consisting of a single state becomes a physical reality. In a sense, the phenomenon is akin to that of the charge fractionalization, or the Jackiw–Rebbi phenomenon [8]. The essence of this well-known phenomenon is as follows: in models with *complex* fermions, where the fermion number is defined, it takes integer values only provided one includes in the measurement the box boundaries. Local measurements on the kink will yield a fractional charge.

3   Domain Walls in (3+1)-Dimensional Theories

Kinks are topological defects in (1+1)-dimensional theories. Topological defects of a similar nature in 1+3 dimensions are domain walls. The corresponding geometry is depicted in Fig. 1. Just like kinks, domain walls interpolate (in the transverse direction, to be denoted as z) between distinct degenerate vacua of the theory. Unlike kinks, domain walls are not localized objects – they extend into the longitudinal directions (x and y in Fig. 1). Therefore, the mass (energy) of the domain wall is infinite and the relevant parameter is the wall tension – the mass per unit area. In (1+3)-dimensional theories the wall tension has dimension m^3.

In this section I will discuss supersymmetric critical (BPS-saturated) domain walls. Before I will be able to proceed, I have to describe the simplest (1+3)-dimensional supersymmetric theory in which such walls exist. Unlike in two dimensions, where field theories with minimal supersymmetry possess two supercharges, in four dimensions the minimal set contains four supercharges,

$$\{Q_\alpha, \bar{Q}_{\dot\alpha}\}, \qquad \alpha, \dot\alpha = 1, 2.$$

Q_α and $\bar{Q}_{\dot\alpha}$ are spinors with respect to the Lorentz group.

3.1   Superspace and Superfields

The four-dimensional space x^μ (with Lorentz vectorial indices $\mu = 0, ..., 3$) can be promoted to superspace by adding four Grassmann coordinates θ_α and $\bar\theta_{\dot\alpha}$,

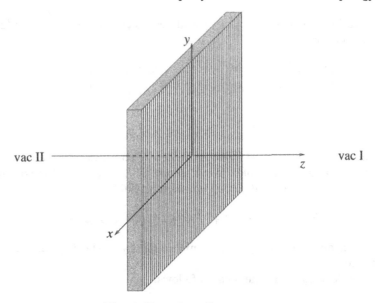

Fig. 1. Domain wall geometry.

(with spinorial indices α, $\dot{\alpha} = 1, 2$). The coordinate transformations

$$\{x^\mu, \theta_\alpha, \bar{\theta}_{\dot{\alpha}}\}: \qquad \delta\theta_\alpha = \varepsilon_\alpha, \qquad \delta\bar{\theta}_{\dot{\alpha}} = \bar{\varepsilon}_{\dot{\alpha}}, \qquad \delta x_{\alpha\dot{\alpha}} = -2i\,\theta_\alpha\bar{\varepsilon}_{\dot{\alpha}} - 2i\,\bar{\theta}_{\dot{\alpha}}\varepsilon_\alpha \quad (3.1)$$

add SUSY to the translational and Lorentz transformations.[5]

Here the Lorentz vectorial indices are transformed into spinorial ones according to the standard rule

$$A_{\beta\dot{\beta}} = A_\mu(\sigma^\mu)_{\beta\dot{\beta}}, \qquad A^\mu = \frac{1}{2}A_{\alpha\dot{\beta}}(\bar{\sigma}^\mu)^{\dot{\beta}\alpha}, \qquad (3.2)$$

where

$$(\sigma^\mu)_{\alpha\dot{\beta}} = \{1, \boldsymbol{\tau}\}_{\alpha\dot{\beta}}, \qquad (\bar{\sigma}^\mu)_{\dot{\beta}\alpha} = (\sigma^\mu)_{\alpha\dot{\beta}}. \qquad (3.3)$$

We use the notation $\boldsymbol{\tau}$ for the Pauli matrices throughout these lecture notes. The lowering and raising of the spinorial indices is performed by virtue of the $\epsilon^{\alpha\beta}$ symbol ($\epsilon^{\alpha\beta} = i(\tau_2)_{\alpha\beta}$, $\epsilon^{12} = 1$). For instance,

$$(\bar{\sigma}^\mu)^{\dot{\beta}\alpha} = \epsilon^{\dot{\beta}\dot{\rho}}\,\epsilon^{\alpha\gamma}\,(\bar{\sigma}^\mu)_{\dot{\rho}\gamma} = \{1, -\boldsymbol{\tau}\}^{\dot{\beta}\alpha}. \qquad (3.4)$$

---

[5] My notation is close but not identical to that of Bagger and Wess [9]. The main distinction is the conventional choice of the metric tensor $g_{\mu\nu} = \mathrm{diag}(+ - --)$ as opposed to the $\mathrm{diag}(- + ++)$ version of Bagger and Wess. For further details see Appendix in [10]. Both, the spinorial and vectorial indices will be denoted by Greek letters. To differentiate between them we will use the letters from the beginning of the alphabet for the spinorial indices (e.g. α, β etc.) reserving those from the end of the alphabet (e.g. μ, ν, etc.) for the vectorial indices.

Two invariant subspaces $\{x_L^\mu, \theta_\alpha\}$ and $\{x_R^\mu, \bar\theta_{\dot\alpha}\}$ are spanned on $1/2$ of the Grassmann coordinates,

$$\begin{aligned} \{x_L^\mu, \theta_\alpha\}: \qquad & \delta\theta_\alpha = \varepsilon_\alpha, \quad \delta(x_L)_{\alpha\dot\alpha} = -4i\,\theta_\alpha\bar\varepsilon_{\dot\alpha}; \\ \{x_R^\mu, \bar\theta_{\dot\alpha}\}: \qquad & \delta\bar\theta_{\dot\alpha} = \bar\varepsilon_{\dot\alpha}, \quad \delta(x_R)_{\alpha\dot\alpha} = -4i\,\bar\theta_{\dot\alpha}\varepsilon_\alpha, \end{aligned} \qquad (3.5)$$

where

$$(x_{L,R})_{\alpha\dot\alpha} = x_{\alpha\dot\alpha} \mp 2i\,\theta_\alpha\bar\theta_{\dot\alpha}. \qquad (3.6)$$

The minimal supermultiplet of fields includes one complex scalar field $\phi(x)$ (two bosonic states) and one complex Weyl spinor $\psi^\alpha(x)$, $\alpha = 1, 2$ (two fermionic states). Both fields are united in one *chiral superfield*,

$$\Phi(x_L, \theta) = \phi(x_L) + \sqrt{2}\theta^\alpha \psi_\alpha(x_L) + \theta^2 F(x_L), \qquad (3.7)$$

where F is an auxiliary component, which appears in the Lagrangian without the kinetic term.

The *superderivatives* are defined as follows:

$$D_\alpha = \frac{\partial}{\partial\theta^\alpha} - i\partial_{\alpha\dot\alpha}\bar\theta^{\dot\alpha}, \quad \bar{D}_{\dot\alpha} = -\frac{\partial}{\partial\bar\theta^{\dot\alpha}} + i\theta^\alpha\partial_{\alpha\dot\alpha}, \quad \{D_\alpha, \bar{D}_{\dot\alpha}\} = 2i\partial_{\alpha\dot\alpha}. \quad (3.8)$$

3.2   Wess–Zumino Models

The Wess–Zumino model describes interactions of an arbitrary number of chiral superfields. We will consider the simplest original Wess–Zumino model [11] (sometimes referred to as the minimal model).

The model contains one chiral superfield $\Phi(x_L, \theta)$ and its complex conjugate $\bar\Phi(x_R, \bar\theta)$, which is anti-chiral. The action of the model is

$$S = \frac{1}{4}\int \mathrm{d}^4x\,\mathrm{d}^4\theta\,\Phi\bar\Phi + \frac{1}{2}\int \mathrm{d}^4x\,\mathrm{d}^2\theta\,\mathcal{W}(\Phi) + \frac{1}{2}\int \mathrm{d}^4x\,\mathrm{d}^2\bar\theta\,\bar{\mathcal{W}}(\bar\Phi). \qquad (3.9)$$

Note that the first term is the integral over the full superspace, while the second and the third run over the chiral subspaces. The holomorphic function $\mathcal{W}(\Phi)$ is called the *superpotential*. In components the Lagrangian has the form

$$\mathcal{L} = (\partial^\mu\bar\phi)(\partial_\mu\phi) + \psi^\alpha i\partial_{\alpha\dot\alpha}\bar\psi^{\dot\alpha} + \bar{F}F + \left\{FW'(\phi) - \frac{1}{2}W''(\phi)\psi^2 + \text{h.c.}\right\}. \quad (3.10)$$

From (3.10) it is obvious that F can be eliminated by virtue of the classical equation of motion,

$$\bar{F} = -\frac{\partial\mathcal{W}(\phi)}{\partial\phi}, \qquad (3.11)$$

so that the *scalar potential* describing the self-interaction of the field ϕ is

$$V(\phi, \bar\phi) = \left|\frac{\partial\mathcal{W}(\phi)}{\partial\phi}\right|^2. \qquad (3.12)$$

In what follows we will often denote the chiral superfield and its lowest (bosonic) component by one and the same letter, making no distinction between capital and small ϕ. Usually it is clear from the context what is meant in each particular case.

If one limits oneself to renormalizable theories, the superpotential $\mathcal{W}$ must be a polynomial function of Φ of power not higher than three. In the model at hand, with one chiral superfield, the generic superpotential then can be always reduced to the following "standard" form

$$\mathcal{W}(\Phi) = \frac{m^2}{\lambda}\Phi - \frac{\lambda}{3}\Phi^3 . \qquad (3.13)$$

The quadratic term can be always eliminated by a redefinition of the field Φ. Moreover, by using the symmetries of the model, one can always choose the phases of the constants m and λ at will. (Note that generically the parameters m and λ are complex.)

Let us study the set of classical vacua of the theory, *the vacuum manifold*. In the simplest case of the vanishing superpotential, $\mathcal{W} = 0$, any coordinate-independent field $\Phi_{\text{vac}} = \phi_0$ can serve as a vacuum. The vacuum manifold is then the one-dimensional (complex) manifold $C^1 = \{\phi_0\}$. The continuous degeneracy is due to the absence of the potential energy, while the kinetic energy vanishes for any constant ϕ_0.

This continuous degeneracy is lifted in the case of a non-vanishing superpotential. In particular, the superpotential (3.13) implies two degenerate classical vacua,

$$\phi_{\text{vac}} = \pm\frac{m}{\lambda} . \qquad (3.14)$$

Thus, the continuous manifold of vacua C^1 reduces to two points. Both vacua are physically equivalent. This equivalence could be explained by the spontaneous breaking of the Z_2 symmetry, $\Phi \to -\Phi$, present in the action (3.9) with the superpotential (3.13).

The determination of the conserved supercharges in this model is a straightforward procedure. We have

$$Q_\alpha = \int d^3x J_\alpha^0 , \qquad \bar{Q}_{\dot{\alpha}} = \int d^3x \bar{J}_{\dot{\alpha}}^0 , \qquad (3.15)$$

where J_α^μ is the conserved supercurrent,

$$J_\alpha^\mu = \frac{1}{2}(\bar{\sigma}^\mu)^{\dot{\beta}\beta} J_{\alpha\beta\dot{\beta}} ,$$

$$J_{\alpha\beta\dot{\beta}} = 2\sqrt{2}\left\{(\partial_{\alpha\dot{\beta}}\phi^+)\psi_\beta - i\,\epsilon_{\beta\alpha}F\bar{\psi}_{\dot{\beta}}\right\} . \qquad (3.16)$$

The Golfand–Likhtman superalgebra in the spinorial notation takes the form

$$\{Q_\alpha , \bar{Q}_{\dot{\alpha}}\} = 2P_{\alpha\dot{\alpha}} , \qquad (3.17)$$

where P is the energy-momentum operator.

3.3   Critical Domain Walls

The minimal Wess–Zumino model has two degenerate vacua (3.14). Field configurations interpolating between two degenerate vacua are called *domain walls*. They have the following properties: (i) the corresponding solutions are static and depend only on one spatial coordinate; (ii) they are topologically stable and indestructible – once a wall is created it cannot disappear. Assume for definiteness that the wall lies in the xy plane. This is the geometry we will always keep in mind. Then the wall solution ϕ_w will depend only on z. Since the wall extends indefinitely in the xy plane, its energy E_w is infinite. However, the wall tension T_w (the energy per unit area $T_w = E_w/A$) is finite, in principle measurable, and has a clear-cut physical meaning.

The wall solution of the classical equations of motion superficially looks very similar to the kink solution in the SPM discussed in Sect. 2,

$$\phi_w = \frac{m}{\lambda}\,\tanh(|m|z)\,. \tag{3.18}$$

Note, however, that the parameters m and λ are not assumed to be real; the field ϕ is complex in the Wess–Zumino model. A remarkable feature of this solution is that it preserves one half of supersymmetry, much in the same way as the critical kinks in Sect. 2. The difference is that $1/2$ in the two-dimensional model meant one supercharge, now it means two supercharges.

Let us now show the preservation of $1/2$ of SUSY explicitly. The SUSY transformations (3.1) generate the following transformation of fields,

$$\delta\phi = \sqrt{2}\varepsilon\psi\,, \qquad \delta\psi^\alpha = \sqrt{2}\left[\varepsilon^\alpha F + i\,\partial_\mu\phi\,(\sigma^\mu)^{\alpha\dot\alpha}\,\bar\varepsilon_{\dot\alpha}\right]\,. \tag{3.19}$$

The domain wall we consider is purely bosonic, $\psi = 0$. Moreover, let us impose the following condition on the domain wall solution (the BPS equation):

$$F\,|_{\bar\phi=\phi_w^*} = -e^{-i\eta}\,\partial_z\phi_w(z)\,, \tag{3.20}$$

where

$$\eta = \arg\frac{m^3}{\lambda^2}\,, \tag{3.21}$$

and, I remind, $F = -\partial\bar{\mathcal{W}}/\partial\bar\phi$, see (3.11). This is a first-order differential equation. The solution quoted above satisfies this condition. The reason for the occurrence of the phase factor $\exp(-i\eta)$ on the right-hand side of (3.20) will become clear shortly. Note that no analog of this phase factor exists in the two-dimensional $\mathcal{N} = 1$ problem on which we dwelled in Sect. 2. There was only a sign ambiguity: two possible choices of signs corresponded respectively to kink and anti-kink.

The first-order BPS equations are, generally speaking, a stronger constraint than the classical equations of motion.[6] If the BPS equation is satisfied, then the second supertransformation in (3.19) reduces to

$$\delta\psi_\alpha \propto \varepsilon_\alpha + i\,e^{i\eta}\,(\sigma^z)_{\alpha\dot\alpha}\,\bar\varepsilon^{\dot\alpha}\,. \tag{3.22}$$

---

[6] I hasten to add that, in the particular problem under consideration, the BPS equation follows from the equation of motion; this is explained in Sect. 3.5.

The right-hand side vanishes provided that

$$\varepsilon_\alpha = -i\, e^{i\eta}\, (\sigma^z)_{\alpha\dot\alpha}\, \bar\varepsilon^{\dot\alpha}\,. \qquad (3.23)$$

This picks up two supertransformations (out of four) which do not act on the domain wall (alternatively people often say that they act trivially). *Quod erat demonstrandum.*

Now, let us calculate the wall tension. To this end we rewrite the expression for the energy functional as follows

$$\mathcal{E} = \int_{-\infty}^{+\infty} dz\, \left[\partial_z\bar\phi\,\partial_z\phi + \bar F F\right]$$

$$\equiv \int_{-\infty}^{+\infty} dz\, \left\{\left[e^{-i\eta}\,\partial_z\mathcal{W} + \text{h.c.}\right] + \left|\partial_z\phi + e^{i\eta}\, F\right|^2\right\}, \qquad (3.24)$$

where ϕ is assumed to depend only on z. In the literature this procedure is called the *Bogomol'nyi completion*. The second term on the right-hand side is non-negative – its minimal value is zero. The first term, being full derivative, depends only on the boundary conditions on ϕ at $z = \pm\infty$.

Equation (3.24) implies that $\mathcal{E} \geq 2\,\text{Re}\left(e^{-i\eta}\,\Delta\mathcal{W}\right)$. The Bogomol'nyi completion can be performed with any η. However, the strongest bound is achieved provided $e^{-i\eta}\,\Delta\mathcal{W}$ is real. This explains the emergence of the phase factor in the BPS equations. In the model at hand, to make $e^{-i\eta}\,\Delta\mathcal{W}$ real, we have to choose η according to (3.21).

When the energy functional is written in the form (3.24), it is perfectly obvious that the absolute minimum is achieved provided the BPS equation (3.20) is satisfied. In fact, the Bogomol'nyi completion provides us with an alternative derivation of the BPS equations. Then, for the minimum of the energy functional – the wall tension T_w – we get

$$T_w = |\mathcal{Z}|\,. \qquad (3.25)$$

Here $\mathcal{Z}$ is the topological charge defined as

$$\mathcal{Z} = 2\left\{\mathcal{W}(\phi(z = \infty)) - \mathcal{W}(\phi(z = -\infty))\right\} = \frac{8\,m^3}{3\,\lambda^2}\,. \qquad (3.26)$$

How come that we got a nonvanishing energy for the state which is annihilated by two supercharges? This is because the original Golfand–Likhtman superalgebra (3.17) gets supplemented by a central extension,

$$\{Q_\alpha, Q_\beta\} = -4\,\Sigma_{\alpha\beta}\,\bar{\mathcal{Z}}\,, \qquad \left\{\bar Q_{\dot\alpha}, \bar Q_{\dot\beta}\right\} = -4\,\bar\Sigma_{\dot\alpha\dot\beta}\,\mathcal{Z}\,, \qquad (3.27)$$

where

$$\Sigma_{\alpha\beta} = -\frac{1}{2}\int dx_{[\mu}dx_{\nu]}\,(\sigma^\mu)_{\alpha\dot\alpha}(\bar\sigma^\nu)^{\dot\alpha}_\beta \qquad (3.28)$$

is the wall area tensor. The particular form of the centrally extended algebra is somewhat different from the one we have discussed in Sect. 2. The central charge is no longer a scalar. Now it is a tensor. However, the structural essence remains the same.

As was mentioned, the general connection between the BPS saturation and the central extension of the superalgebra was noted long ago by Olive and Witten [2] shortly after the advent of supersymmetry. In the context of supersymmetric domain walls, the topic was revisited and extensively discussed in [10] and [12] which I closely follow in my presentation.

Now let us consider representations of the centrally extended superalgebra (with four supercharges). We will be interested not in a generic representation but, rather, in a special one where one half of the supercharges annihilates all states (the famous *short representations*). The existence of such supercharges was demonstrated above at the classical level. The covariant expressions for the residual supercharges $\tilde{Q}_\alpha$ are

$$\tilde{Q}_\alpha = e^{i\eta/2} Q_\alpha - \frac{2}{A} e^{-i\eta/2} \Sigma_{\alpha\beta} n_{\dot\alpha}^\beta \bar{Q}^{\dot\alpha}, \qquad (3.29)$$

where A is the wall area ($A \to \infty$) and

$$n_{\alpha\dot\alpha} = \frac{P_{\alpha\dot\alpha}}{T_w A} \qquad (3.30)$$

is the unit vector proportional to the wall four-momentum $P_{\alpha\dot\alpha}$; it has only the time component in the rest frame. The subalgebra of these residual supercharges in the rest frame is

$$\left\{ \tilde{Q}_\alpha, \tilde{Q}_\beta \right\} = 8 \Sigma_{\alpha\beta} \left\{ T_w - |\mathcal{Z}| \right\}. \qquad (3.31)$$

The existence of the subalgebra (3.31) immediately proves that the wall tension T_w is equal to the central charge $\mathcal{Z}$. Indeed, $\tilde{Q}|\text{wall}\rangle = 0$ implies that $T_w - |\mathcal{Z}| = 0$. This equality is valid both to any order in perturbation theory and non-perturbatively.

From the non-renormalization theorem for the superpotential [13] we additionally infer that the central charge $\mathcal{Z}$ is not renormalized. This is in contradistinction with the situation in the two-dimensional model [7] of Sect. 2. The fact that there are more conserved supercharges in four dimensions than in two turns out crucial. As a consequence, the result

$$T_w = \frac{8}{3} \left| \frac{m^3}{\lambda^2} \right| \qquad (3.32)$$

for the wall tension is *exact* [12,10].

The wall tension T_w is a physical parameter and, as such, should be expressible in terms of the physical (renormalized) parameters m_{ren} and λ_{ren}. One can easily verify that this is compatible with the statement of non-renormalization of T_w. Indeed,

$$m = Z m_{\text{ren}}, \qquad \lambda = Z^{3/2} \lambda_{\text{ren}},$$

---

[7] There one has to deal with the fact that Z *is* renormalized and, moreover, a quantum anomaly was found in the central charge. See Sect. 2.6. What stays exact is the relation $M - \mathcal{Z} = 0$.

where Z is the Z factor coming from the kinetic term. Consequently,

$$\frac{m^3}{\lambda^2} = \frac{m_{ren}^3}{\lambda_{ren}^2}.$$

Thus, the absence of the quantum corrections to (3.32), the renormalizability of the theory, and the non-renormalization theorem for superpotentials – all these three elements are intertwined with each other. In fact, every two elements taken separately imply the third one.

What lessons have we drawn from the example of the domain walls? In the centrally extended SUSY algebras the exact relation $E_{vac} = 0$ is replaced by the exact relation $T_w - |Z| = 0$. Although this statement is valid both perturbatively and non-perturbatively, it is very instructive to visualize it as an explicit cancelation between bosonic and fermionic modes in perturbation theory. The non-renormalization of Z is a specific feature of four dimensions. We have seen previously that it does *not* take place in minimally supersymmetric models in two dimensions.

3.4   Finding the Solution to the BPS Equation

In the two-dimensional theory the integration of the first-order BPS equation (2.26) was trivial. Now the BPS equation (3.20) presents in fact two equations – one for the real part and one for the imaginary part. Nevertheless, it is still trivial to find the solution. This is due to the existence of an "integral of motion,"

$$\frac{\partial}{\partial z}\left(\operatorname{Im} e^{-i\eta}\mathcal{W}\right) = 0.\tag{3.33}$$

The proof is straightforward and is valid in the generic Wess–Zumino model with an arbitrary number of fields. Indeed, differentiating $\mathcal{W}$ and using the BPS equations we get

$$\frac{\partial}{\partial z}\left(e^{-i\eta}\mathcal{W}\right) = \left|\frac{\partial\mathcal{W}}{\partial\phi}\right|^2,\tag{3.34}$$

which immediately entails (3.33).

If we deal with more than one field ϕ, the above "integral of motion" is of limited help. However, for a single field ϕ it solves the problem: our boundary conditions fix $e^{-i\eta}\mathcal{W}$ to be real along the wall trajectory, which allows one to find the trajectory immediately. In this way we arrive at (3.18).

The constraint

$$\operatorname{Im} e^{-i\eta}\mathcal{W} = \text{const}\tag{3.35}$$

can be interpreted as follows: in the complex $\mathcal{W}$ plane the domain wall trajectory is a straight line.

3.5   Does the BPS Equation Follow from the Second Order Equation of Motion?

As we already know, every solution of the BPS equations is automatically a solution of the second-order equations of motion. The inverse is certainly not true in the general case. However, in the minimal Wess–Zumino model under consideration, given the boundary conditions appropriate for the domain walls, this is true, much in the same way as in the minimal two-dimensional model with which we began. Namely, every solution of the equations of motion with the appropriate boundary conditions is simultaneously the solution of the BPS equation (3.20).

The proof of this statement is rather straightforward [14]. Indeed, we start from the equations of motion

$$\partial_z^2 \phi = W' \bar{W}'' , \qquad \partial_z^2 \bar{\phi} = \bar{W}' W'' , \qquad (3.36)$$

where the prime denotes differentiation with respect to the corresponding argument, and use them to show that

$$\frac{\partial}{\partial z} |W'|^2 = \frac{\partial}{\partial z} \left| \frac{\partial \phi}{\partial z} \right|^2 . \qquad (3.37)$$

This implies, in turn, that

$$|W'|^2 - \left| \frac{\partial \phi}{\partial z} \right|^2 = z \text{ independent const.} \qquad (3.38)$$

From the domain wall boundary conditions, one immediately concludes that this constant must vanish, so that in fact

$$|W'|^2 - \left| \frac{\partial \phi}{\partial z} \right|^2 = 0 . \qquad (3.39)$$

If z is interpreted as "time" this equation is nothing but "energy" conservation along the wall trajectory.

Now, let us introduce the ratio

$$R \equiv \left(\bar{W}' \right)^{-1} \frac{\partial \phi}{\partial z} . \qquad (3.40)$$

Please, observe that its absolute value is unity – this is an immediate consequence of (3.39). Our task is to show that the phase of R is z independent. To this end we perform differentiation (again exploiting (3.20)) to arrive at

$$\frac{\partial R}{\partial z} = \left(\bar{W}' \right)^{-2} \left(\bar{W}'' \right) \left(|W'|^2 - \left| \frac{\partial \phi}{\partial z} \right|^2 \right) = 0 . \qquad (3.41)$$

The statement that R reduces to a z independent phase factor is equivalent to the BPS equation (3.20), *quod erat demonstrandum*.

3.6   Living on a Wall

This section could have been entitled "The fate of two broken supercharges." As we already know, two out of four supercharges annihilate the wall – these supersymmetries are preserved in the given wall background. The two other super-charges are broken: being applied to the wall solution, they create two fermion zero modes. these zero modes correspond to a (2+1)-dimensional (massless) Majorana spinor field $\psi(t, x, y)$ localized on the wall.

To elucidate the above assertion it is convenient to turn first to the fate of another symmetry of the original theory, which is spontaneously broken for each given wall, namely, translational invariance in the z direction.

Indeed, each wall solution, e.g. (3.18), breaks this invariance. This means that in fact we must deal with a family of solutions: if $\phi(z)$ is a solution, so is $\phi(z - z_0)$. The parameter z_0 is a collective coordinate – the wall center. People also refer to it as a *modulus* (in plural, moduli). For the static wall, z_0 is a fixed constant.

Assume, however, that the wall is slightly bent. The bending should be negligible compared to the wall thickness (which is of the order of m^{-1}). The bending can be described as an adiabatically slow dependence of the wall center z_0 on t, x, and y. We will write this slightly bent wall field configuration as

$$\phi(t, x, y, z) = \phi_{\rm w}(z - \zeta(t, x, y)) . \tag{3.42}$$

Substituting this field in the original action, we arrive at the following effective (2+1)-dimensional action for the field $\zeta(t, x, y)$:

$$S^\zeta_{2+1} = \frac{T_{\rm w}}{2} \int d^3x \, (\partial^m \zeta)(\partial_m \zeta) , \qquad m = 0, 1, 2 . \tag{3.43}$$

It is clear that $\zeta(t, x, y)$ can be viewed as a massless scalar field (called the *translational modulus*) which lives on the wall. It is nothing but a Goldstone field corresponding to the spontaneous breaking of the translational invariance.

Returning to the two broken supercharges, they generate a Majorana (2+1)-dimensional Goldstino field $\psi_\alpha(t, x, y)$, $(\alpha = 1, 2)$ localized on the wall. The total (2+1)-dimensional effective action on the wall world volume takes the form

$$S_{2+1} = \frac{T_{\rm w}}{2} \int d^3x \left\{ (\partial^m \zeta)(\partial_m \zeta) + \bar{\psi} i \partial_m \gamma^m \psi \right\} \tag{3.44}$$

where γ^m are three-dimensional gamma matrices in the Majorana representation, e.g.

$$\gamma_0 = \sigma_2, \quad \gamma_1 = i\sigma_3, \quad \gamma_2 = i\sigma_1,$$

with the Pauli matrices $\sigma_{1,2,3}$.

The effective theory of the moduli fields on the wall world volume is supersymmetric, with two conserved supercharges. This is the minimal supersymmetry in 2+1 dimensions. It corresponds to the fact that two out of four supercharges are conserved.

4 Extended Supersymmetry in Two Dimensions: The Supersymmetric CP(1) Model

In this part I will return to kinks in two dimensions. The reason is three-fold. First, I will get you acquainted with a very interesting supersymmetric model which is routinely used in a large variety of applications and as a theoretical laboratory. It is called, rather awkwardly, O(3) sigma model. It also goes under the name of CP(1) sigma model. Initial data for this model, which will be useful in what follows, are collected in Appendix A. Second, supersymmetry of this model is extended (it is more than minimal). It has four conserved supercharges rather than two, as was the case in Sect. 2. Since the number of supercharges is twice as large as in the minimal case, people call it $\mathcal{N} = 2$ supersymmetry. So, we will get familiar with extended supersymmetries. Finally, solitons in the $\mathcal{N} = 2$ sigma model present a showcase for a variety of intriguing dynamical phenomena. One of them is charge "irrationalization:" in the presence of the θ term (topological term) the U(1) charge of the soliton acquires an extra $\theta/(2\pi)$. This phenomenon was first discovered by Witten [15] in the 't Hooft–Polyakov monopoles [16,17]. The kinks in the CP(1) sigma model are subject to charge irrationalization too. Since they are simpler than the 't Hooft–Polyakov monopoles, it makes sense to elucidate the rather unexpected addition of $\theta/(2\pi)$ in the CP(1) kink example.

The Lagrangian of the original CP(1) model is [18]

$$\mathcal{L}_{\mathrm{CP}(1)} = G \left\{ \partial_\mu \bar{\phi} \partial^\mu \phi + \frac{i}{2} \left(\bar{\Psi}_L \overset{\leftrightarrow}{\partial}_R \Psi_L + \bar{\Psi}_R \overset{\leftrightarrow}{\partial}_L \Psi_R \right) \right.$$

$$\left. - \frac{i}{\chi} \left[\bar{\Psi}_L \Psi_L \left(\bar{\phi} \overset{\leftrightarrow}{\partial}_R \phi \right) + \bar{\Psi}_R \Psi_R \left(\bar{\phi} \overset{\leftrightarrow}{\partial}_L \phi \right) \right] - \frac{2}{\chi^2} \bar{\Psi}_L \Psi_L \bar{\Psi}_R \Psi_R \right\}$$

$$+ \frac{i\theta}{2\pi} \frac{1}{\chi^2} \varepsilon^{\mu\nu} \partial_\mu \bar{\phi} \partial_\nu \phi \,, \tag{4.1}$$

where G is the metric on the target space,

$$G \equiv \frac{2}{g^2} \frac{1}{\left(1 + \phi\bar{\phi}\right)^2} \,, \tag{4.2}$$

and $\chi \equiv 1 + \phi\bar{\phi}$. (It is useful to note that $R = 2\chi^{-2}$ is the Ricci tensor.) The derivatives $\partial_{R,L}$ are defined as

$$\partial_R = \frac{\partial}{\partial t} - \frac{\partial}{\partial z} \,, \qquad \partial_L = \frac{\partial}{\partial t} + \frac{\partial}{\partial z} \,. \tag{4.3}$$

The target space in the case at hand is the two-dimensional sphere S_2 with radius
$$\mathcal{R}_{S_2} = g^{-1} \,.$$

As is well-known, one can introduce complex coordinates $\bar{\phi}, \phi$ on S_2. The choice of coordinates in (4.1) corresponds to the stereographic projection of the sphere. The term in the last line of (4.1) is the θ term. It can be represented as an integral over a total derivative. Moreover, the fermion field is a two-component Dirac spinor

$$\Psi = \begin{pmatrix} \Psi_R \\ \Psi_L \end{pmatrix} \,. \tag{4.4}$$

Bars over ϕ and $\Psi_{L,R}$ denote Hermitean conjugation.

This model has the extended $\mathcal{N} = 2$ supersymmetry since the Lagrangian (4.1) is invariant (up to total derivatives) under the following supertransformations (see e.g. the review paper [19])

$$\delta\phi = -i\bar{\varepsilon}_R\Psi_L + i\bar{\varepsilon}_L\Psi_R\,,$$

$$\delta\Psi_R = -i\left(\partial_R\phi\right)\varepsilon_L - 2i\frac{\bar{\phi}}{\chi}\left(\bar{\varepsilon}_R\Psi_L - \bar{\varepsilon}_L\Psi_R\right)\Psi_R\,,$$

$$\delta\Psi_L = i\left(\partial_L\phi\right)\varepsilon_R - 2i\frac{\bar{\phi}}{\chi}\left(\bar{\varepsilon}_R\Psi_L - \bar{\varepsilon}_L\Psi_R\right)\Psi_L\,, \qquad (4.5)$$

with complex parameters $\varepsilon_{R,L}$. The corresponding conserved supercurrent is

$$J^\mu = G\left(\partial_\lambda\bar{\varphi}\right)\gamma^\lambda\gamma^\mu\Psi\,. \qquad (4.6)$$

Since the fermion sector is most conveniently formulated in terms of the chiral components, it makes sense to rewrite the supercurrent (4.6) accordingly,

$$J_R^+ = G\left(\partial_R\bar{\phi}\right)\Psi_R\,, \quad J_R^- = 0\,;$$
$$J_L^- = G\left(\partial_L\bar{\phi}\right)\Psi_L\,, \quad J_L^+ = 0\,. \qquad (4.7)$$

where

$$J^\pm = \frac{1}{2}\left(J^0 \pm J^1\right)\,.$$

The current conservation law takes the form

$$\partial_L J^+ + \partial_R J^- = 0\,. \qquad (4.8)$$

The superalgebra induced by the four supercharges

$$Q = \int dz\, J^0(t, x) \qquad (4.9)$$

is as follows:

$$\{\bar{Q}_L, Q_L\} = (H + P)\,, \quad \{\bar{Q}_R, Q_R\} = (H - P)\,; \qquad (4.10)$$

$$\{Q_L, Q_R\} = 0\,, \quad \{Q_R, Q_L\} = 0\,; \qquad (4.11)$$

$$\{Q_R, Q_R\} = 0\,, \quad \{\bar{Q}_R, \bar{Q}_R\} = 0\,; \qquad (4.12)$$

$$\{Q_L, Q_L\} = 0\,, \quad \{\bar{Q}_L, \bar{Q}_L\} = 0\,; \qquad (4.13)$$

$$\{\bar{Q}_R, Q_L\} = \frac{i}{\pi}\int dz\, \partial_z\left(\chi^{-2}\,\bar{\Psi}_R\Psi_L\right)\,, \qquad (4.14)$$

$$\{\bar{Q}_L, Q_R\} = -\frac{i}{\pi}\int dz\, \partial_z\left(\chi^{-2}\,\bar{\Psi}_L\Psi_R\right)\,. \qquad (4.15)$$

where (H, P) is the energy-momentum operator,

$$(H, P) = \int dz \theta^{0i} , \qquad i = 0, 1 ,$$

and $\theta^{\mu\nu}$ is the energy-momentum tensor. Equations (4.14) and (4.15) present a quantum anomaly – these anticommutators vanish at the classical level. These anomalies will not be used in what follows. I quote them here only for the sake of completeness.

As is well-known, the model (4.1) is asymptotically free [20]. The coupling constant defined in (4.2) runs according to the law

$$\frac{1}{g^2(\mu)} = \frac{1}{g_0^2} - \frac{1}{4\pi} \ln \frac{M_{uv}^2}{\mu^2} , \tag{4.16}$$

where M_{uv} is the ultraviolet cut-off and g_0^2 is the coupling constant at this cut-off. At small momenta the theory becomes strongly coupled. The scale parameter of the model is

$$\Lambda^2 = M_{uv}^2 \exp\left(-\frac{4\pi}{g_0^2}\right) . \tag{4.17}$$

Our task is to study solitons in a pedagogical setting, which means, by default, that the theory must be weakly coupled. One can make the CP(1) model (4.1) weakly coupled, still preserving $\mathcal{N} = 2$ supersymmetry, by introducing the so-called *twisted mass* [21].

4.1   Twisted Mass

I will explain here neither genesis of twisted masses nor the origin of the name. Crucial is the fact that the target space of the CP(1) model has isometries. It was noted by Alvarez-Gaumé and Freedman that one can exploit these isometries to introduce supersymmetric mass terms, namely,

$$\Delta_m \mathcal{L}_{CP(1)} = G\left\{ -|m|^2 \phi\bar\phi - \frac{1 - \bar\phi\phi}{\chi}\left(m\bar\Psi_L\Psi_R + \bar m\bar\Psi_R\Psi_L\right)\right\} . \tag{4.18}$$

Here m is a complex parameter. Certainly, one can always eliminate the phase of m by a chiral rotation of the fermion fields. Due to the chiral anomaly, this will lead to a shift of the vacuum angle θ. In fact, it is the combination $\theta_{eff} = \theta + 2\arg m$ on which physics depends.

With the mass term included, the symmetry of the model is reduced to a global U(1) symmetry,

$$\phi \to e^{i\alpha}\phi, \qquad \bar\phi \to e^{-i\alpha}\bar\phi,$$

$$\Psi \to e^{i\alpha}\Psi, \qquad \bar\Psi \to e^{-i\alpha}\bar\Psi. \tag{4.19}$$

Needless to say that in order to get the conserved supercurrent, one must modify (4.7) appropriately,

$$J_R^+ = G\left(\partial_R\bar\phi\right)\Psi_R, \qquad J_R^- = -iG\,\bar m\bar\phi\Psi_L ;$$

$$J_L^- = G\left(\partial_L\bar\phi\right)\Psi_L, \qquad J_L^+ = iG\,m\phi\Psi_R. \tag{4.20}$$

The only change twisted mass terms introduce in the superalgebra is that (4.14) and (4.15) are to be replaced by

$$\{Q_L, \bar{Q}_R\} = m q_{U(1)} - im \int dz\, \partial_z\, h + \text{anom.}\,,$$

$$\{Q_R, \bar{Q}_L\} = \bar{m} q_{U(1)} + i\bar{m} \int dz\, \partial_z\, h + \text{anom.}\,, \tag{4.21}$$

where $q_{U(1)}$ is the conserved U(1) charge,

$$q_{U(1)} \equiv \int dz\, \mathcal{J}^0_{U(1)}\,,$$

$$\mathcal{J}^\mu_{U(1)} = G\left(\bar{\phi}\, i \overset{\leftrightarrow}{\partial}{}^\mu \phi + \bar{\Psi}\gamma^\mu\Psi - 2\,\frac{\phi\bar{\phi}}{\chi}\,\bar{\Psi}\gamma^\mu\Psi\right)\,, \tag{4.22}$$

and

$$h = -\frac{2}{g^2}\frac{1}{\chi}\,. \tag{4.23}$$

(Remember, χ is defined after (4.2).) As already mentioned, in what follows, the anomaly in (4.2) will be neglected. Equation (4.21) clearly demonstrates that the very possibility of introducing twisted mass terms is due to the U(1) symmetry.

Most important for our purposes is the fact that the model at hand is weakly coupled provided that $m \gg \Lambda$. Indeed, in this case the running of $g^2(\mu)$ is frozen at $\mu = m$. Consequently, the solitons emerging in this model can be treated quasiclassically.

4.2   BPS Solitons at the Classical Level

As already mentioned, the target space of the CP(1) model is S_2. The U(1) invariant scalar potential term

$$V = |m|^2\, G\,\bar{\phi}\phi \tag{4.24}$$

lifts the vacuum degeneracy leaving us with two discrete vacua: at the south and north poles of the sphere (Fig. 2) i.e. $\phi = 0$ and $\phi = \infty$.

The kink solutions interpolate between these two vacua. Let us focus, for definiteness, on the kink with the boundary conditions

$$\phi \to 0 \quad \text{at} \quad z \to -\infty\,, \qquad \phi \to \infty \quad \text{at} \quad z \to \infty\,. \tag{4.25}$$

Consider the following linear combinations of supercharges

$$q = Q_R - i\,e^{-i\beta}Q_L\,, \qquad \bar{q} = \bar{Q}_R + i\,e^{i\beta}\bar{Q}_L\,, \tag{4.26}$$

where β is the argument of the mass parameter,

$$m = |m|\,e^{i\beta}\,. \tag{4.27}$$

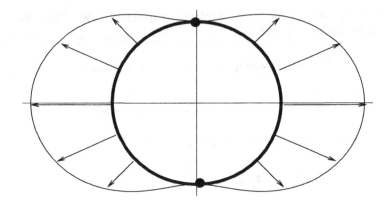

Fig. 2. Meridian slice of the target space sphere (thick solid line). The arrows present the scalar potential (4.24), their length being the strength of the potential. The two vacua of the model are denoted by the closed circles at the north and south pole.

Then

$$\{q, \bar{q}\} = 2H - 2|m| \int dz\, \partial_z h\,, \qquad \{q, q\} = \{\bar{q}, \bar{q}\} = 0\,. \tag{4.28}$$

Now, let us require q and $\bar{q}$ to vanish on the classical solution. Since for static field configurations

$$q = -\left(\partial_z \bar{\phi} - |m|\bar{\phi}\right) \left(\Psi_R + ie^{-i\beta}\Psi_L\right),$$

the vanishing of these two supercharges implies

$$\partial_z \bar{\phi} = |m|\bar{\phi} \quad \text{or} \quad \partial_z \phi = |m|\phi\,. \tag{4.29}$$

This is the BPS equation in the sigma model with twisted mass.

The BPS equation (4.29) has a number of peculiarities compared to those in more familiar Landau–Ginzburg $\mathcal{N} = 2$ models. The most important feature is its complexification, i.e. the fact that (4.29) is holomorphic in ϕ. The solution of this equation is, of course, trivial and can be written as

$$\phi(z) = e^{|m|(z-z_0)-i\alpha}\,. \tag{4.30}$$

Here z_0 is the kink center while α is an arbitrary phase. In fact, these two parameters enter only in the combination $|m|z_0 + i\alpha$. We see that the notion of the kink center also gets complexified.

The physical meaning of the modulus α is obvious: there is a continuous family of solitons interpolating between the north and south poles of the target space sphere. This is due to the U(1) symmetry. The soliton trajectory can follow

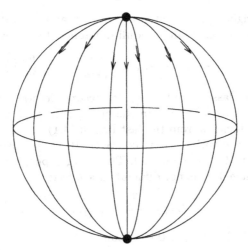

Fig. 3. The soliton solution family. The collective coordinate α in (4.30) spans the interval $0 \le \alpha \le 2\pi$. For given α the soliton trajectory on the target space sphere follows a meridian, so that when α varies from 0 to 2π all meridians are covered.

any meridian (Fig. 3). It is instructive to derive the BPS equation directly from the (bosonic part of the) Lagrangian, performing the Bogomol'nyi completion,

$$\int d^2x\, \mathcal{L} = \int d^2x\, G\, \{\partial_\mu \bar\phi \partial^\mu \phi - |m|^2 \bar\phi \phi\}$$

$$\rightarrow - \Big\{ \int dz\, G\, (\partial_z \bar\phi - |m|\bar\phi)\, (\partial_z \phi - |m|\phi)$$

$$+ \, |m| \int dz\, \partial_z h \Big\}, \qquad (4.31)$$

where I assumed ϕ to be time-independent and the following identity has been used

$$\partial_z h \equiv G(\phi \partial_z \bar\phi + \bar\phi \partial_z \phi)\,.$$

Equation (4.29) ensues immediately. In addition, (4.31) implies that (classically) the kink mass is

$$M_0 = |m|\, (h(\infty) - h(0)) = \frac{2|m|}{g^2}\,. \qquad (4.32)$$

The subscript 0 emphasizes that this result is obtained at the classical level. Quantum corrections will be considered below.

4.3   Quantization of the Bosonic Moduli

To carry out conventional quasiclassical quantization we, as usual, assume the moduli z_0 and α in (4.30) to be (weakly) time-dependent, substitute (4.30) in the bosonic Lagrangian (4.31), integrate over z and thus derive a quantum-mechanical Lagrangian describing moduli dynamics. In this way we obtain

$$\mathcal{L}_{\mathrm{QM}} = -M_0 + \frac{M_0}{2}\dot{z}_0^2 + \left\{\frac{1}{g^2|m|}\dot{\alpha}^2 - \frac{\theta}{2\pi}\dot{\alpha}\right\}. \tag{4.33}$$

The first term is the classical kink mass, the second describes the free motion of the kink along the z axis. The term in the braces is most interesting (I included the θ term which originates from the last line in (4.1)).

Remember that the variable α is compact. Its very existence is related to the exact $U(1)$ symmetry of the model. The energy spectrum corresponding to α dynamics is quantized. It is not difficult to see that

$$E_{[\alpha]} = \frac{g^2|m|}{4}\, q_{U(1)}^2, \tag{4.34}$$

where $q_{U(1)}$ is the $U(1)$ charge of the soliton,

$$q_{U(1)} = k + \frac{\theta}{2\pi}, \qquad k = \text{ an integer}. \tag{4.35}$$

This is the same effect as the occurrence of an irrational electric charge $\theta/(2\pi)$ on the magnetic monopole, a phenomenon first noted by Witten [15]. Objects which carry both magnetic and electric charges are called dyons. The standard four-dimensional magnetic monopole becomes a dyon in the presence of the θ term if $\theta \neq 0$. The $q_{U(1)} \neq 0$ kinks in the CP(1) model are sometimes referred to as Q-kinks.

A brief comment regarding (4.34) and (4.35) is in order here. The dynamics of the compact modulus α is described by the Hamiltonian

$$H_{\mathrm{QM}} = \frac{1}{g^2|m|}\dot{\alpha}^2 \tag{4.36}$$

while the canonic momentum conjugated to α is

$$p_{[\alpha]} = \frac{\delta \mathcal{L}_{\mathrm{QM}}}{\delta \dot{\alpha}} = \frac{2}{g^2|m|}\dot{\alpha} - \frac{\theta}{2\pi}. \tag{4.37}$$

In terms of the canonic momentum the Hamiltonian takes the form

$$H_{\mathrm{QM}} = \frac{g^2|m|}{4}\left(p_{[\alpha]} + \frac{\theta}{2\pi}\right)^2 \tag{4.38}$$

The eigenfunctions obviously are

$$\Psi_k(\alpha) = e^{ik\alpha}, \qquad k = \text{ an integer}, \tag{4.39}$$

which immediately leads to $E_{[\alpha]} = (g^2|m|/4)(k + \theta(2\pi)^{-1})^2$.

Let us now calculate the $U(1)$ charge of the k-th state. Starting from (4.22) we arrive at

$$q_{U(1)} = \frac{2}{g^2|m|}\dot{\alpha} = p_{[\alpha]} + \frac{\theta}{2\pi} \rightarrow k + \frac{\theta}{2\pi}, \tag{4.40}$$

quod erat demonstrandum, cf. (4.35).

4.4   The Soliton Mass and Holomorphy

Taking account of $E_{[\alpha]}$ – the energy of an "internal motion" – the kink mass can be written as

$$M = \frac{2|m|}{g^2} + \frac{g^2|m|}{4}\left(k + \frac{\theta}{2\pi}\right)^2$$

$$= \frac{2|m|}{g^2}\left\{1 + \frac{g^4}{4}\left(k + \frac{\theta}{2\pi}\right)^2\right\}^{1/2}$$

$$= 2|m|\left|\frac{1}{g^2} + i\,\frac{\theta + 2\pi k}{4\pi}\right|. \qquad (4.41)$$

The transition from the first to the second line is approximate, valid to the leading order in the coupling constant. The quantization procedure and derivation of (4.34) presented in Sect. 4.3 are also valid to the leading order in the coupling constant. At the same time, the expressions in the second and last lines in (4.41) are valid to all orders and, in this sense, are more general. They will be derived below from the consideration of the relevant central charge.

The important circumstance to be stressed is that the kink mass depends on a special combination of the coupling constant and θ, namely,

$$\tau = \frac{1}{g^2} + i\,\frac{\theta}{4\pi} \qquad (4.42)$$

In other words, it is the complexified coupling constant that enters.

It is instructive to make a pause here to examine the issue of the kink mass from a slightly different angle. Equation (4.21) tells us that there is a central charge $Z_{L\bar{R}}$ in the anticommutator $\{Q_{Lc}\bar{Q}_R\}$,

$$Z_{L\bar{R}} = -i\,m\left\{\int dz\,\partial_z\,h + i\,q_{U(1)}\right\}, \qquad (4.43)$$

where the anomalous term is omitted, as previously, which is fully justified at weak coupling. If the soliton under consideration is critical – and it is – its mass *must* be equal to the absolute value $|Z_{L\bar{R}}|$. This leads us directly to (4.41). However, one can say more.

Indeed, g^2 in (4.41) is the bare coupling constant. It is quite clear that the kink mass, being a physical parameter, should contain the renormalized constant $g^2(m)$, after taking account of radiative corrections. In other words, switching on radiative corrections in $Z_{L\bar{R}}$ one *must* replace the bare $1/g^2$ by the renormalized $1/g^2(m)$. We will see now how it comes out, verifying *en route* a very important assertion – the dependence of $Z_{L\bar{R}}$ on all relevant parameters, τ and m, being holomorphic.

I will perform the one-loop calculation in two steps. First, I will rotate the mass parameter m in such a way as to make it real, $m \to |m|$. Simultaneously,

$$\langle \delta\bar{\phi}\, \delta\phi \rangle$$

Fig. 4. h renormalization.

the θ angle will be replaced by an effective θ,

$$\theta \to \theta_{\text{eff}} = \theta + 2\beta, \qquad (4.44)$$

where the phase β is defined in (4.26). Next, I decompose the field ϕ into a classical and a quantum part,

$$\phi \to \phi + \delta\phi.$$

Then the h part of the central charge $Z_{L\bar{R}}$ becomes

$$h \to h + \frac{2}{g^2} \frac{1 - \bar{\phi}\phi}{\left(1 - \bar{\phi}\phi\right)^3} \delta\bar{\phi}\,\delta\phi. \qquad (4.45)$$

Contracting $\delta\bar{\phi}\,\delta\phi$ into a loop (Fig. 4) and calculating this loop – quite a trivial exercise – we find with ease that

$$h \to h + -\frac{2}{g^2} \frac{1}{\chi} + \frac{1}{\chi} \frac{2}{4\pi} \ln \frac{M_{\text{uv}}^2}{|m|^2}. \qquad (4.46)$$

Combining this result with (4.40) and (4.42), we arrive at

$$Z_{L\bar{R}} = 2im \left\{ \tau - \frac{1}{4\pi} \ln \frac{M_{\text{uv}}^2}{m^2} - i\frac{k}{2} \right\} \qquad (4.47)$$

(remember, the kink mass $M = |Z_{L\bar{R}}|$). A salient feature of this formula, to be noted, is the holomorphic dependence of $Z_{L\bar{R}}$ on m and τ. Such a holomorphic dependence would be impossible if two and more loops contributed to h renormalization. Thus, h renormalization beyond one loop must cancel, and it does.[8] Note also that the bare coupling in (4.47) conspires with the logarithm in such a way as to replace the bare coupling by that renormalized at $|m|$, as was expected.

---

[8] Fermions are important for this cancelation.

The analysis carried out above is quasiclassical. It tells us nothing about the possible occurrence of non-perturbative terms in $Z_{L\bar{R}}$. In fact, all terms of the type

$$\left\{ \frac{M_{uv}^2}{m^2} \exp\left(-4\pi\tau\right) \right\}^{\ell}, \qquad \ell = \text{integer}$$

are fully compatible with holomorphy; they can and do emerge from instantons. An indirect calculation of non-perturbative terms was performed in [22]. I will skip it altogether referring the interested reader to the above publication.

4.5   Switching On Fermions

Fermion non-zero modes are irrelevant for our consideration since, being combined with the boson non-zero modes, they cancel for critical solitons, a usual story. Thus, for our purposes it is sufficient to focus on the (static) zero modes in the kink background (4.30). The coefficients in front of the fermion zero modes will become (time-dependent) fermion moduli, for which we are going to build the corresponding quantum mechanics. There are two such moduli, $\bar{\eta}$ and η.

The equations for the fermion zero modes are

$$\partial_z \Psi_L - \frac{2}{\chi}\left(\bar{\phi}\partial_z\phi\right)\Psi_L - i\frac{1-\bar{\phi}\phi}{\chi}|m|e^{i\beta}\Psi_R = 0\,,$$

$$\partial_z \Psi_R - \frac{2}{\chi}\left(\bar{\phi}\partial_z\phi\right)\Psi_R + i\frac{1-\bar{\phi}\phi}{\chi}|m|e^{-i\beta}\Psi_L = 0 \qquad (4.48)$$

(plus similar equations for $\bar{\Psi}$; since our operator is Hermitean we do not need to consider them separately.)

It is not difficult to find solutions to these equations, either directly or by using supersymmetry. Indeed, if we know the bosonic solution (4.30), its fermionic superpartner – and the fermion zero modes are such superpartners – is obtained from the bosonic one by those two supertransformations which act on $\bar{\phi}$, ϕ nontrivially. In this way we conclude that the functional form of the fermion zero mode must coincide with the functional form of the boson solution (4.30). Concretely,

$$\begin{pmatrix} \Psi_R \\ \Psi_L \end{pmatrix} = \eta \left(\frac{g^2|m|}{2}\right)^{1/2} \begin{pmatrix} -ie^{-i\beta} \\ 1 \end{pmatrix} e^{|m|(z-z_0)} \qquad (4.49)$$

and

$$\begin{pmatrix} \bar{\Psi}_R \\ \bar{\Psi}_L \end{pmatrix} = \bar{\eta} \left(\frac{g^2|m|}{2}\right)^{1/2} \begin{pmatrix} ie^{i\beta} \\ 1 \end{pmatrix} e^{|m|(z-z_0)}\,, \qquad (4.50)$$

where the numerical factor is introduced to ensure the proper normalization of the quantum-mechanical Lagrangian. Another solution which asymptotically, at large z, behaves as $e^{3|m|(z-z_0)}$ must be discarded as non-normalizable.

Now, to perform the quasiclassical quantization we follow the standard route: the moduli are assumed to be time-dependent, and we derive the quantum mechanics of the moduli starting from the original Lagrangian (4.1) with the twisted mass terms (4.18). Substituting the kink solution and the fermion zero modes for Ψ, one gets

$$\mathcal{L}'_{\text{QM}} = i\,\bar{\eta}\dot{\eta}. \tag{4.51}$$

In the Hamiltonian approach the only remnants of the fermion moduli are the anticommutation relations

$$\{\bar{\eta}, \eta\} = 1, \quad \{\bar{\eta}, \bar{\eta}\} = 0, \quad \{\eta, \eta\} = 0, \tag{4.52}$$

which tell us that the wave function is *two-component* (i.e. the kink supermultiplet is two-dimensional). One can implement (4.52) by choosing, e.g., $\bar{\eta} = \sigma^+$, $\eta = \sigma^-$, where $\sigma^P m = (\sigma_1 \pm \sigma_2)/2$.

The fact that there are two critical kink states in the supermultiplet is consistent with the multiplet shortening in $\mathcal{N} = 2$. Indeed, in two dimensions the full $\mathcal{N} = 2$ supermultiplet must consist of four states: two bosonic and two fermionic. 1/2 BPS multiplets are shortened – they contain twice less states than the full supermultiplets, one bosonic and one fermionic. This is to be contrasted with the single-state kink supermultiplet in the minimal supersymmetric model of Sect. 2. The notion of the fermion parity remains well-defined in the kink sector of the CP(1) model.

4.6   Combining Bosonic and Fermionic Moduli

Quantum dynamics of the kink at hand is summarized by the Hamiltonian

$$H_{\text{QM}} = \frac{M_0}{2}\,\dot{\bar{\zeta}}\dot{\zeta} \tag{4.53}$$

acting in the space of *two-component wave functions*. The variable ζ here is a complexified kink center,

$$\zeta = z_0 + \frac{i}{|m|}\,\alpha. \tag{4.54}$$

For simplicity, I set the vacuum angle $\theta = 0$ for the time being (it will be reinstated later).

The original field theory we deal with has four conserved supercharges. Two of them, q and $\bar{q}$, see (4.26), act trivially in the critical kink sector. In moduli quantum mechanics they take the form

$$q = \sqrt{M_0}\,\dot{\zeta}\eta, \qquad \bar{q} = \sqrt{M_0}\,\dot{\bar{\zeta}}\bar{\eta}; \tag{4.55}$$

they do indeed vanish provided that the kink is at rest. The superalgebra describing kink quantum mechanics is $\{\bar{q}, q\} = 2H_{\text{QM}}$. This is nothing but Witten's $\mathcal{N} = 1$ supersymmetric quantum mechanics [23] (two supercharges). The realization we deal with is peculiar and distinct from that of Witten. Indeed, the standard supersymmetric quantum mechanics of Witten includes one (real) bosonic degree of freedom and two fermionic ones, while we have two bosonic degrees of freedom, x_0 and α. Nevertheless, the superalgebra remains the same due to the fact that the bosonic coordinate is complexified.

Finally, to conclude this section, let us calculate the U(1) charge of the kink states. We start from (4.22), substitute the fermion zero modes and get [9]

$$\Delta q_{U(1)} = \frac{1}{2}[\bar{\eta}\eta] \qquad (4.56)$$

(this is to be added to the bosonic part given in (4.40)). Given that $\bar{\eta} = \sigma^+$ and $\eta = \sigma^-$ we arrive at $\Delta q_{U(1)} = \frac{1}{2}\sigma_3$. This means that the U(1) charges of two kink states in the supermultiplet split from the value given in (4.40): one has the U(1) charges

$$k + \frac{1}{2} + \frac{\theta}{2\pi},$$

and

$$k - \frac{1}{2} + \frac{\theta}{2\pi}.$$

5 Conclusions

Supersymmetric solitons is a vast topic, with a wide range of applications in field and string theories. In spite of almost thirty years of development, the review literature on this subject is scarce. Needless to say, I was unable to cover this topic in an exhaustive manner. No attempt at such coverage was made. Instead, I focused on basic notions and on pedagogical aspects in the hope of providing a solid introduction, allowing the interested reader to navigate themselves in the ocean of the original literature.

Appendix A.
CP(1) Model = O(3) Model ($\mathcal{N} = 1$ Superfields N)

In this Appendix we follow the review paper [24]. One introduces a (real) superfield

$$N^a(x, \theta) = \sigma^a(x) + \bar{\theta}\psi^a(x) + \frac{1}{2}\bar{\theta}\theta F^a, \qquad a = 1, 2, 3, \qquad (A.1)$$

where σ is a scalar field, ψ is a Majorana two-component spinor,

$$\bar{\psi} \equiv \psi\gamma^0, \qquad \bar{\theta} \equiv \theta\gamma^0,$$

and F is the auxiliary component (without kinetic term in the action). A convenient choice of gamma matrices is the following:

$$\gamma^0 = \sigma_2, \quad \gamma^1 = i\sigma_3, \quad \gamma^5 = \gamma^0\gamma^1 = -\sigma_1, \qquad \sigma_i \text{ are Pauli matrices.} \quad (A.2)$$

---

[9] To set the scale properly, so that the U(1) charge of the vacuum state vanishes, one must antisymmetrize the fermion current, $\bar{\Psi}\gamma^\mu\Psi \to (1/2)\left(\bar{\Psi}\gamma^\mu\Psi - \bar{\Psi}^c\gamma^\mu\Psi^c\right)$ where the superscript c denotes C conjugation.

In terms of the superfield N^a the action of the original O(3) sigma model can be written as follows:

$$S = \frac{1}{2g^2} \int d^2x\, d^2\theta\, (\bar{D}_\alpha N^a)(D_\alpha N^a) \tag{A.3}$$

with the constraint

$$N^a(x,\theta)N^a(x,\theta) = 1. \tag{A.4}$$

Here g^2 is the coupling constant, integration over the Grassmann parameters is normalized as

$$\int d^2\theta\, \frac{i}{2}\bar{\theta}\theta = 1,$$

while the spinorial derivatives are

$$D_\alpha = \frac{\partial}{\partial\bar{\theta}_\alpha} - i(\gamma^\mu\theta)_\alpha\partial_\mu, \qquad \bar{D}_\alpha = -\frac{\partial}{\partial\theta_\alpha} + i(\bar{\theta}\gamma^\mu)_\alpha\partial_\mu. \tag{A.5}$$

The mass deformation of (A.3) that preserves $\mathcal{N} = 2$ but breaks O(3) down to U(1) is

$$S = \frac{1}{2g^2} \int d^2x\, d^2\theta\, \left\{(\bar{D}_\alpha N^a)(D_\alpha N^a) + 4imN^3\right\} \tag{A.6}$$

where m is a mass parameter. Note that $\mathcal{N} = 2$ is preserved only because the added term is very special – linear in the third ($a = 3$) component of the superfield N.

In components the Lagrangian in (A.6) has the form

$$\begin{aligned}
L &= \frac{1}{2g^2}\left\{(\partial_\mu\sigma^a)^2 + \bar{\psi}^a i\,\slashed{\partial}\psi^a + F^2 + 2m\,F^3\right\} \\
&= \frac{1}{2g^2}\left\{(\partial_\mu\sigma^a)^2 + \bar{\psi}^a i\,\slashed{\partial}\psi^a + \frac{1}{4}\left(\bar{\psi}\psi\right)^2\right. \\
&\quad \left. + m\sigma^3\,\bar{\psi}\psi - m^2\left[(\sigma^1)^2 + (\sigma^2)^2\right]\right\} \\
&\quad + \frac{i\theta}{8\pi}\,\varepsilon^{\mu\nu}\,\varepsilon^{abc}\,\sigma^a\left(\partial_\mu\sigma^b\right)\left(\partial_\nu\sigma^c\right).
\end{aligned} \tag{A.7}$$

I added the θ term in the last line. The constraint (A.4) is equivalent to

$$\sigma^2 = 1, \qquad \sigma\psi = 0, \qquad \sigma F = \frac{1}{2}(\bar{\psi}\psi) \tag{A.8}$$

while the auxiliary F term was eliminated through the equation of motion

$$F^a = \frac{1}{2}(\bar{\psi}\psi + 2m\sigma^3)\sigma^a - m\delta^{3a}. \tag{A.9}$$

The equations of motion for σ and ψ have the form

$$\left(-\delta^{ab} + \sigma^a \sigma^b\right) \partial^2 \sigma^b - \sigma^b \bar{\psi}^a i \not{\partial} \psi^b$$

$$-\left\{\frac{1}{2}m\sigma^3 \left(\bar{\psi}\psi\right) + m^2 \left(\sigma^3\right)^2\right\} \sigma^a + \left\{\frac{1}{2}m\delta^{3a} \left(\bar{\psi}\psi\right) + m^2 \sigma^3 \delta^{3a}\right\} = 0,$$

$$\left(\delta^{ab} - \sigma^a \sigma^b\right) i \not{\partial} \psi^b + \frac{1}{2} \left(\bar{\psi}\psi\right) \psi^a + m\sigma^3 \psi^a = 0. \tag{A.10}$$

The first conserved supercurrent is

$$S_{(1)}^{\mu} = \frac{1}{g^2} \left\{ \left(\partial_\lambda \sigma^a\right) \gamma^\lambda \gamma^\mu \psi^a + im\, \gamma^\mu \psi^3 \right\}. \tag{A.11}$$

The second conserved supercurrent (remember that we deal with $\mathcal{N} = 2$) is

$$S_{(2)}^{\mu} = \frac{1}{g^2} \left\{ \varepsilon^{abc} \sigma^a \left(\partial_\lambda \sigma^b\right) \gamma^\lambda \gamma^\mu \psi^c - im\, \varepsilon^{3ab} \sigma^a \gamma^\mu \psi^b \right\}. \tag{A.12}$$

In this form the model is usually called O(3) sigma model. The conversion to the complex representation used in Sect. 4, in which form the model is usually referred to as CP(1) sigma model, can be carried out by virtue of the well-known formulae given, for example, in (67) and (69) of [24].

Appendix B.
Getting Started (Supersymmetry for Beginners)

To visualize conventional (non-supersymmetric) field theory one usually thinks of a space filled with a large number of coupled anharmonic oscillators. For instance, in the case of 1+1 dimensional field theory, with a single spatial dimension, one can imagine an infinite chain of penduli connected by springs (Fig. 5). Each pendulum represents an anharmonic oscillator. One can think of it as of a massive ball in a gravitational field. Each spring works in the harmonic regime, i.e. the corresponding force grows linearly with the displacement between the penduli. Letting the density of penduli per unit length tend to infinity, we return to field theory.

If a pendulum is pushed aside, it starts oscillating and initiates a wave which propagates along the chain. After quantization one interprets this wave as a scalar particle.

Can one present a fermion in this picture? The answer is yes. Imagine that each pendulum acquires a spin degree of freedom (i.e. each ball can rotate, see Fig. 6). Spins are coupled to their neighbors. Now, in addition to the wave that propagates in Fig. 5, one can imagine a spin wave propagating in Fig. 6. If one perturbs a single spin, this perturbation will propagate along the chain.

Our world is 1+3 dimensional, one time and three space coordinates. In this world bosons manifest themselves as particles with integer spins. For instance,

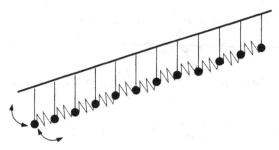

Fig. 5. A mechanical analogy for the scalar field theory.

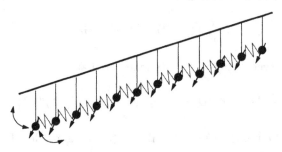

Fig. 6. A mechanical analogy for the spinor field theory.

the scalar (spin-0) particle from which we started is a boson. The photon (spin-1) particle) is a boson too. On the other hand, particles with semi-integer spins – electrons, protons, etc. – are fermions.

Conventional symmetries, such as isotopic invariance, do not mix bosons with fermions. Isosymmetry tells us that the proton and neutron masses are the same. It also tells us that the masses of π^0 and π^+ are the same. However, no prediction for the ratio of the pion to proton masses emerges.

Supersymmetry is a very unusual symmetry. It connects masses and other properties of bosons with those of fermions. Thus, each known particle acquires a superpartner: the superpartner of the photon (spin 1) is the *photino* (spin 1/2), the superpartner of the electron (spin 1/2) is the *selectron* (spin 0). Since spin is involved, which is related to geometry of space-time, it is clear that supersymmetry has a deep geometric nature. Unfortunately, I have no time to dwell on further explanations. Instead, I would like to present here a quotation from Witten which nicely summarizes the importance of this concept for modern physics. Witten writes [25]:

"... One of the biggest adventures of all is the search for supersymmetry. Supersymmetry is the framework in which theoretical physicists have sought to answer some of the questions left open by the Standard Model of particle physics.

Supersymmetry, if it holds in nature, is part of the quantum structure of space and time. In everyday life, we measure space and time by numbers, "It is now three o'clock, the elevation is two hundred meters above sea

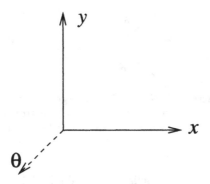

Fig. 7. Superspace.

level," and so on. Numbers are classical concepts, known to humans since long before Quantum Mechanics was developed in the early twentieth century. The discovery of Quantum Mechanics changed our understanding of almost everything in physics, but our basic way of thinking about space and time has not yet been affected.

Showing that nature is supersymmetric would change that, by revealing a quantum dimension of space and time, not measurable by ordinary numbers. .... Discovery of supersymmetry would be one of the real milestones in physics."

I have tried to depict "a quantum dimension of space and time" in Fig. 7. Two coordinates, x and y represent the conventional space-time. I should have drawn four coordinates, x, y, z and t, but this is impossible – we should try to imagine them.

The axis depicted by a dashed line (going in the perpendicular direction) is labeled by θ (again, one should try to imagine four distinct θ's rather than one). The dimensions along these directions cannot be measured in meters, the coordinates along these directions are very unusual, they anticommute,

$$\theta_1\theta_2 = -\theta_2\theta_1 \,, \tag{B.1}$$

and, as a result, $\theta^2 = 0$. This is in sharp contrast with ordinary coordinates for which 5 meters × 3 meters is, certainly, the same as 3 meters × 5 meters. In mathematics the θ's are known as Grassmann numbers, the square of every given Grassmann number vanishes. These extra θ directions are pure quantum structures. In our world they would manifest themselves through the fact that every integer spin particle has a half-integer spin superpartner.

A necessary condition for any theory to be supersymmetric is the balance between the number of the bosonic and fermionic degrees of freedom, having the same mass and the same "external" quantum numbers, e.g. electric charge. To give you an idea of supersymmetric field theories, let us turn to the most familiar and simplest gauge theory, quantum electrodynamics (QED). This theory describes electrons and positrons (one Dirac spinor with four degrees of freedom)

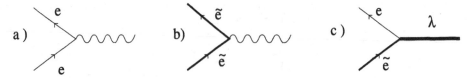

Fig. 8. Interaction vertices in QED and its supergeneralization, SQED. (a) $\bar{e}e\gamma$ vertex; (b) selectron coupling to photon; (c) electron–selectron–photino vertex. All vertices have the same coupling constant. The quartic self-interaction of selectrons is also present but not shown.

interacting with photons (an Abelian gauge field with two physical degrees of freedom). Correspondingly, in its supersymmetric version, SQED, one has to add one massless Majorana spinor, the photino (two degrees of freedom), and two complex scalar fields, the selectrons (four degrees of freedom).

Balancing the number of degrees of freedom is a necessary but not sufficient condition for supersymmetry in dynamically nontrivial theories, of course. All interaction vertices must be supersymmetric too. This means that each line in every vertex can be replaced by that of a superpartner. Say, we start from the electron–electron–photon coupling (Fig. 8a). Now, as we already know, in SQED the electron is accompanied by two selectrons. Thus, supersymmetry requires the selectron–selectron–photon vertices (Fig. 8b) with the same coupling constant. Moreover, the photon can be replaced by its superpartner, photino, which generates the electron–selectron–photino vertex (Fig. 8c) with the same coupling.

With the above set of vertices one can show that the theory is supersymmetric at the level of trilinear interactions, provided that the electrons and selectrons are degenerate in mass, while the photon and photino fields are both massless. To make it fully supersymmetric, one should also add some quartic terms, which describe the self-interactions of the selectron fields. Historically, SQED was the first supersymmetric theory discovered in four dimensions [1].

B.1   Promises of Supersymmetry

Supersymmetry has yet to be discovered experimentally. In spite of the absence of direct experimental evidence, immense theoretical effort was invested in this subject in the last thirty years; over 30,000 papers are published. The so-called Minimal Supersymmetric Standard Model (MSSM) became a generally accepted paradigm in high-energy physics. In this respect the phenomenon is rather unprecedented in the history of physics. Einstein's general relativity, the closest possible analogy one can give, was experimentally confirmed within several years after its creation. Only in one or two occasions, theoretical predictions of comparable magnitude had to wait for experimental confirmation that long. For example, the neutrino had a time lag of 27 years. A natural question arises: why do we believe that this concept is so fundamental?

Supersymmetry may help us to solve two of the the deepest mysteries of nature – the cosmological term problem and the hierarchy problem.

B.2   Cosmological Term

An additional term in the Einstein action of the form

$$\Delta S = \int d^4 x \sqrt{g}\, \Lambda \tag{B.2}$$

goes under the name of the cosmological term. It is compatible with general covariance and, therefore, can be added freely; this fact was known to Einstein. Empirically Λ is very small, see below. In classical theory there is no problem with fine-tuning Λ to any value.

The problem arises at the quantum level. In conventional (non-supersymmetric) quantum field theory it is practically inevitable that

$$\Lambda \sim M_{\rm Pl}^4, \tag{B.3}$$

where $M_{\rm Pl}$ is the Planck scale, $M_{\rm Pl} \sim 10^{19}$ GeV. This is to be confronted with the experimental value of the cosmological term,

$$\Lambda_{\rm exp} \sim (10^{-12}\,{\rm GeV})^4 . \tag{B.4}$$

The divergence between theoretical expectations and experiment is 124 orders of magnitude! This is probably the largest discrepancy in the history of physics.

Why may supersymmetry help? In supersymmetric theories Λ is strictly forbidden by supersymmetry, $\Lambda \equiv 0$. Of course, supersymmetry, even if it is there, must be broken in nature. People hope that the breaking occurs in a way ensuring splittings between the superpartners' masses in the ball-park of 100 GeV, with the cosmological term in the ball-park of the experimental value (B.4).

B.3   Hierarchy Problem

The masses of the spinor particles (electrons, quarks) are protected against large quantum corrections by chirality ("handedness"). For scalar particles the only natural mass scale is $M_{\rm Pl}$. Even if originally you choose this mass in the "human" range of, say, 100 GeV, quantum loops will inevitably drag it to $M_{\rm Pl}$. A crucial element of the Standard Model of electroweak interactions is the Higgs boson (not yet discovered). Its mass has to be in the ball-park of 100 GeV. If you let its mass to be $\sim M_{\rm Pl}$, this will drag, in turn, the masses of the W bosons. Thus, you would expect $(M_W)_{\rm theor} \sim 10^{19}$ GeV while $(M_W)_{\rm exp} \sim 10^2$ GeV. The discrepancy is 17 orders of magnitude.

Again, supersymmetry comes to rescue. In supersymmetry the notion of chirality extends to bosons, through their fermion superpartners. There are no quadratic divergences in the boson masses, at most they are logarithmic, just like in the fermion case. Thus, the Higgs boson mass gets protected against large quantum corrections.

Having explained that supersymmetry may help to solve two of the most challenging problems in high-energy physics, I hasten to add that it does a lot of other good things already right now. It proved to be a remarkable tool in

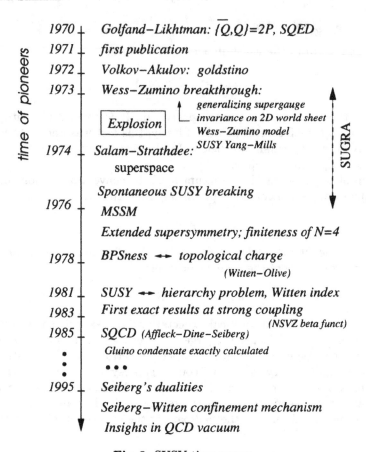

Fig. 9. SUSY time arrow.

dealing with previously "uncrackable" issues in gauge theories at strong coupling. Let me give a brief list of achievements: (i) first finite four-dimensional field theories; (ii) first exact results in four-dimensional gauge theories [26]; (iii) first fully dynamical (albeit toy) theory of confinement [27]; (iv) dualities in gauge theories [28]. The latter finding was almost immediately generalized to strings which gave rise to the breakthrough discovery of *string dualities*.

To conclude my mini-introduction, I present an arrow of time in supersymmetry (Fig. 9).

References

1. Yu. A. Golfand and E. P. Likhtman, *JETP Lett.* **13**, 323 (1971) [Reprinted in *Supersymmetry*, Ed. S. Ferrara, (North-Holland/World Scientific, Amsterdam – Singapore, 1987), Vol. 1, page 7].
2. E. Witten and D. I. Olive, Phys. Lett. B **78**, 97 (1978).
3. R. Haag, J. T. Lopuszanski, and M. Sohnius, Nucl. Phys. B **88**, 257 (1975).

4. E. B. Bogomol'nyi, Sov. J. Nucl. Phys. **24**, 449 (1976) [Reprinted in *Solitons and Particles*, Eds. C. Rebbi and G. Soliani (World Scientific, Singapore, 1984) p. 389]; M. K. Prasad and C. M. Sommerfield, Phys. Rev. Lett. **35**, 760 (1975) [Reprinted in *Solitons and Particles*, Eds. C. Rebbi and G. Soliani (World Scientific, Singapore, 1984) p. 530].

5. J. Milnor, *Morse theory* (Princeton University Press, 1973).

6. M. Shifman, A. Vainshtein, and M. Voloshin, Phys. Rev. D **59**, 045016 (1999) [hep-th/9810068].

7. A. Losev, M. A. Shifman, and A. I. Vainshtein, Phys. Lett. B **522**, 327 (2001) [hep-th/0108153]; New J. Phys. **4**, 21 (2002) [hep-th/0011027], reprinted in *Multiple Facets of Quantization and Supersymmetry*, the Michael Marinov Memorial Volume, Eds. M. Olshanetsky and A. Vainshtein (World Scientific, Singapore, 2002), p. 585–625.

8. R. Jackiw and C. Rebbi, Phys. Rev. D **13**, 3398 (1976), reprinted in *Solitons and Particles*, Eds. C. Rebbi and G. Soliani, (World Scientific, Singapore, 1984), p. 331.

9. J. Bagger and J. Wess, *Supersymmetry and Supergravity*, (Princeton University Press, 1990).

10. B. Chibisov and M. A. Shifman, Phys. Rev. D **56**, 7990 (1997) (E) **D58** (1998) 109901. [hep-th/9706141].

11. J. Wess and B. Zumino, *Phys. Lett.* **B49** (1974) 52 [Reprinted in *Supersymmetry*, Ed. S. Ferrara, (North-Holland/World Scientific, Amsterdam – Singapore, 1987), Vol. 1, page 77].

12. G. R. Dvali and M. A. Shifman, Nucl. Phys. B **504**, 127 (1997) [hep-th/9611213]; Phys. Lett. B **396**, 64 (1997) (E) **407**, 452 (1997) [hep-th/9612128].

13. J. Wess and B. Zumino, *Phys. Lett.* **B49** (1974) 52;
J. Iliopoulos and B. Zumino, *Nucl. Phys.* **B76** (1974) 310;
P. West, *Nucl. Phys.* **B106** (1976) 219;
M. Grisaru, M. Roček, and W. Siegel, *Nucl. Phys.* **B159** (1979) 429.

14. D. Bazeia, J. Menezes and M. M. Santos, Phys. Lett. B **521**, 418 (2001) [hep-th/0110111]; Nucl. Phys. B **636**, 132 (2002) [hep-th/0103041].

15. E. Witten, Phys. Lett. B **86**, 283 (1979) [Reprinted in *Solitons and Particles*, Eds. C. Rebbi and G. Soliani, (World Scinetific, Singapore, 1984) p. 777].

16. G. 't Hooft, Nucl. Phys. B **79**, 276 (1974).

17. A. M. Polyakov, Pisma Zh. Eksp. Teor. Fiz. **20**, 430 (1974) [Engl. transl. JETP Lett. **20**, 194 (1974), reprinted in *Solitons and Particles*, Eds. C. Rebbi and G. Soliani, (World Scientific, Singapore, 1984), p. 522].

18. B. Zumino, Phys. Lett. B **87**, 203 (1979).

19. J. Bagger, *Supersymmetric Sigma Models*, Report SLAC-PUB-3461, published in *Supersymmetry*, Proc. NATO Advanced Study Institute on Supersymmetry, Bonn, Germany, August 1984, Eds. K. Dietz, R. Flume, G. von Gehlen, and V. Rittenberg (Plenum Press, New York 1985) pp. 45-87, and in *Supergravities in Diverse Dimensions*, Eds. A. Salam and E. Sezgin (World Scientific, Singapore, 1989), Vol. 1, pp. 569-611.

20. A. M. Polyakov, Phys. Lett. B **59**, 79 (1975).

21. L. Alvarez-Gaume and D. Z. Freedman, Commun. Math. Phys. **91**, 87 (1983).

22. N. Dorey, JHEP **9811**, 005 (1998) [hep-th/9806056].

23. E. Witten, Nucl. Phys. B **202**, 253 (1982).

24. V. A. Novikov *et al.*, Phys. Rept. **116**, 103 (1984).

25. E. Witten, in G. Kane, *Supersymmetry: Unveiling the Ultimate Laws of Nature* (Perseus Books, 2000).

26. V. A. Novikov, M. A. Shifman, A. I. Vainshtein, and V. I. Zakharov, Nucl. Phys. B **229**, 381 (1983); Phys. Lett. B **166**, 329 (1986).

27. N. Seiberg and E. Witten, Nucl. Phys. B **426**, 19 (1994), (E) B **430**, 485 (1994); [hep-th/9407087]. Nucl. Phys. B **431**, 484 (1994) [hep-th/9408099].

28. N. Seiberg, Proceedings 4th Int. Symposium on Particles, Strings, and Cosmology (PASCOS 94), Syracuse, New York, May 1994, Ed. K. C. Wali (World Scientific, Singapore, 1995), p. 183 [hep-th/9408013]; Int. J. Mod. Phys. A **12** (1997) 5171 [hep-th/9506077].

Forces from Connes' Geometry

T. Schücker

Centre de Physique Théorique, CNRS – Luminy, Case 907, 13288 Marseille Cedex 9, France

Abstract. Einstein derived general relativity from Riemannian geometry. Connes extends this derivation to noncommutative geometry and obtains electro–magnetic, weak, and strong forces. These are pseudo forces, that accompany the gravitational force just as in Minkowskian geometry the magnetic force accompanies the electric force. The main physical input of Connes' derivation is parity violation. His main output is the Higgs boson which breaks the gauge symmetry spontaneously and gives masses to gauge and Higgs bosons.

1 Introduction

Still today one of the major summits in physics is the understanding of the spectrum of the hydrogen atom. The phenomenological formula by Balmer and Rydberg was a remarkable pre-summit on the way up. The true summit was reached by deriving this formula from quantum mechanics. We would like to compare the standard model of electro–magnetic, weak, and strong forces with the Balmer–Rydberg formula [1] and review the present status of Connes' derivation of this model from noncommutative geometry, see Table 1. This geometry extends Riemannian geometry, and Connes' derivation is a natural extension of another major summit in physics: Einstein's derivation of general relativity from Riemannian geometry. Indeed, Connes' derivation unifies gravity with the other three forces.

Let us briefly recall four nested, analytic geometries and their impact on our understanding of forces and time, see Table 2. *Euclidean geometry* is underlying Newton's mechanics as space of positions. Forces are described by vectors living in the same space and the Euclidean scalar product is needed to define work and potential energy. Time is not part of geometry, it is absolute. This point of view is abandoned in special relativity unifying space and time into *Minkowskian geometry*. This new point of view allows to derive the magnetic field from the electric field as a pseudo force associated to a Lorentz boost. Although time has become relative, one can still imagine a grid of synchronized clocks, i.e. a universal time. The next generalization is *Riemannian geometry =* curved spacetime. Here gravity can be viewed as the pseudo force associated to a uniformly accelerated coordinate transformation. At the same time, universal time loses all meaning and we must content ourselves with proper time. With today's precision in time measurement, this complication of life becomes a bare necessity, e.g. the global positioning system (GPS).

T. Schücker, Forces from Connes' Geometry, Lect. Notes Phys. **659**, 285–350 (2005)
http://www.springerlink.com/

Table 1. An analogy

atoms	particles and forces
Balmer–Rydberg formula	standard model
quantum mechanics	noncommutative geometry

Table 2. Four nested analytic geometries

geometry	force	time
Euclidean	$E = \int \boldsymbol{F} \cdot \mathrm{d}\boldsymbol{x}$	absolute
Minkowskian	$\boldsymbol{E}, \epsilon_0 \Rightarrow \boldsymbol{B}, \mu_0 = \frac{1}{\epsilon_0 c^2}$	universal
Riemannian	Coriolis $\leftrightarrow$ gravity	proper, τ
noncommutative	gravity $\Rightarrow$ YMH, $\lambda = \frac{1}{3}g_2^2$	$\Delta\tau \sim 10^{-40}$ s

Our last generalization is to Connes' *noncommutative geometry* = curved space(time) with uncertainty. It allows to understand some Yang–Mills and some Higgs forces as pseudo forces associated to transformations that extend the two coordinate transformations above to the new geometry without points. Also, proper time comes with an uncertainty. This uncertainty of some hundred Planck times might be accessible to experiments through gravitational wave detectors within the next ten years [2].

Prerequisites

On the physical side, the reader is supposed to be acquainted with general relativity, e.g. [3], Dirac spinors at the level of e.g. the first few chapters in [4] and Yang–Mills theory with spontaneous symmetry break-down, for example the standard model, e.g. [5]. I am not ashamed to adhere to the minimax principle: a maximum of pleasure with a minimum of effort. The effort is to do a calculation, the pleasure is when its result coincides with an experiment result. Consequently our mathematical treatment is as low-tech as possible. We do need *local* differential and Riemannian geometry at the level of e.g. the first few chapters in [6]. Local means that our spaces or manifolds can be thought of as open subsets of $\mathbb{R}^4$. Nevertheless, we sometimes use compact spaces like the torus: only to simplify some integrals. We do need some group theory, e.g. [7], mostly matrix groups and their representations. We also need a few basic facts on associative algebras. Most of them are recalled as we go along and can be found for instance in [8]. For the reader's convenience, a few simple definitions from groups and algebras are collected in the Appendix. And, of course, we need some chapters of noncommutative geometry which are developped in the text. For a more detailed presentation still with particular care for the physicist see Refs. [9,10].

2   Gravity from Riemannian Geometry

In this section we briefly review Einstein's derivation of general relativity from Riemannian geometry. His derivation is in two strokes, kinematics and dynamics.

2.1   First Stroke: Kinematics

Consider flat space(time) M in inertial or Cartesian coordinates $\tilde{x}^{\tilde{\lambda}}$. Take as 'matter a free, classical point particle. Its dynamics, Newton's free equation, fixes the trajectory $\tilde{x}^{\tilde{\lambda}}(p)$:

$$\frac{\mathrm{d}^2 \tilde{x}^{\tilde{\lambda}}}{\mathrm{d}p^2} = 0. \tag{1}$$

After a general coordinate transformation, $x^{\lambda} = \sigma^{\lambda}(\tilde{x})$, Newton's equation reads

$$\frac{\mathrm{d}^2 x^{\lambda}}{\mathrm{d}p^2} + \Gamma^{\lambda}{}_{\mu\nu}(g)\, \frac{\mathrm{d}x^{\mu}}{\mathrm{d}p}\, \frac{\mathrm{d}x^{\nu}}{\mathrm{d}p} = 0. \tag{2}$$

Pseudo forces have appeared. They are coded in the Levi–Civita connection

$$\Gamma^{\lambda}{}_{\mu\nu}(g) = \tfrac{1}{2} g^{\lambda\kappa} \left[\frac{\partial}{\partial x^{\mu}} g_{\kappa\nu} + \frac{\partial}{\partial x^{\nu}} g_{\kappa\mu} - \frac{\partial}{\partial x^{\kappa}} g_{\mu\nu} \right], \tag{3}$$

where $g_{\mu\nu}$ is obtained by 'fluctuating' the flat metric $\tilde{\eta}_{\tilde{\mu}\tilde{\nu}} = \mathrm{diag}(1, -1, -1, -1,)$ with the Jacobian of the coordinate transformation σ:

$$g_{\mu\nu}(x) = J(x)^{-1\tilde{\mu}}{}_{\mu}\, \tilde{\eta}_{\tilde{\mu}\tilde{\nu}}\, J(x)^{-1\tilde{\nu}}{}_{\nu}, \quad J(\tilde{x})^{\mu}{}_{\tilde{\mu}} := \partial \sigma^{\mu}(\tilde{x}) / \partial\, \tilde{x}^{\tilde{\mu}}. \tag{4}$$

For the coordinates of the rotating disk, the pseudo forces are precisely the centrifugal and Coriolis forces. Einstein takes uniformly accelerated coordinates, $ct = c\tilde{t}$, $z = \tilde{z} + \tfrac{1}{2}\frac{g}{c^2}(c\tilde{t})^2$ with $g = 9.81$ m/s^2. Then the geodesic equation (2) reduces to $\mathrm{d}^2 z / \mathrm{d}t^2 = -g$. So far this gravity is still a pseudo force which means that the curvature of its Levi–Civita connection vanishes. This constraint is relaxed by the equivalence principle: pseudo forces and true gravitational forces are coded together in a not necessarily flat connection Γ, that derives from a potential, the not necessarily flat metric g. The kinematical variable to describe gravity is therefore the Riemannian metric. By construction the dynamics of matter, the geodesic equation, is now covariant under general coordinate transformations.

2.2   Second Stroke: Dynamics

Now that we know the kinematics of gravity let us see how Einstein obtains its dynamics, i.e. differential equations for the metric tensor $g_{\mu\nu}$. Of course Einstein wants these equations to be covariant under general coordinate transformations and he wants the energy-momentum tensor $T_{\mu\nu}$ to be the source of gravity. From

Riemannian geometry he knew that there is no covariant, first order differential operator for the metric. But there are second order ones:

Theorem: The most general tensor of degree 2 that can be constructed from the metric tensor $g_{\mu\nu}(x)$ with at most two partial derivatives is

$$\alpha R_{\mu\nu} + \beta R g_{\mu\nu} + \Lambda g_{\mu\nu}, \quad \alpha, \beta, \Lambda \in \mathbb{R}.. \tag{5}$$

Here are our conventions for the curvature tensors:

Riemann tensor : $R^\lambda{}_{\mu\nu\kappa} = \partial_\nu \Gamma^\lambda{}_{\mu\kappa} - \partial_\kappa \Gamma^\lambda{}_{\mu\nu} + \Gamma^\eta{}_{\mu\kappa}\Gamma^\lambda{}_{\nu\eta} - \Gamma^\eta{}_{\mu\nu}\Gamma^\lambda{}_{\kappa\eta},$ (6)

Ricci tensor : $R_{\mu\kappa} = R^\lambda{}_{\mu\lambda\kappa},$ (7)

curvature scalar : $R = R_{\mu\nu}g^{\mu\nu}.$ (8)

The miracle is that the tensor (5) is symmetric just as the energy-momentum tensor. However, the latter is covariantly conserved, $D^\mu T_{\mu\nu} = 0$, while the former one is conserved if and only if $\beta = -\frac{1}{2}\alpha$. Consequently, Einstein puts his equation

$$R_{\mu\nu} - \tfrac{1}{2} R g_{\mu\nu} - \Lambda_c g_{\mu\nu} = \tfrac{8\pi G}{c^4} T_{\mu\nu}. \tag{9}$$

He chooses a vanishing cosmological constant, $\Lambda_c = 0$. Then for small static mass density T_{00}, his equation reproduces Newton's universal law of gravity with G the Newton constant. However for not so small masses there are corrections to Newton's law like precession of perihelia. Also Einstein's theory applies to massless matter and produces the curvature of light. Einstein's equation has an agreeable formal property, it derives via the Euler–Lagrange variational principle from an action, the famous Einstein–Hilbert action:

$$S_{\mathrm{EH}}[g] = \frac{-1}{16\pi G} \int_M R\, dV - \frac{2\Lambda_c}{16\pi G} \int_M dV, \tag{10}$$

with the invariant volume element $dV := |\det g_{..}|^{1/2}\, d^4x$.

General relativity has a precise geometric origin: the left-hand side of Einstein's equation is a sum of some 80 000 terms in first and second partial derivatives of $g_{\mu\nu}$ and its matrix inverse $g^{\mu\nu}$. All of these terms are completely fixed by the requirement of covariance under general coordinate transformations. General relativity is verified experimentally to an extraordinary accuracy, even more, it has become a cornerstone of today's technology. Indeed length measurements had to be abandoned in favour of proper time measurements, e.g. the GPS. Nevertheless, the theory still leaves a few questions unanswered:

- Einstein's equation is nonlinear and therefore does not allow point masses as source, in contrast to Maxwell's equation that does allow point charges as source. From this point of view it is not satisfying to consider point-like matter.
- The gravitational force is coded in the connection Γ. Nevertheless we have accepted its potential, the metric g, as kinematical variable.

- The equivalence principle states that locally, i.e. on the trajectory of a point-like particle, one cannot distinguish gravity from a pseudo force. In other words, there is always a coordinate system, 'the freely falling lift', in which gravity is absent. This is not true for electro–magnetism and we would like to derive this force (as well as the weak and strong forces) as a pseudo force coming from a geometric transformation.
- So far general relativity has resisted all attempts to reconcile it with quantum mechanics.

3   Slot Machines and the Standard Model

Today we have a very precise phenomenological description of electro–magnetic, weak, and strong forces. This description, the standard model, works on a pertur-bative quantum level and, as classical gravity, it derives from an action principle. Let us introduce this action by analogy with the Balmer–Rydberg formula.

One of the new features of atomic physics was the appearance of discrete frequencies and the measurement of atomic spectra became a highly developed art. It was natural to label the discrete frequencies ν by natural numbers n. To fit the spectrum of a given atom, say hydrogen, let us try the ansatz

$$\nu = g_1 n_1^{q_1} + g_2 n_2^{q_2}. \tag{11}$$

We view this ansatz as a slot machine. You input two bills, the integers q_1, q_2 and two coins, the two real numbers g_1, g_2, and compare the output with the measured spectrum. (See Fig. 1.) If you are rich enough, you play and replay on the slot machine until you win. The winner is the Balmer–Rydberg formula, i.e., $q_1 = q_2 = -2$ and $g_1 = -g_2 = 3.289 \ 10^{15}$ Hz, which is the famous Rydberg constant R. Then came quantum mechanics. It explained why the spectrum of the hydrogen atom was discrete in the first place and derived the exponents and the Rydberg constant,

$$R = \frac{m_e}{4\pi \hbar^3} \frac{e^4}{(4\pi \epsilon_0)^2}, \tag{12}$$

from a noncommutativity, $[x, p] = i\hbar 1$.

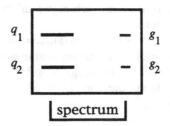

q_1 ———— g_1
q_2 ———— g_2

spectrum

Fig. 1. A slot machine for atomic spectra

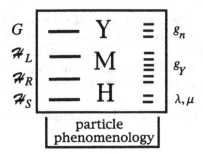

Fig. 2. The Yang–Mills–Higgs slot machine

To cut short its long and complicated history we introduce the standard model as the winner of a particular slot machine. This machine, which has become popular under the names Yang, Mills and Higgs, has four slots for four bills. Once you have decided which bills you choose and entered them, a certain number of small slots will open for coins. Their number depends on the choice of bills. You make your choice of coins, feed them in, and the machine starts working. It produces as output a Lagrange density. From this density, perturbative quantum field theory allows you to compute a complete particle phenomenology: the particle spectrum with the particles' quantum numbers, cross sections, life times, and branching ratios. (See Fig. 2.) You compare the phenomenology to experiment to find out whether your input wins or loses.

3.1 Input

The first bill is a finite dimensional, real, compact Lie group G. The gauge bosons, spin 1, will live in its adjoint representation whose Hilbert space is the complexification of the Lie algebra $\mathfrak{g}$ (cf. Appendix).

The remaining bills are three unitary representations of G, ρ_L, ρ_R, ρ_S, defined on the complex Hilbert spaces, $\mathcal{H}_L$, $\mathcal{H}_R$, $\mathcal{H}_S$. They classify the left- and right-handed fermions, spin $\frac{1}{2}$, and the scalars, spin 0. The group G is chosen compact to ensure that the unitary representations are finite dimensional, we want a finite number of 'elementary particles' according to the credo of particle physics that particles are orthonormal basis vectors of the Hilbert spaces which carry the representations. More generally, we might also admit multi-valued representations, 'spin representations', which would open the debate on charge quantization. More on this later.

The coins are numbers, coupling constants, more precisely coefficients of invariant polynomials. We need an invariant scalar product on $\mathfrak{g}$. The set of all these scalar products is a cone and the gauge couplings are particular coordinates of this cone. If the group is simple, say $G = SU(n)$, then the most general, invariant scalar product is

$$(X, X') = \tfrac{2}{g_n^2} \mathrm{tr}\,[X^* X'], \quad X, X' \in su(n). \tag{13}$$

If $G = U(1)$, we have

$$(Y, Y') = \tfrac{1}{g_1^2} \bar{Y} Y', \quad Y, Y' \in u(1). \tag{14}$$

We denote by $\bar{}$ the complex conjugate and by $\cdot^*$ the Hermitean conjugate. Mind the different normalizations, they are conventional. The g_n are positive numbers, *the gauge couplings.* For every simple factor of G there is one gauge coupling.

Then we need the Higgs potential $V(\varphi)$. It is an invariant, fourth order, stable polynomial on $\mathcal{H}_S \ni \varphi$. Invariant means $V(\rho_S(u)\varphi) = V(\varphi)$ for all $u \in G$. Stable means bounded from below. For $G = U(2)$ and the Higgs scalar in the fundamental or defining representation, $\varphi \in \mathcal{H}_S = \mathbb{C}^2$, $\rho_S(u) = u$, we have

$$V(\varphi) = \lambda \, (\varphi^* \varphi)^2 - \tfrac{1}{2}\mu^2 \, \varphi^* \varphi. \tag{15}$$

The coefficients of the Higgs potential are the Higgs couplings, λ must be positive for stability. We say that the potential breaks G spontaneously if no minimum of the potential is a trivial orbit under G. In our example, if μ is positive, the minima of $V(\varphi)$ lie on the 3-sphere $|\varphi| = v := \tfrac{1}{2}\mu/\sqrt{\lambda}$. v is called vacuum expectation value and $U(2)$ is said to break down spontaneously to its little group

$$U(1) \ni \begin{pmatrix} 1 & 0 \\ 0 & e^{i\alpha} \end{pmatrix}. \tag{16}$$

The little group leaves invariant any given point of the minimum, e.g. $\varphi = (v, 0)^T$. On the other hand, if μ is purely imaginary, then the minimum of the potential is the origin, no spontaneous symmetry breaking and the little group is all of G.

Finally, we need the Yukawa couplings g_Y. They are the coefficients of the most general, real, trilinear invariant on $\mathcal{H}_L^* \otimes \mathcal{H}_R \otimes (\mathcal{H}_S \oplus \mathcal{H}_S^*)$. For every 1-dimensional invariant subspace in the reduction of this tensor representation, we have one complex Yukawa coupling. For example $G = U(2)$, $\mathcal{H}_L = \mathbb{C}^2$, $\rho_L(u)\psi_L = (\det u)^{q_L} u \psi_L$, $\mathcal{H}_R = \mathbb{C}$, $\rho_R(u)\psi_R = (\det u)^{q_R}\psi_R$, $\mathcal{H}_S = \mathbb{C}^2$, $\rho_S(u)\varphi = (\det u)^{q_S} u \varphi$. If $-q_L + q_R + q_S \neq 0$ there is no Yukawa coupling, otherwise there is one: $(\psi_L, \psi_R, \varphi) = \mathrm{Re}(g_Y \, \psi_L^* \psi_R \varphi)$.

If the symmetry is broken spontaneously, gauge and Higgs bosons acquire masses related to gauge and Higgs couplings, fermions acquire masses equal to the 'vacuum expectation value' v times the Yukawa couplings.

As explained in Jan-Willem van Holten's and Jean Zinn-Justin's lectures at this School [11,12], one must require for consistency of the quantum theory that the fermionic representations be free of Yang–Mills anomalies,

$$\mathrm{tr}\,((\tilde{\rho}_L(X))^3) - \mathrm{tr}\,((\tilde{\rho}_R(X))^3) = 0, \quad \text{for all } X \in \mathfrak{g}. \tag{17}$$

We denote by $\tilde{\rho}$ the Lie algebra representation of the group representation ρ. Sometimes one also wants the mixed Yang–Mills–gravitational anomalies to vanish:

$$\mathrm{tr}\,\tilde{\rho}_L(X) - \mathrm{tr}\,\tilde{\rho}_R(X) = 0, \quad \text{for all } X \in \mathfrak{g}. \tag{18}$$

3.2 Rules

It is time to open the slot machine and to see how it works. Its mechanism has five pieces:

The Yang–Mills Action. The actor in this piece is $A = A_\mu dx^\mu$, called connection, gauge potential, gauge boson or Yang–Mills field. It is a 1-form on spacetime $M \ni x$ with values in the Lie algebra $\mathfrak{g}$, $A \in \Omega^1(M, \mathfrak{g})$. We define its curvature or field strength,

$$F := dA + \tfrac{1}{2}[A, A] = \tfrac{1}{2}F_{\mu\nu}dx^\mu dx^\nu \ \in \Omega^2(M, \mathfrak{g}), \tag{19}$$

and the Yang–Mills action,

$$S_{\mathrm{YM}}[A] = -\tfrac{1}{2}\int_M (F, *F) = \frac{-1}{2g_n^2}\int_M \operatorname{tr} F_{\mu\nu}^* F^{\mu\nu}dV. \tag{20}$$

The gauge group $^M G$ is the infinite dimensional group of differentiable functions $g : M \to G$ with pointwise multiplication. $\cdot^*$ is the Hermitean conjugate of matrices, $*\cdot$ is the Hodge star of differential forms. The space of all connections carries an affine representation (cf. Appendix) ρ_V of the gauge group:

$$\rho_V(g)A = gAg^{-1} + gdg^{-1}. \tag{21}$$

Restricted to x-independent ('rigid') gauge transformation, the representation is linear, the adjoint one. The field strength transforms homogeneously even under x-dependent ('local') gauge transformations, $g : M \to G$ differentiable,

$$\rho_V(g)F = gFg^{-1}, \tag{22}$$

and, as the scalar product $(\cdot, \cdot)$ is invariant, the Yang–Mills action is gauge invariant,

$$S_{\mathrm{YM}}[\rho_V(g)A] = S_{\mathrm{YM}}[A] \quad \text{for all } g \in \ ^M G. \tag{23}$$

Note that a mass term for the gauge bosons,

$$\tfrac{1}{2}\int_M m_A^2(A, *A) = \frac{1}{g_n^2}\int_M m_A^2 \operatorname{tr} A_\mu^* A^\mu dV, \tag{24}$$

is not gauge invariant because of the inhomogeneous term in the transformation law of a connection (21). Gauge invariance forces the gauge bosons to be massless.

In the Abelian case $G = U(1)$, the Yang–Mills Lagrangian is nothing but Maxwell's Lagrangian, the gauge boson A is the photon and its coupling constant g is $e/\sqrt{\epsilon_0}$. Note however, that the Lie algebra of $U(1)$ is $i\mathbb{R}$ and the vector potential is purely imaginary, while conventionally, in Maxwell's theory it is chosen real. Its quantum version is QED, quantum electro-dynamics. For $G = SU(3)$ and $\mathcal{H}_L = \mathcal{H}_R = \mathbb{C}^3$ we have today's theory of strong interaction, quantum chromo-dynamics, QCD.

The Dirac Action. Schrödinger's action is non-relativistic. Dirac generalized it to be Lorentz invariant, e.g. [4]. The price to be paid is twofold. His generalization only works for spin $\frac{1}{2}$ particles and requires that for every such particle there must be an antiparticle with same mass and opposite charges. Therefore, Dirac's wave function $\psi(x)$ takes values in $\mathbb{C}^4$, spin up, spin down, particle, antiparticle. antiparticles have been discovered and Dirac's theory was celebrated. Here it is in short for (flat) Minkowski space of signature $+---$, $\eta_{\mu\nu} = \eta^{\mu\nu} = \mathrm{diag}(+1,-1,-1,-1)$. Define the four Dirac matrices,

$$\gamma^0 = \begin{pmatrix} 0 & -1_2 \\ -1_2 & 0 \end{pmatrix}, \quad \gamma^j = \begin{pmatrix} 0 & \sigma_j \\ -\sigma_j & 0 \end{pmatrix}, \tag{25}$$

for $j = 1,2,3$ with the three Pauli matrices,

$$\sigma_1 = \begin{pmatrix} 0 & 1 \\ 1 & 0 \end{pmatrix}, \quad \sigma_2 = \begin{pmatrix} 0 & -i \\ i & 0 \end{pmatrix}, \quad \sigma_3 = \begin{pmatrix} 1 & 0 \\ 0 & -1 \end{pmatrix}. \tag{26}$$

They satisfy the anticommutation relations,

$$\gamma^\mu \gamma^\nu + \gamma^\nu \gamma^\mu = 2\eta^{\mu\nu} 1_4. \tag{27}$$

In even spacetime dimensions, the chirality,

$$\gamma_5 := -\tfrac{i}{4!}\epsilon_{\mu\nu\rho\sigma}\gamma^\mu\gamma^\nu\gamma^\rho\gamma^\sigma = -i\gamma^0\gamma^1\gamma^2\gamma^3 = \begin{pmatrix} -1_2 & 0 \\ 0 & 1_2 \end{pmatrix} \tag{28}$$

is a natural operator and it paves the way to an understanding of parity violation in weak interactions. The chirality is a unitary matrix of unit square, which anticommutes with all four Dirac matrices. $(1 - \gamma_5)/2$ projects a Dirac spinor onto its left-handed part, $(1 + \gamma_5)/2$ projects onto the right-handed part. The two parts are called Weyl spinors. A massless left-handed (right-handed) spinor, has its spin parallel (anti-parallel) to its direction of propagation. The chirality maps a left-handed spinor to a right-handed spinor. A space reflection or parity transformation changes the sign of the velocity vector and leaves the spin vector unchanged. It therefore has the same effect on Weyl spinors as the chirality operator. Similarly, there is the charge conjugation, an anti-unitary operator (cf. Appendix) of unit square, that applied on a particle ψ produces its antiparticle

$$J = \tfrac{1}{i}\gamma^0\gamma^2 \circ \text{complex conjugation} = \begin{pmatrix} 0 & -1 & 0 & 0 \\ 1 & 0 & 0 & 0 \\ 0 & 0 & 0 & 1 \\ 0 & 0 & -1 & 0 \end{pmatrix} \circ \mathrm{c\,c}, \tag{29}$$

i.e. $J\psi = \tfrac{1}{i}\gamma^0\gamma^2\,\bar{\psi}$. Attention, here and for the last time $\bar{\psi}$ stands for the complex conjugate of ψ. In a few lines we will adopt a different more popular convention. The charge conjugation commutes with all four Dirac matrices. In flat spacetime, the free Dirac operator is simply defined by,

$$\partial\!\!\!/ := i\hbar\gamma^\mu\partial_\mu. \tag{30}$$

It is sometimes referred to as square root of the wave operator because $\partial\!\!\!/^2 = -\Box$. The coupling of the Dirac spinor to the gauge potential $A = A_\mu dx^\mu$ is done via the covariant derivative, and called Minimal coupling. In order to break parity, we write left- and right-handed parts independently:

$$S_D[A, \psi_L, \psi_R] = \int_M \bar\psi_L \left[\partial\!\!\!/ + i\hbar\gamma^\mu \tilde\rho_L(A_\mu)\right] \frac{1 - \gamma_5}{2} \psi_L \, dV$$
$$+ \int_M \bar\psi_R \left[\partial\!\!\!/ + i\hbar\gamma^\mu \tilde\rho_R(A_\mu)\right] \frac{1 + \gamma_5}{2} \psi_R \, dV. \tag{31}$$

The new actors in this piece are ψ_L and ψ_R, two multiplets of Dirac spinors or fermions, that is with values in $\mathcal{H}_L$ and $\mathcal{H}_R$. We use the notations, $\bar\psi := \psi^* \gamma^0$, where $\cdot^*$ denotes the Hermitean conjugate with respect to the four spinor components and the dual with respect to the scalar product in the (internal) Hilbert space $\mathcal{H}_L$ or $\mathcal{H}_R$. The γ^0 is needed for energy reasons and for invariance of the pseudo–scalar product of spinors under lifted Lorentz transformations. The γ^0 is absent if spacetime is Euclidean. Then we have a genuine scalar product and the square integrable spinors form a Hilbert space $\mathcal{L}^2(\mathcal{S}) = \mathcal{L}^2(\mathbb{R}^4) \otimes \mathbb{C}^4$, the infinite dimensional brother of the internal one. The Dirac operator is then self adjoint in this Hilbert space. We denote by $\tilde\rho_L$ the Lie algebra representation in $\mathcal{H}_L$. The covariant derivative, $D_\mu := \partial_\mu + \tilde\rho_L(A_\mu)$, deserves its name,

$$\left[\partial_\mu + \tilde\rho_L(\rho_V(g)A_\mu)\right] (\rho_L(g)\psi_L) = \rho_L(g) \left[\partial_\mu + \tilde\rho_L(A_\mu)\right] \psi_L, \tag{32}$$

for all gauge transformations $g \in {}^M G$. This ensures that the Dirac action (31) is gauge invariant.

If parity is conserved, $\mathcal{H}_L = \mathcal{H}_R$, we may add a mass term

$$-c \int_M \bar\psi_R \, m_\psi \frac{1 - \gamma_5}{2} \psi_L \, dV \; - \; c \int_M \bar\psi_L \, m_\psi \frac{1 + \gamma_5}{2} \psi_R \, dV \; =$$
$$-c \int_M \bar\psi \, m_\psi \, \psi \, dV \tag{33}$$

to the Dirac action. It gives identical masses to all members of the multiplet. The fermion masses are gauge invariant if all fermions in $\mathcal{H}_L = \mathcal{H}_R$ have the same mass. For instance QED preserves parity, $\mathcal{H}_L = \mathcal{H}_R = \mathbb{C}$, the representation being characterized by the electric charge, -1 for both the left- and right handed electron. Remember that gauge invariance forces gauge bosons to be massless. For fermions, it is parity *non*-invariance that forces them to be massless.

Let us conclude by reviewing briefly why the Dirac equation is the Lorentz invariant generalization of the Schrödinger equation. Take the free Schrödinger equation on (flat) $\mathbb{R}^4$. It is a linear differential equation with constant coefficients,

$$\left(\frac{2m}{i\hbar} \frac{\partial}{\partial t} - \Delta\right) \psi = 0. \tag{34}$$

We compute its polynomial following Fourier and de Broglie,

$$-\frac{2m}{\hbar} \omega + k^2 = -\frac{2m}{\hbar^2} \left[E - \frac{p^2}{2m}\right]. \tag{35}$$

Energy conservation in Newtonian mechanics is equivalent to the vanishing of the polynomial. Likewise, the polynomial of the free, massive Dirac equation $(\partial\!\!\!/ - cm_\psi)\psi = 0$ is

$$\frac{\hbar}{c}\,\omega\gamma^0 + \hbar\,k_j\gamma^j - c\,m\mathbf{1}. \tag{36}$$

Putting it to zero implies energy conservation in special relativity,

$$(\tfrac{\hbar}{c})^2\,\omega^2 - \hbar^2\,\mathbf{k}^2 - c^2\,m^2 = 0. \tag{37}$$

In this sense, Dirac's equation generalizes Schrödinger's to special relativity. To see that Dirac's equation is really Lorentz invariant we must lift the Lorentz transformations to the space of spinors. We will come back to this lift.

So far we have seen the two noble pieces by Yang–Mills and Dirac. The remaining three pieces are cheap copies of the two noble ones with the gauge boson A replaced by a scalar φ. We need these three pieces to cure only one problem, give masses to some gauge bosons and to some fermions. These masses are forbidden by gauge invariance and parity violation. To simplify the notation we will work from now on in units with $c = \hbar = 1$.

The Klein–Gordon Action. The Yang–Mills action contains the kinetic term for the gauge boson. This is simply the quadratic term, $(\mathrm{d}A, \mathrm{d}A)$, which by Euler–Lagrange produces linear field equations. We copy this for our new actor, a multiplet of scalar fields or Higgs bosons,

$$\varphi \in \Omega^0(M, \mathcal{H}_S), \tag{38}$$

by writing the Klein–Gordon action,

$$S_{\mathrm{KG}}[A, \varphi] = \tfrac{1}{2}\int_M (\mathrm{D}\varphi)^* * \mathrm{D}\varphi = \tfrac{1}{2}\int_M (\mathrm{D}_\mu\varphi)^* \mathrm{D}^\mu\varphi\,\mathrm{d}V, \tag{39}$$

with the covariant derivative here defined with respect to the scalar representation,

$$\mathrm{D}\varphi := \mathrm{d}\varphi + \tilde{\rho}_S(A)\varphi. \tag{40}$$

Again we need this Minimal coupling $\varphi^* A\varphi$ for gauge invariance.

The Higgs Potential. The non-Abelian Yang–Mills action contains interaction terms for the gauge bosons, an invariant, fourth order polynomial, $2(\mathrm{d}A, [A, A]) + ([A, A], [A, A])$. We mimic these interactions for scalar bosons by adding the integrated Higgs potential $\int_M *V(\varphi)$ to the action.

The Yukawa Terms. We also mimic the (minimal) coupling of the gauge boson to the fermions $\psi^* A\psi$ by writing all possible trilinear invariants,

$$S_{\mathrm{Y}}[\psi_L, \psi_R, \varphi] :=$$

$$\mathrm{Re}\int_M * \left(\sum_{j=1}^n g_{Yj}\,(\psi_L^*, \psi_R, \varphi)_j + \sum_{j=n+1}^m g_{Yj}\,(\psi_L^*, \psi_R, \varphi^*)_j \right). \tag{41}$$

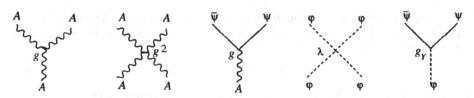

Fig. 3. Tri- and quadrilinear gauge couplings, minimal gauge coupling to fermions, Higgs self-coupling and Yukawa coupling

In the standard model, there are 27 complex Yukawa couplings, $m = 27$.

The Yang–Mills and Dirac actions, contain three types of couplings, a trilinear self coupling AAA, a quadrilinear self coupling $AAAA$ and the trilinear Minimal coupling $\psi^* A\psi$. The gauge self couplings are absent if the group G is Abelian, the photon has no electric charge, Maxwell's equations are linear. The beauty of gauge invariance is that if G is simple, all these couplings are fixed in terms of one positive number, the gauge coupling g. To see this, take an orthonormal basis T_b, $b = 1, 2, \ldots \dim G$ of the complexification $\mathfrak{g}^{\mathbb{C}}$ of the Lie algebra with respect to the invariant scalar product and an orthonormal basis F_k, $k = 1, 2, \ldots \dim \mathcal{H}_L$, of the fermionic Hilbert space, say $\mathcal{H}_L$, and expand the actors,

$$A =: A_\mu^b T_b \mathrm{d}x^\mu, \quad \psi =: \psi^k F_k. \tag{42}$$

Insert these expressions into the Yang–Mills and Dirac actions, then you get the following interaction terms, see Fig. 3,

$$g\, \partial_\rho A_\mu^a A_\nu^b A_\sigma^c\, f_{abc}\, \epsilon^{\rho\mu\nu\sigma}, \quad g^2\, A_\mu^a A_\nu^b A_\rho^c A_\sigma^d\, f_{ab}{}^e f_{ecd}\, \epsilon^{\rho\mu\nu\sigma},$$
$$g\, \psi^{k*} A_\mu^b \gamma^\mu \psi_\ell\, t_{bk}{}^\ell, \tag{43}$$

with the structure constants $f_{ab}{}^e$,

$$[T_a, T_b] =: f_{ab}{}^e T_e. \tag{44}$$

The indices of the structure constants are raised and lowered with the matrix of the invariant scalar product in the basis T_b, that is the identity matrix. The $t_{bk}{}^\ell$ is the matrix of the operator $\tilde{\rho}_L(T_b)$ with respect to the basis F_k. The difference between the noble and the cheap actions is that the Higgs couplings, λ and μ in the standard model, and the Yukawa couplings g_{Yj} are arbitrary, are neither connected among themselves nor connected to the gauge couplings g_i.

3.3   The Winner

Physicists have spent some thirty years and billions of Swiss Francs playing on the slot machine by Yang, Mills and Higgs. There is a winner, the standard model of electro–weak and strong forces. Its bills are

$$G = SU(2) \times U(1) \times SU(3)/(\mathbb{Z}_2 \times \mathbb{Z}_3), \tag{45}$$

$$\mathcal{H}_L = \bigoplus_1^3 \left[(2, \tfrac{1}{6}, 3) \oplus (2, -\tfrac{1}{2}, 1) \right], \tag{46}$$

$$\mathcal{H}_R = \bigoplus_1^3 \left[(1, \tfrac{2}{3}, 3) \oplus (1, -\tfrac{1}{3}, 3) \oplus (1, -1, 1) \right], \tag{47}$$

$$\mathcal{H}_S = (2, -\tfrac{1}{2}, 1), \tag{48}$$

where (n_2, y, n_3) denotes the tensor product of an n_2 dimensional representation of $SU(2)$, an n_3 dimensional representation of $SU(3)$ and the one dimensional representation of $U(1)$ with hypercharge y: $\rho(\exp(i\theta)) = \exp(iy\theta)$. For historical reasons the hypercharge is an integer multiple of $\tfrac{1}{6}$. This is irrelevant: only the product of the hypercharge with its gauge coupling is measurable and we do not need multi-valued representations, which are characterized by non-integer, rational hypercharges. In the direct sum, we recognize the three generations of fermions, the quarks are $SU(3)$ colour triplets, the leptons colour singlets. The basis of the fermion representation space is

$$\begin{pmatrix} u \\ d \end{pmatrix}_L, \begin{pmatrix} c \\ s \end{pmatrix}_L, \begin{pmatrix} t \\ b \end{pmatrix}_L, \begin{pmatrix} \nu_e \\ e \end{pmatrix}_L, \begin{pmatrix} \nu_\mu \\ \mu \end{pmatrix}_L, \begin{pmatrix} \nu_\tau \\ \tau \end{pmatrix}_L$$

$$\begin{matrix} u_R, & c_R, & t_R, \\ d_R, & s_R, & b_R, \end{matrix} \quad e_R, \quad \mu_R, \quad \tau_R$$

The parentheses indicate isospin doublets.

The eight gauge bosons associated to $su(3)$ are called gluons. Attention, the $U(1)$ is not the one of electric charge, it is called hypercharge, the electric charge is a linear combination of hypercharge and weak isospin, parameterized by the weak mixing angle θ_w to be introduced below. This mixing is necessary to give electric charges to the W bosons. The W^+ and W^- are pure isospin states, while the Z^0 and the photon are (orthogonal) mixtures of the third isospin generator and hypercharge.

Because of the high degree of reducibility in the bills, there are many coins, among them 27 complex Yukawa couplings. Not all Yukawa couplings have a physical meaning and we only remain with 18 physically significant, positive numbers [13], three gauge couplings at energies corresponding to the Z mass,

$$g_1 = 0.3574 \pm 0.0001, \ g_2 = 0.6518 \pm 0.0003, \ g_3 = 1.218 \pm 0.01, \tag{49}$$

two Higgs couplings, λ and μ, and 13 positive parameters from the Yukawa couplings. The Higgs couplings are related to the boson masses:

$$m_W = \tfrac{1}{2} g_2 v = 80.419 \pm 0.056 \text{ GeV}, \tag{50}$$

$$m_Z = \tfrac{1}{2} \sqrt{g_1^2 + g_2^2} \, v = m_W / \cos \theta_w = 91.1882 \pm 0.0022 \text{ GeV}, \tag{51}$$

$$m_H = 2\sqrt{2}\sqrt{\lambda} v > 98 \text{ GeV}, \tag{52}$$

with the vacuum expectation value $v := \frac{1}{2}\mu/\sqrt{\lambda}$ and the weak mixing angle θ_w defined by

$$\sin^2\theta_w := g_2^{-2}/(g_2^{-2}+g_1^{-2}) = 0.23117 \pm 0.00016. \tag{53}$$

For the standard model, there is a one–to–one correspondence between the physically relevant part of the Yukawa couplings and the fermion masses and mixings,

$$m_e = 0.510998902 \pm 0.000000021 \text{ MeV},$$
$$m_\mu = 0.105658357 \pm 0.000000005 \text{ GeV},$$
$$m_\tau = 1.77703 \pm 0.00003 \text{ GeV},$$

$$m_u = 3 \pm 2 \text{ MeV}, \qquad m_d = 6 \pm 3 \text{ MeV},$$
$$m_c = 1.25 \pm 0.1 \text{ GeV}, \quad m_s = 0.125 \pm 0.05 \text{ GeV},$$
$$m_t = 174.3 \pm 5.1 \text{ GeV}, \quad m_b = 4.2 \pm 0.2 \text{ GeV}.$$

For simplicity, we take massless neutrinos. Then mixing only occurs for quarks and is given by a unitary matrix, the Cabibbo–Kobayashi–Maskawa matrix

$$C_{\text{KM}} := \begin{pmatrix} V_{ud} & V_{us} & V_{ub} \\ V_{cd} & V_{cs} & V_{cb} \\ V_{td} & V_{ts} & V_{tb} \end{pmatrix}. \tag{54}$$

For physical purposes it can be parameterized by three angles θ_{12}, θ_{23}, θ_{13} and one CP violating phase δ:

$$C_{\text{KM}} = \begin{pmatrix} c_{12}c_{13} & s_{12}c_{13} & s_{13}e^{-i\delta} \\ -s_{12}c_{23}-c_{12}s_{23}s_{13}e^{i\delta} & c_{12}c_{23}-s_{12}s_{23}s_{13}e^{i\delta} & s_{23}c_{13} \\ s_{12}s_{23}-c_{12}c_{23}s_{13}e^{i\delta} & -c_{12}s_{23}-s_{12}c_{23}s_{13}e^{i\delta} & c_{23}c_{13} \end{pmatrix}, \tag{55}$$

with $c_{kl} := \cos\theta_{kl}$, $s_{kl} := \sin\theta_{kl}$. The absolute values of the matrix elements in C_{KM} are:

$$\begin{pmatrix} 0.9750 \pm 0.0008 & 0.223 \pm 0.004 & 0.004 \pm 0.002 \\ 0.222 \pm 0.003 & 0.9742 \pm 0.0008 & 0.040 \pm 0.003 \\ 0.009 \pm 0.005 & 0.039 \pm 0.004 & 0.9992 \pm 0.0003 \end{pmatrix}. \tag{56}$$

The physical meaning of the quark mixings is the following: when a sufficiently energetic W^+ decays into a u quark, this u quark is produced together with a $\bar{d}$ quark with probability $|V_{ud}|^2$, together with a $\bar{s}$ quark with probability $|V_{us}|^2$, together with a $\bar{b}$ quark with probability $|V_{ub}|^2$. The fermion masses and mixings together are an entity, the fermionic mass matrix or the matrix of Yukawa couplings multiplied by the vacuum expectation value.

Let us note six intriguing properties of the standard model.

- The gluons couple in the same way to left- and right-handed fermions, the gluon coupling is vectorial, the strong interaction does not break parity.

- The fermionic mass matrix commutes with $SU(3)$, the three colours of a given quark have the same mass.
- The scalar is a colour singlet, the $SU(3)$ part of G does not suffer spontaneous symmetry break down, the gluons remain massless.
- The $SU(2)$ couples only to left-handed fermions, its coupling is chiral, the weak interaction breaks parity maximally.
- The scalar is an isospin doublet, the $SU(2)$ part suffers spontaneous symmetry break down, the $W^{\pm}$ and the Z^0 are massive.
- The remaining colourless and neutral gauge boson, the photon, is massless and couples vectorially. This is certainly the most ad-hoc feature of the standard model. Indeed the photon is a linear combination of isospin, which couples only to left-handed fermions, and of a $U(1)$ generator, which may couple to both chiralities. Therefore only the careful fine tuning of the hypercharges in the three input representations (46-48) can save parity conservation and gauge invariance of electro–magnetism,

$$y_{u_R} = y_{q_L} - y_{\ell_L} \quad y_{d_R} = y_{q_L} + y_{\ell_L}, \quad y_{e_R} = 2y_{\ell_L}, \quad y_{\varphi} = y_{\ell_L}, \quad (57)$$

The subscripts label the multiplets, qL for the left-handed quarks, ℓL for the left-handed leptons, uR for the right-handed up-quarks and so forth and φ for the scalar.

Nevertheless the phenomenological success of the standard model is phenomenal: with only a handful of parameters, it reproduces correctly some millions of experimental numbers. Most of these numbers are measured with an accuracy of a few percent and they can be reproduced by classical field theory, no $\hbar$ needed. However, the experimental precision has become so good that quantum corrections cannot be ignored anymore. At this point it is important to note that the fermionic representations of the standard model are free of Yang–Mills (and mixed) anomalies. Today the standard model stands uncontradicted.

Let us come back to our analogy between the Balmer–Rydberg formula and the standard model. One might object that the ansatz for the spectrum, equation (11), is completely ad hoc, while the class of all (anomaly free) s is distinguished by perturbative renormalizability. This is true, but this property was proved [14] only years after the electro–weak part of the standard model was published [15].

By placing the hydrogen atom in an electric or magnetic field, we know experimentally that every frequency 'state' n, $n = 1, 2, 3, ...$, comes with n irreducible unitary representations of the rotation group $SO(3)$. These representations are labelled by ℓ, $\ell = 0, 1, 2, ...n - 1$, of dimensions $2\ell + 1$. An orthonormal basis of each representation ℓ is labelled by another integer m, $m = -\ell, -\ell + 1, ...\ell$. This experimental fact has motivated the credo that particles are orthonormal basis vectors of unitary representations of compact groups. This credo is also behind the standard model. While $SO(3)$ has a clear geometric interpretation, we are still looking for such an interpretation of $SU(2) \times U(1) \times SU(3)/[\mathbb{Z}_2 \times \mathbb{Z}_3]$.

We close this subsection with Iliopoulos' joke [16] from 1976:

Do-It-Yourself Kit for Gauge Models:

1) Choose a gauge group G.
2) Choose the fields of the "elementary particles" you want to introduce, and their representations. Do not forget to include enough fields to allow for the Higgs mechanism.
3) Write the most general renormalizable Lagrangian invariant under G. At this stage gauge invariance is still exact and all vector bosons are massless.
4) Choose the parameters of the Higgs scalars so that spontaneous symmetry breaking occurs. In practice, this often means to choose a negative value [positive in our notations] for the parameter μ^2.
5) Translate the scalars and rewrite the Lagrangian in terms of the translated fields. Choose a suitable gauge and quantize the theory.
6) Look at the properties of the resulting model. If it resembles physics, even remotely, publish it.
7) GO TO 1.

Meanwhile his joke has become experimental reality.

3.4   Wick Rotation

Euclidean signature is technically easier to handle than Minkowskian. What is more, in Connes' geometry it will be vital that the spinors form a Hilbert space with a true scalar product and that the Dirac action takes the form of a scalar product. We therefore put together the Einstein–Hilbert and Yang–Mills–Higgs actions with emphasis on the relative signs and indicate the changes necessary to pass from Minkowskian to Euclidean signature.

In 1983 the meter disappeared as fundamental unit of science and technology. The conceptual revolution of general relativity, the abandon of length in favour of time, had made its way up to the domain of technology. Said differently, general relativity is not really geo-metry, but chrono-metry. Hence our choice of Minkowskian signature is $+ - --$.

With this choice the combined Lagrangian reads,

$$\{-\frac{2\Lambda_c}{16\pi G} - \frac{1}{16\pi G} R - \frac{1}{2g^2} \operatorname{tr}(F^*_{\mu\nu}F^{\mu\nu}) + \frac{1}{g^2}m_A^2 \operatorname{tr}(A^*_\mu A^\mu)$$
$$+ \frac{1}{2}(D_\mu\varphi)^*D^\mu\varphi - \frac{1}{2}m_\varphi^2|\varphi|^2 + \frac{1}{2}\mu^2|\varphi|^2 - \lambda|\varphi|^4$$
$$+ \psi^*\gamma^0 [i\gamma^\mu D_\mu - m_\psi 1_4]\psi\} \, |\det g..|^{1/2}. \tag{58}$$

This Lagrangian is real if we suppose that all fields vanish at infinity. The relative coefficients between kinetic terms and mass terms are chosen as to reproduce the correct energy momentum relations from the free field equations using Fourier transform and the de Broglie relations as explained after equation (34). With the chiral decomposition

$$\psi_L = \frac{1-\gamma_5}{2}\psi, \quad \psi_R = \frac{1+\gamma_5}{2}\psi, \tag{59}$$

the Dirac Lagrangian reads

$$\psi^* \gamma^0 \left[i\gamma^\mu D_\mu - m_\psi 1_4 \right] \psi$$
$$= \psi_L^* \gamma^0 \, i\gamma^\mu D_\mu \, \psi_L + \psi_R^* \gamma^0 \, i\gamma^\mu D_\mu \, \psi_R - m_\psi \psi_L^* \gamma^0 \psi_R - m_\psi \psi_R^* \gamma^0 \psi_L. \quad (60)$$

The relativistic energy momentum relations are quadratic in the masses. Therefore the sign of the fermion mass m_ψ is conventional and merely reflects the choice: who is particle and who is antiparticle. We can even adopt one choice for the left-handed fermions and the opposite choice for the right-handed fermions. Formally this can be seen by the change of field variable (chiral transformation):

$$\psi := \exp(i\alpha\gamma_5) \, \psi'. \quad (61)$$

It leaves invariant the kinetic term and the mass term transforms as,

$$-m_\psi \psi'^* \gamma^0 [\cos(2\alpha) \, 1_4 + i \sin(2\alpha) \, \gamma_5] \psi'. \quad (62)$$

With $\alpha = -\pi/4$ the Dirac Lagrangian becomes:

$$\psi'^* \gamma^0 [i\gamma^\mu D_\mu + im_\psi \gamma_5] \psi' \quad (63)$$
$$= \psi_L'^* \gamma^0 \, i\gamma^\mu D_\mu \, \psi_L' + \psi_R'^* \gamma^0 \, i\gamma^\mu D_\mu \, \psi'_R + m_\psi \psi_L'^* \gamma^0 i\gamma_5 \psi'_R$$
$$+ m_\psi \psi_R'^* \gamma^0 i\gamma_5 \psi'_L$$
$$= \psi_L'^* \gamma^0 \, i\gamma^\mu D_\mu \, \psi_L' + \psi_R'^* \gamma^0 \, i\gamma^\mu D_\mu \, \psi'_R + im_\psi \psi_L'^* \gamma^0 \psi'_R - im_\psi \psi_R'^* \gamma^0 \psi'_L.$$

We have seen that gauge invariance forbids massive gauge bosons, $m_A = 0$, and that parity violation forbids massive fermions, $m_\psi = 0$. This is fixed by spontaneous symmetry breaking, where we take the scalar mass term with wrong sign, $m_\varphi = 0$, $\mu > 0$. The shift of the scalar then induces masses for the gauge bosons, the fermions and the physical scalars. These masses are calculable in terms of the gauge, Yukawa, and Higgs couplings.

The other relative signs in the combined Lagrangian are fixed by the requirement that the energy density of the non-gravitational part T_{00} be positive (up to a cosmological constant) and that gravity in the Newtonian limit be attractive. In particular this implies that the Higgs potential must be bounded from below, $\lambda > 0$. The sign of the Einstein–Hilbert action may also be obtained from an asymptotically flat space of weak curvature, where we can define gravitational energy density. Then the requirement is that the kinetic terms of all physical bosons, spin 0, 1, and 2, be of the same sign. Take the metric of the form

$$g_{\mu\nu} = \eta_{\mu\nu} + h_{\mu\nu}, \quad (64)$$

$h_{\mu\nu}$ small. Then the Einstein–Hilbert Lagrangian becomes [17],

$$-\tfrac{1}{16\pi G} R \, |\det g_{..}|^{1/2} = \tfrac{1}{16\pi G} \{ \tfrac{1}{4} \partial_\mu h_{\alpha\beta} \partial^\mu h^{\alpha\beta} - \tfrac{1}{8} \partial_\mu h_\alpha{}^\alpha \partial^\mu h_\beta{}^\beta \quad (65)$$
$$- [\partial_\nu h_\mu{}^\nu - \tfrac{1}{2} \partial_\mu h_\nu{}^\nu][\partial_{\nu'} h^{\mu\nu'} - \tfrac{1}{2} \partial^\mu h_{\nu'}{}^{\nu'}] + O(h^3) \}.$$

Here indices are raised with $\eta^{..}$. After an appropriate choice of coordinates, 'harmonic coordinats', the bracket $[\partial_\nu h_\mu{}^\nu - \tfrac{1}{2} \partial_\mu h_\nu{}^\nu]$ vanishes and only two independent components of $h_{\mu\nu}$ remain, $h_{11} = -h_{22}$ and h_{12}. They represent the two physical states of the graviton, helicity ± 2. Their kinetic terms are both positive, e.g.:

$$+\tfrac{1}{16\pi G}\tfrac{1}{4}\partial_\mu h_{12}\partial^\mu h_{12}. \tag{66}$$

Likewise, by an appropriate gauge transformation, we can achieve $\partial_\mu A^\mu = 0$, 'Lorentz gauge', and remain with only two 'transverse' components A_1, A_2 of helicity ± 1. They have positive kinetic terms, e.g.:

$$+\tfrac{1}{2g^2}\operatorname{tr}\left(\partial_\mu A_1^*\partial^\mu A_1\right). \tag{67}$$

Finally, the kinetic term of the scalar is positive:

$$+\tfrac{1}{2}\partial_\mu\varphi^*\partial^\mu\varphi. \tag{68}$$

An old recipe from quantum field theory, 'Wick rotation', amounts to replacing spacetime by a Riemannian manifold with Euclidean signature. Then certain calculations become feasible or easier. One of the reasons for this is that Euclidean quantum field theory resembles statistical mechanics, the imaginary time playing formally the role of the inverse temperature. Only at the end of the calculation the result is 'rotated back' to real time. In some cases, this recipe can be justified rigorously. The precise formulation of the recipe is that the n-point functions computed from the Euclidean Lagrangian be the analytic continuations in the complex time plane of the Minkowskian n-point functions. We shall indicate a hand waving formulation of the recipe, that is sufficient for our purpose: In a first stroke we pass to the signature $-+++$. In a second stroke we replace t by it and replace all Minkowskian scalar products by the corresponding Euclidean ones.

The first stroke amounts simply to replacing the metric by its negative. This leaves invariant the Christoffel symbols, the Riemann and Ricci tensors, but reverses the sign of the curvature scalar. Likewise, in the other terms of the Lagrangian we get a minus sign for every contraction of indices, e.g.: $\partial_\mu\varphi^*\partial^\mu\varphi = \partial_\mu\varphi^*\partial_{\mu'}\varphi g^{\mu\mu'}$ becomes $\partial_\mu\varphi^*\partial_{\mu'}\varphi(-g^{\mu\mu'}) = -\partial_\mu\varphi^*\partial^\mu\varphi$. After multiplication by a conventional overall minus sign the combined Lagrangian reads now,

$$\begin{aligned} \{ \tfrac{2\Lambda_c}{16\pi G} &- \tfrac{1}{16\pi G}\,R + \tfrac{1}{2g^2}\operatorname{tr}\left(F_{\mu\nu}^*F^{\mu\nu}\right) + \tfrac{1}{g^2}m_A^2\operatorname{tr}\left(A_\mu^*A^\mu\right) \\ &+ \tfrac{1}{2}\left(D_\mu\varphi\right)^*D^\mu\varphi + \tfrac{1}{2}m_\varphi^2|\varphi|^2 - \tfrac{1}{2}\mu^2|\varphi|^2 + \lambda|\varphi|^4 \\ &+ \psi^*\gamma^0[i\gamma^\mu D_\mu + m_\psi 1_4]\psi \} \,|\det g_{..}|^{1/2}. \end{aligned} \tag{69}$$

To pass to the Euclidean signature, we multiply time, energy and mass by i. This amounts to $\eta^{\mu\nu} = \delta^{\mu\nu}$ in the scalar product. In order to have the Euclidean anticommutation relations,

$$\gamma^\mu\gamma^\nu + \gamma^\nu\gamma^\mu = 2\delta^{\mu\nu}1_4, \tag{70}$$

we change the Dirac matrices to the Euclidean ones,

$$\gamma^0 = \begin{pmatrix} 0 & -1_2 \\ -1_2 & 0 \end{pmatrix}, \quad \gamma^j = \tfrac{1}{i}\begin{pmatrix} 0 & \sigma_j \\ -\sigma_j & 0 \end{pmatrix}, \tag{71}$$

All four are now self adjoint. For the chirality we take

$$\gamma_5 := \gamma^0\gamma^1\gamma^2\gamma^3 = \begin{pmatrix} -1_2 & 0 \\ 0 & 1_2 \end{pmatrix}. \tag{72}$$

The Minkowskian scalar product for spinors has a γ^0. This γ^0 is needed for the correct physical interpretation of the energy of antiparticles and for invariance under lifted Lorentz transformations, $Spin(1,3)$. In the Euclidean, there is no physical interpretation and we can only retain the requirement of a $Spin(4)$ invariant scalar product. This scalar product has no γ^0. But then we have a problem if we want to write the Dirac Lagrangian in terms of chiral spinors as above. For instance, for a purely left-handed neutrino, $\psi_R = 0$ and $\psi_L^* i\gamma^\mu D_\mu \psi_L$ vanishes identically because γ_5 anticommutes with the four γ^μ. The standard trick of Euclidean field theoreticians [12] is fermion doubling, ψ_L and ψ_R are treated as two *independent*, four component spinors. They are not chiral projections of one four component spinor as in the Minkowskian, equation (59). The spurious degrees of freedom in the Euclidean are kept all the way through the calculation. They are projected out only after the Wick rotation back to Minkowskian, by imposing $\gamma_5\psi_L = -\psi_L, \gamma_5\psi_R = \psi_R$.

In noncommutative geometry the Dirac operator must be self adjoint, which is not the case for the Euclidean Dirac operator $i\gamma^\mu D_\mu + im_\psi 1_4$ we get from the Lagrangian (69) after multiplication of the mass by i. We therefore prefer the primed spinor variables ψ' producing the self adjoint Euclidean Dirac operator $i\gamma^\mu D_\mu + m_\psi \gamma_5$. Dropping the prime, the combined Lagrangian in the Euclidean then reads:

$$\{ \frac{2\Lambda_c}{16\pi G} - \frac{1}{16\pi G} R + \frac{1}{2g^2}\text{tr}\,(F_{\mu\nu}^* F^{\mu\nu}) + \frac{1}{g^2}m_A^2\text{tr}\,(A_\mu^* A^\mu) \tag{73}$$
$$+ \frac{1}{2}(D_\mu\varphi)^* D^\mu\varphi + \frac{1}{2}m_\varphi^2|\varphi|^2 - \frac{1}{2}\mu^2|\varphi|^2 + \lambda|\varphi|^4$$
$$+ \psi_L^* i\gamma^\mu D_\mu \psi_L + \psi_R^* i\gamma^\mu D_\mu \psi_R + m_\psi\psi_L^*\gamma_5\psi_R + m_\psi\psi_R^*\gamma_5\psi_L \} \,(\det g_{..})^{1/2}.$$

4 Connes' Noncommutative Geometry

Connes equips Riemannian spaces with an uncertainty principle. As in quantum mechanics, this uncertainty principle is derived from noncommutativity.

4.1 Motivation: Quantum Mechanics

Consider the classical harmonic oscillator. Its phase space is $\mathbb{R}^2$ with points labelled by position x and momentum p. A classical observable is a differentiable function on phase space such as the total energy $p^2/(2m) + kx^2$. Observables can be added and multiplied, they form the algebra $\mathcal{C}^\infty(\mathbb{R}^2)$, which is associative and commutative. To pass to quantum mechanics, this algebra is rendered noncommutative by means of the following noncommutation relation for the generators x and p,

$$[x, p] = i\hbar 1. \tag{74}$$

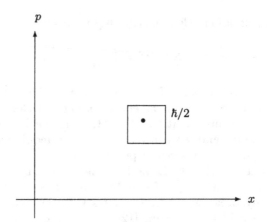

Fig. 4. The first example of noncommutative geometry

Let us call $\mathcal{A}$ the resulting algebra 'of quantum observables'. It is still associative, has an involution $\cdot^*$ (the adjoint or Hermitean conjugation) and a unit 1. Let us briefly recall the defining properties of an involution: it is a linear map from the *real* algebra into itself that reverses the product, $(ab)^* = b^*a^*$, respects the unit, $1^* = 1$, and is such that $a^{**} = a$.

Of course, there is no space anymore of which $\mathcal{A}$ is the algebra of functions. Nevertheless, we talk about such a 'quantum phase space' as a space that has no points or a space with an uncertainty relation. Indeed, the noncommutation relation (74) implies

$$\Delta x \Delta p \geq \hbar/2 \tag{75}$$

and tells us that points in phase space lose all meaning, we can only resolve cells in phase space of volume $\hbar/2$, see Fig. 4. To define the uncertainty Δa for an observable $a \in \mathcal{A}$, we need a faithful representation of the algebra on a Hilbert space, i.e. an injective homomorphism $\rho : \mathcal{A} \to \mathrm{End}(\mathcal{H})$ (cf. Appendix). For the harmonic oscillator, this Hilbert space is $\mathcal{H} = \mathcal{L}^2(\mathbb{R})$. Its elements are the wave functions $\psi(x)$, square integrable functions on configuration space. Finally, the dynamics is defined by a self adjoint observable $H = H^* \in \mathcal{A}$ via Schrödinger's equation

$$\left(i\hbar \frac{\partial}{\partial t} - \rho(H) \right) \psi(t, x) = 0. \tag{76}$$

Usually the representation is not written explicitly. Since it is faithful, no confusion should arise from this abuse. Here time is considered an external parameter, in particular, time is not considered an observable. This is different in the special relativistic setting where Schrödinger's equation is replaced by Dirac's equation,

$$\partial\!\!\!/\psi = 0. \tag{77}$$

Now the wave function ψ is the four-component spinor consisting of left- and right-handed, particle and antiparticle wave functions. The Dirac operator is not in $\mathcal{A}$ anymore, but $\partial\!\!\!/ \in \text{End}(\mathcal{H})$. The Dirac operator is only formally self adjoint because there is no *positive definite* scalar product, whereas in Euclidean spacetime it is truly self adjoint, $\partial\!\!\!/^* = \partial\!\!\!/$.

Connes' geometries are described by these three purely algebraic items, $(\mathcal{A}, \mathcal{H}, \partial\!\!\!/)$, with $\mathcal{A}$ a real, associative, possibly noncommutative involution algebra with unit, faithfully represented on a complex Hilbert space $\mathcal{H}$, and $\partial\!\!\!/$ is a self adjoint operator on $\mathcal{H}$.

4.2   The Calibrating Example: Riemannian Spin Geometry

Connes' geometry [18] does to spacetime what quantum mechanics does to phase space. Of course, the first thing we have to learn is how to reconstruct the Riemannian geometry from the algebraic data $(\mathcal{A}, \mathcal{H}, \partial\!\!\!/)$ in the case where the algebra is commutative. We start the easy way and construct the triple $(\mathcal{A}, \mathcal{H}, \partial\!\!\!/)$ given a four dimensional, compact, Euclidean spacetime M. As before $\mathcal{A} = C^\infty(M)$ is the real algebra of complex valued differentiable functions on spacetime and $\mathcal{H} = \mathcal{L}^2(\mathcal{S})$ is the Hilbert space of complex, square integrable spinors ψ on M. Locally, in any coordinate neighborhood, we write the spinor as a column vector, $\psi(x) \in \mathbb{C}^4$, $x \in M$. The scalar product of two spinors is defined by

$$(\psi, \psi') = \int_M \psi^*(x)\psi'(x)\, dV, \tag{78}$$

with the invariant volume form $dV := |\det g_{..}|^{1/2}\, d^4x$ defined with the metric tensor,

$$g_{\mu\nu} = g\left(\frac{\partial}{\partial x^\mu}, \frac{\partial}{\partial x^\nu}\right), \tag{79}$$

that is the matrix of the Riemannian metric g with respect to the coordinates x^μ, $\mu = 0, 1, 2, 3$. Note – and this is important – that with Euclidean signature the Dirac action is simply a scalar product, $S_\text{D} = (\psi, \partial\!\!\!/\psi)$. The representation is defined by pointwise multiplication, $(\rho(a)\,\psi)(x) := a(x)\psi(x)$, $a \in \mathcal{A}$. For a start, it is sufficient to know the Dirac operator on a flat manifold M and with respect to inertial or Cartesian coordinates $\tilde{x}^\mu$ such that $\tilde{g}_{\tilde\mu\tilde\nu} = \delta^{\tilde\mu}_{\tilde\nu}$. Then we use Dirac's original definition,

$$\mathcal{D} = \partial\!\!\!/ = i\gamma^{\tilde\mu}\partial/\partial\tilde{x}^{\tilde\mu}, \tag{80}$$

with the self adjoint γ-matrices

$$\gamma^0 = \begin{pmatrix} 0 & -1_2 \\ -1_2 & 0 \end{pmatrix}, \quad \gamma^j = \frac{1}{i}\begin{pmatrix} 0 & \sigma_j \\ -\sigma_j & 0 \end{pmatrix}, \tag{81}$$

with the Pauli matrices

$$\sigma_1 = \begin{pmatrix} 0 & 1 \\ 1 & 0 \end{pmatrix}, \quad \sigma_2 = \begin{pmatrix} 0 & -i \\ i & 0 \end{pmatrix}, \quad \sigma_3 = \begin{pmatrix} 1 & 0 \\ 0 & -1 \end{pmatrix}. \tag{82}$$

We will construct the general curved Dirac operator later.

When the dimension of the manifold is even like in our case, the representation ρ is reducible. Its Hilbert space decomposes into left- and right-handed spaces,

$$\mathcal{H} = \mathcal{H}_L \oplus \mathcal{H}_R, \quad \mathcal{H}_L = \frac{1-\chi}{2}\mathcal{H}, \quad \mathcal{H}_R = \frac{1+\chi}{2}\mathcal{H}. \tag{83}$$

Again we make use of the unitary chirality operator,

$$\chi = \gamma_5 := \gamma^0\gamma^1\gamma^2\gamma^3 = \begin{pmatrix} -1_2 & 0 \\ 0 & 1_2 \end{pmatrix}. \tag{84}$$

We will also need the charge conjugation or real structure, the anti-unitary operator:

$$J = C := \gamma^0\gamma^2 \circ \text{complex conjugation} = \begin{pmatrix} 0 & -1 & 0 & 0 \\ 1 & 0 & 0 & 0 \\ 0 & 0 & 0 & 1 \\ 0 & 0 & -1 & 0 \end{pmatrix} \circ c\,c, \tag{85}$$

that permutes particles and antiparticles.

The five items $(\mathcal{A}, \mathcal{H}, \mathcal{D}, J, \chi)$ form what Connes calls an even, real spectral triple [19].

$\mathcal{A}$ is a real, associative involution algebra with unit, represented faithfully by bounded operators on the Hilbert space $\mathcal{H}$.

$\mathcal{D}$ is an unbounded self adjoint operator on $\mathcal{H}$.

J is an anti-unitary operator,

χ a unitary one.

They enjoy the following properties:

- $J^2 = -1$ in four dimensions ($J^2 = 1$ in zero dimensions).
- $[\rho(a), J\rho(\tilde{a})J^{-1}] = 0$ for all $a, \tilde{a} \in \mathcal{A}$.
- $\mathcal{D}J = J\mathcal{D}$, particles and antiparticles have the same dynamics.
- $[\mathcal{D}, \rho(a)]$ is bounded for all $a \in \mathcal{A}$ and $[[\mathcal{D}, \rho(a)], J\rho(\tilde{a})J^{-1}] = 0$ for all $a, \tilde{a} \in \mathcal{A}$. This property is called first order condition because in the calibrating example it states that the genuine Dirac operator is a first order differential operator.
- $\chi^2 = 1$ and $[\chi, \rho(a)] = 0$ for all $a \in \mathcal{A}$. These properties allow the decomposition $\mathcal{H} = \mathcal{H}_L \oplus \mathcal{H}_R$.
- $J\chi = \chi J$.
- $\mathcal{D}\chi = -\chi\mathcal{D}$, chirality does not change under time evolution.
- There are three more properties, that we do not spell out, orientability, which relates the chirality to the volume form, Poincaré duality and regularity, which states that our functions $a \in \mathcal{A}$ are differentiable.

Connes promotes these properties to the axioms defining an even, real spectral triple. These axioms are justified by his

Reconstruction theorem (Connes 1996 [20]): Consider an (even) spectral triple $(\mathcal{A}, \mathcal{H}, \mathcal{D}, J, (\chi))$ whose algebra $\mathcal{A}$ is commutative. Then here exists a compact, Riemannian spin manifold M (of even dimensions), whose spectral triple $(\mathcal{C}^\infty(M), \mathcal{L}^2(S), \partial\!\!\!/, C, (\gamma_5))$ coincides with $(\mathcal{A}, \mathcal{H}, \mathcal{D}, J, (\chi))$.

For details on this theorem and noncommutative geometry in general, I warmly recommend the Costa Rica book [10]. Let us try to get a feeling of the *local* information contained in this theorem. Besides describing the dynamics of the spinor field ψ, the Dirac operator $\partial\!\!\!/$ encodes the dimension of spacetime, its Riemannian metric, its differential forms and its integration, that is all the tools that we need to define a . In Minkowskian signature, the square of the Dirac operator is the wave operator, which in 1+2 dimensions governs the dynamics of a drum. The deep question: 'Can you hear the shape of a drum?' has been raised. This question concerns a global property of spacetime, the boundary. Can you reconstruct it from the spectrum of the wave operator?

The dimension of spacetime is a local property. It can be retrieved from the asymptotic behaviour of the spectrum of the Dirac operator for large eigenvalues. Since M is compact, the spectrum is discrete. Let us order the eigenvalues, $...\lambda_{n-1} \leq \lambda_n \leq \lambda_{n+1}...$ Then states that the eigenvalues grow asymptotically as $n^{1/\dim M}$. To explore a local property of spacetime we only need the high energy part of the spectrum. This is in nice agreement with our intuition from quantum mechanics and motivates the name 'spectral triple'.

The metric can be reconstructed from the commutative spectral triple by Connes distance formula (86) below. In the commutative case a point $x \in M$ is reconstructed as the pure state. The general definition of a pure state of course does not use the commutativity. A state δ of the algebra $\mathcal{A}$ is a linear form on $\mathcal{A}$, that is normalized, $\delta(1) = 1$, and positive, $\delta(a^*a) \geq 0$ for all $a \in \mathcal{A}$. A state is pure if it cannot be written as a linear combination of two states. For the calibrating example, there is a one–to–one correspondence between points $x \in M$ and pure states δ_x defined by the Dirac distribution, $\delta_x(a) := a(x) = \int_M \delta_x(y)a(y)\mathrm{d}^4y$. The geodesic distance between two points x and y is reconstructed from the triple as:

$$\sup\{|\delta_x(a) - \delta_y(a)|;\ a \in \mathcal{C}^\infty(M) \text{ such that } ||[\partial\!\!\!/, \rho(a)]|| \leq 1\}. \qquad (86)$$

For the calibrating example, $[\partial\!\!\!/, \rho(a)]$ is a bounded operator. Indeed, $[\partial\!\!\!/, \rho(a)]$ $\psi = i\gamma^\mu\partial_\mu(a\psi) - ia\gamma^\mu\partial_\mu\psi = i\gamma^\mu(\partial_\mu a)\psi$, and $\partial_\mu a$ is bounded as a differentiable function on a compact space.

For a general spectral triple this operator is bounded by axiom. In any case, the operator norm $||[\partial\!\!\!/, \rho(a)]||$ in the distance formula is finite.

Consider the circle, $M = S^1$, of circumference 2π with Dirac operator $\partial\!\!\!/ = i\,\mathrm{d}/\mathrm{d}x$. A function $a \in \mathcal{C}^\infty(S^1)$ is represented faithfully on a wave-function $\psi \in \mathcal{L}^2(S^1)$ by pointwise multiplication, $(\rho(a)\psi)(x) = a(x)\psi(x)$. The commutator $[\partial\!\!\!/, \rho(a)] = i\rho(a')$ is familiar from quantum mechanics. Its operator norm is $||[\partial\!\!\!/, \rho(a)]|| := \sup_\psi ||[\partial\!\!\!/, \rho(a)]\psi|/|\psi| = \sup_x |a'(x)|$, with $|\psi|^2 = \int_0^{2\pi} \bar\psi(x)\psi(x)\,\mathrm{d}x$. Therefore, the distance between two points x and y on the circle is

$$\sup_a\{|a(x) - a(y)|;\ \sup_x |a'(x)| \leq 1\} = |x - y|. \qquad (87)$$

Note that Connes' distance formula continues to make sense for non-connected manifolds, like discrete spaces of dimension zero, i.e. collections of points.

Differential forms, for example of degree one like da for a function $a \in \mathcal{A}$, are reconstructed as $(-i)[\partial\!\!\!/, \rho(a)]$. This is again motivated from quantum mechanics. Indeed in a 1+0 dimensional spacetime da is just the time derivative of the 'observable' a and is associated with the commutator of the Hamilton operator with a.

Motivated from quantum mechanics, we define a noncommutative geometry by a real spectral triple with noncommutative algebra $\mathcal{A}$.

4.3   Spin Groups

Let us go back to quantum mechanics of spin and recall how a space rotation acts on a spin $\frac{1}{2}$ particle. For this we need group homomorphisms between the rotation group $SO(3)$ and the probability preserving unitary group $SU(2)$. We construct first the group homomorphism

$$\begin{aligned} p: SU(2) &\longrightarrow SO(3) \\ U &\longmapsto p(U). \end{aligned}$$

With the help of the auxiliary function

$$f : \mathbb{R}^3 \longrightarrow su(2)$$

$$\boldsymbol{x} = \begin{pmatrix} x^1 \\ x^2 \\ x^3 \end{pmatrix} \longmapsto -\tfrac{1}{2} i x^j \sigma_j \,,$$

we define the rotation $p(U)$ by

$$p(U)\boldsymbol{x} := f^{-1}(U f(\boldsymbol{x}) U^{-1}). \tag{88}$$

The conjugation by the unitary U will play an important role and we give it a special name, $i_U(w) := U w U^{-1}$, i for inner. Since $i_{(-U)} = i_U$, the projection p is two to one, $\mathrm{Ker}(p) = \{\pm 1\}$. Therefore the spin lift

$$\begin{aligned} L: SO(3) &\longrightarrow SU(2) \\ R = \exp(\omega) &\longmapsto \exp(\tfrac{1}{8}\omega^{jk}[\sigma_j, \sigma_k]) \end{aligned} \tag{89}$$

is double-valued. It is a local group homomorphism and satisfies $p(L(R)) = R$. Its double-valuedness is accessible to quantum mechanical experiments: neutrons have to be rotated through an angle of 720° before interference patterns repeat [21].

The lift L was generalized by Dirac to the special relativistic setting, e.g. [4], and by E. Cartan [22] to the general relativistic setting. Connes [23] generalizes it to noncommutative geometry, see Fig. 5. The transformations we need to lift are

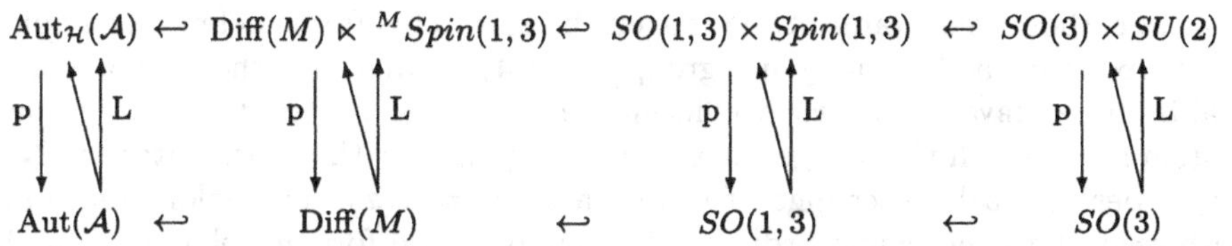

$$\text{Aut}_{\mathcal{H}}(\mathcal{A}) \hookleftarrow \text{Diff}(M) \ltimes {}^{M}Spin(1,3) \hookleftarrow SO(1,3) \times Spin(1,3) \hookleftarrow SO(3) \times SU(2)$$

$$\begin{array}{cccc} p\Big\downarrow \Big\Vert L & p\Big\downarrow \Big\Vert L & p\Big\downarrow \Big\Vert L & p\Big\downarrow \Big\Vert L \end{array}$$

$$\text{Aut}(\mathcal{A}) \hookleftarrow \quad \text{Diff}(M) \quad \hookleftarrow \quad SO(1,3) \quad \hookleftarrow \quad SO(3)$$

Fig. 5. The nested spin lifts of Connes, Cartan, Dirac, and Pauli

Lorentz transformations in special relativity, and general coordinate transformations in general relativity, i.e. our calibrating example. The latter transformations are the local elements of the diffeomorphism group $\text{Diff}(M)$. In the setting of noncommutative geometry, this group is the group of algebra automorphisms $\text{Aut}(\mathcal{A})$. Indeed, in the calibrating example we have $\text{Aut}(\mathcal{A})=\text{Diff}(M)$. In order to generalize the spin group to spectral triples, Connes defines the receptacle of the group of 'lifted automorphisms',

$$\text{Aut}_{\mathcal{H}}(\mathcal{A}) := \{U \in \text{End}(\mathcal{H}), \ UU^* = U^*U = 1, \ UJ = JU, \ U\chi = \chi U,$$
$$i_U \in \text{Aut}(\rho(\mathcal{A}))\}. \tag{90}$$

The first three properties say that a lifted automorphism U preserves probability, charge conjugation, and chirality. The fourth, called *covariance property*, allows to define the projection $p: \ \text{Aut}_{\mathcal{H}}(\mathcal{A}) \longrightarrow \text{Aut}(\mathcal{A})$ by

$$p(U) = \rho^{-1}i_U\rho \tag{91}$$

We will see that the covariance property will protect the locality of field theory. For the calibrating example of a four dimensional spacetime, a local calculation, i.e. in a coordinate patch, that we still denote by M, yields the semi-direct product (cf. Appendix) of diffeomorphisms with local or gauged spin transformations, $\text{Aut}_{\mathcal{L}^2(S)}(\mathcal{C}^\infty(M)) = \text{Diff}(M) \ltimes {}^{M}Spin(4)$. We say receptacle because already in six dimensions, $\text{Aut}_{\mathcal{L}^2(S)}(\mathcal{C}^\infty(M))$ is larger than $\text{Diff}(M) \ltimes {}^{M}Spin(6)$. However we can use the lift L with $p(L(\sigma)) = \sigma$, $\sigma \in \text{Aut}(\mathcal{A})$ to correctly identify the spin group in any dimension of M. Indeed we will see that the spin group is the image of the spin lift $L(\text{Aut}(\mathcal{A}))$, in general a proper subgroup of the receptacle $\text{Aut}_{\mathcal{H}}(\mathcal{A})$.

Let σ be a diffeomorphism close to the identity. We interpret σ as coordinate transformation, all our calculations will be local, M standing for one chart, on which the coordinate systems $\tilde{x}^\mu$ and $x^\mu = (\sigma(\tilde{x}))^\mu$ are defined. We will work out the local expression of a lift of σ to the Hilbert space of spinors. This lift $U = L(\sigma)$ will depend on the metric and on the initial coordinate system $\tilde{x}^\mu$.

In a first step, we construct a group homomorphism $\Lambda : \text{Diff}(M) \to \text{Diff}(M) \ltimes {}^{M}SO(4)$ into the group of local 'Lorentz' transformations, i.e. the group of differentiable functions from spacetime into $SO(4)$ with pointwise multiplication. Let $(\tilde{e}^{-1}(\tilde{x}))^{\tilde{\mu}}{}_a = (\tilde{g}^{-1/2}(\tilde{x}))^{\tilde{\mu}}{}_a$ be the inverse of the square root of the positive matrix $\tilde{g}$ of the metric with respect to the initial coordinate system $\tilde{x}^\mu$. Then the four vector fields $\tilde{e}_a$, $a = 0, 1, 2, 3$, defined by

$$\tilde{e}_a := (\tilde{e}^{-1})^{\tilde{\mu}}{}_a \frac{\partial}{\partial\tilde{x}^{\tilde{\mu}}} \tag{92}$$

give an orthonormal frame of the tangent bundle. This frame defines a complete gauge fixing of the Lorentz gauge group $^M SO(4)$ because it is the only orthonormal frame to have symmetric coefficients $(\tilde{e}^{-1})^{\tilde{\mu}}{}_a$ with respect to the coordinate system $\tilde{x}^{\tilde{\mu}}$. We call this gauge the symmetric gauge for the coordinates $\tilde{x}^{\tilde{\mu}}$. Now let us perform a local change of coordinates, $x = \sigma(\tilde{x})$. The holonomic frame with respect to the new coordinates is related to the former holonomic one by the inverse Jacobian matrix of σ

$$\frac{\partial}{\partial x^\mu} = \frac{\partial \tilde{x}^{\tilde{\mu}}}{\partial x^\mu} \frac{\partial}{\partial \tilde{x}^{\tilde{\mu}}} = (\mathcal{J}^{-1})^{\tilde{\mu}}{}_\mu \frac{\partial}{\partial \tilde{x}^{\tilde{\mu}}}, \quad (\mathcal{J}^{-1}(x))^{\tilde{\mu}}{}_\mu = \frac{\partial \tilde{x}^{\tilde{\mu}}}{\partial x^\mu}. \tag{93}$$

The matrix g of the metric with respect to the new coordinates reads,

$$g_{\mu\nu}(x) := g\left(\frac{\partial}{\partial x^\mu}, \frac{\partial}{\partial x^\nu} \right)\bigg|_x = \left(\mathcal{J}^{-1T}(x)\tilde{g}(\sigma^{-1}(x))\mathcal{J}^{-1}(x) \right)_{\mu\nu}, \tag{94}$$

and the symmetric gauge for the new coordinates x is the new orthonormal frame

$$e_b = e^{-1\mu}{}_b \frac{\partial}{\partial x^\mu} = g^{-1/2\,\mu}{}_b \mathcal{J}^{-1\,\tilde{\mu}}{}_\mu \frac{\partial}{\partial \tilde{x}^{\tilde{\mu}}} = \left(\mathcal{J}^{-1}\sqrt{\mathcal{J}\tilde{g}^{-1}\mathcal{J}^T} \right)^{\tilde{\mu}}{}_b \frac{\partial}{\partial \tilde{x}^{\tilde{\mu}}}. \tag{95}$$

New and old orthonormal frames are related by a Lorentz transformation Λ, $e_b = \Lambda^{-1\,a}{}_b \tilde{e}_a$, with

$$\Lambda(\sigma)|_{\tilde{x}} = \sqrt{\mathcal{J}^{-1T}\tilde{g}\mathcal{J}^{-1}}\bigg|_{\sigma(\tilde{x})} \mathcal{J}|_{\tilde{x}} \sqrt{\tilde{g}^{-1}}\bigg|_{\tilde{x}} = \sqrt{g}\mathcal{J}\sqrt{\tilde{g}^{-1}}. \tag{96}$$

If M is flat and $\tilde{x}^{\tilde{\mu}}$ are 'inertial' coordinates, i.e. $\tilde{g}_{\tilde{\mu}\tilde{\nu}} = \delta^{\tilde{\mu}}{}_{\tilde{\nu}}$, and σ is a local isometry then $\mathcal{J}(\tilde{x}) \in SO(4)$ for all $\tilde{x}$ and $\Lambda(\sigma) = \mathcal{J}$. In special relativity, therefore, the symmetric gauge ties together Lorentz transformations in spacetime with Lorentz transformations in the tangent spaces.

In general, if the coordinate transformation σ is close to the identity, so is its Lorentz transformation $\Lambda(\sigma)$ and it can be lifted to the spin group,

$$S : SO(4) \longrightarrow Spin(4)$$
$$\Lambda = \exp\omega \longmapsto \exp\left[\tfrac{1}{4}\omega_{ab}\gamma^{ab}\right] \tag{97}$$

with $\omega = -\omega^T \in so(4)$ and $\gamma^{ab} := \tfrac{1}{2}[\gamma^a, \gamma^b]$. With our choice (81) for the γ matrices, we have

$$\gamma^{0j} = i\begin{pmatrix} -\sigma_j & 0 \\ 0 & \sigma_j \end{pmatrix}, \quad \gamma^{jk} = i\epsilon^{jk\ell}\begin{pmatrix} \sigma_\ell & 0 \\ 0 & \sigma_\ell \end{pmatrix}, \quad j,k = 1,2,3, \quad \epsilon^{123} = 1. \tag{98}$$

We can write the local expression [24] of the lift $L : \text{Diff}(M) \to \text{Diff}(M) \ltimes {}^M Spin(4)$,

$$(L(\sigma)\psi)(x) = S(\Lambda(\sigma))|_{\sigma^{-1}(x)} \psi(\sigma^{-1}(x)). \tag{99}$$

L is a double-valued group homomorphism. For any σ close to the identity, $L(\sigma)$ is unitary, commutes with charge conjugation and chirality, satisfies the covariance property, and $p(L(\sigma)) = \sigma$. Therefore, we have locally

$$L(\text{Diff}(M)) \subset \text{Diff}(M) \ltimes {}^M Spin(4) = \text{Aut}_{\mathcal{L}^2(S)}(\mathcal{C}^\infty(M)). \tag{100}$$

The symmetric gauge is a complete gauge fixing and this reduction follows Einstein's spirit in the sense that the only arbitrary choice is the one of the initial coordinate system $\tilde{x}^{\mu}$ as will be illustrated in the next section. Our computations are deliberately local. The global picture can be found in reference [25].

5   The Spectral Action

5.1   Repeating Einstein's Derivation in the Commutative Case

We are ready to parallel Einstein's derivation of general relativity in Connes' language of spectral triples. The associative algebra $C^{\infty}(M)$ is commutative, but this property will never be used. As a by-product, the lift L will reconcile Einstein's and Cartan's formulations of general relativity and it will yield a self contained introduction to Dirac's equation in a gravitational field accessible to particle physicists. For a comparison of Einstein's and Cartan's formulations of general relativity see for example [6].

First Stroke: Kinematics. Instead of a point-particle, Connes takes as matter a field, the free, massless Dirac particle $\psi(\tilde{x})$ in the flat spacetime of special relativity. In inertial coordinates $\tilde{x}^{\mu}$, its dynamics is given by the Dirac equation,

$$\tilde{\partial\!\!\!/}\psi = i\delta^{\tilde{\mu}}{}_a \gamma^a \frac{\partial}{\partial \tilde{x}^{\tilde{\mu}}}\, \psi = 0. \tag{101}$$

We have written $\delta^{\tilde{\mu}}{}_a \gamma^a$ instead of $\gamma^{\tilde{\mu}}$ to stress that the γ matrices are $\tilde{x}$-independent. This Dirac equation is covariant under Lorentz transformations. Indeed if σ is a local isometry then

$$L(\sigma)\, \tilde{\partial\!\!\!/} L(\sigma)^{-1} = \partial\!\!\!/ = i\delta^{\mu}{}_a \gamma^a \frac{\partial}{\partial x^{\mu}}. \tag{102}$$

To prove this special relativistic covariance, one needs the identity $S(\Lambda)\gamma^a S(\Lambda)^{-1}$ $= \Lambda^{-1}{}^a{}_b \gamma^b$ for Lorentz transformations $\Lambda \in SO(4)$ close to the identity. Take a general coordinate transformation σ close to the identity. Now comes a long, but straightforward calculation. It is a useful exercise requiring only matrix multiplication and standard calculus, Leibniz and chain rules. Its result is the Dirac operator in curved coordinates,

$$L(\sigma)\, \tilde{\partial\!\!\!/} L(\sigma)^{-1} = \partial\!\!\!/ = ie^{-1\,\mu}{}_a \gamma^a \left[\frac{\partial}{\partial x^{\mu}} + s(\omega_{\mu}) \right], \tag{103}$$

where $e^{-1} = \sqrt{JJ^T}$ is a symmetric matrix,

$$\begin{aligned} s : so(4) &\longrightarrow spin(4) \\ \omega &\longmapsto \tfrac{1}{4}\omega_{ab}\gamma^{ab} \end{aligned} \tag{104}$$

is the Lie algebra isomorphism corresponding to the lift (97) and

$$\omega_{\mu}(x) = \Lambda|_{\sigma^{-1}(x)}\, \partial_{\mu}\, \Lambda^{-1}\big|_x. \tag{105}$$

The 'spin connection' ω is the gauge transform of the Levi–Civita connection Γ, the latter is expressed with respect to the holonomic frame ∂_μ, the former is written with respect to the orthonormal frame $e_a = e^{-1\mu}{}_a \partial_\mu$. The gauge transformation passing between them is $e \in {}^M GL_4$,

$$\omega = e\Gamma e^{-1} + e\,de^{-1}. \tag{106}$$

We recover the well known explicit expression

$$\omega^a{}_{b\mu}(e) = \tfrac{1}{2}\left[(\partial_\beta e^a{}_\mu) - (\partial_\mu e^a{}_\beta) + e^m{}_\mu(\partial_\beta e^m{}_a)e^{-1\,\alpha}{}_a\right]e^{-1\,\beta}{}_b \; - \; [a \leftrightarrow b] \tag{107}$$

of the spin connection in terms of the first derivatives of $e^a{}_\mu = \sqrt{g}^a{}_\mu$. Again the spin connection has zero curvature and the equivalence principle relaxes this constraint. But now equation (103) has an advantage over its analogue (2). Thanks to Connes' distance formula (86), the metric can be read explicitly in (103) from the matrix of functions $e^{-1\mu}{}_a$, while in (2) first derivatives of the metric are present. We are used to this nuance from electro–magnetism, where the classical particle feels the force while the quantum particle feels the potential. In Einstein's approach, the zero connection fluctuates, in Connes' approach, the flat metric fluctuates. This means that the constraint $e^{-1} = \sqrt{JJ^T}$ is relaxed and e^{-1} now is an arbitrary symmetric matrix depending smoothly on x.

Let us mention two experiments with neutrons confirming the 'Minimal coupling' of the Dirac operator to curved coordinates, equation (103). The first takes place in flat spacetime. The neutron interferometer is mounted on a loud speaker and shaken periodically [26]. The resulting pseudo forces coded in the spin connection do shift the interference patterns observed. The second experiment takes place in a true gravitational field in which the neutron interferometer is placed [27]. Here shifts of the interference patterns are observed that do depend on the gravitational *potential*, $e^a{}_\mu$ in equation (103).

Second Stroke: Dynamics. The second stroke, the covariant dynamics for the new class of Dirac operators $\partial\!\!\!/$ is due to Chamseddine & Connes [28]. It is the celebrated spectral action. The beauty of their approach to general relativity is that it works precisely because the Dirac operator $\partial\!\!\!/$ plays two roles simultaneously, it defines the dynamics of matter and the kinematics of gravity. For a discussion of the transformation passing from the metric to the Dirac operator I recommend the article [29] by Landi & Rovelli.

The starting point of Chamseddine & Connes is the simple remark that the spectrum of the Dirac operator is invariant under diffeomorphisms interpreted as general coordinate transformations. From $\partial\!\!\!/\chi = -\chi\,\partial\!\!\!/$ we know that the spectrum of $\partial\!\!\!/$ is even. Indeed, for every eigenvector ψ of $\partial\!\!\!/$ with eigenvalue E, $\chi\psi$ is eigenvector with eigenvalue $-E$. We may therefore consider only the spectrum of the positive operator $\partial\!\!\!/^2/\Lambda^2$ where we have divided by a fixed arbitrary energy scale to make the spectrum dimensionless. If it was not divergent the trace $\operatorname{tr}\partial\!\!\!/^2/\Lambda^2$ would be a general relativistic action functional. To make it convergent, take a differentiable function $f : \mathbb{R}_+ \to \mathbb{R}_+$ of sufficiently fast decrease such that the action

$$S_{CC} := \operatorname{tr} f(\partial\!\!\!/^2/\Lambda^2) \tag{108}$$

converges. It is still a diffeomorphism invariant action. The following theorem, also known as heat kernel expansion, is a local version of an index theorem [30], that as explained in Jean Zinn-Justin's lectures [12] is intimately related to Feynman graphs with one fermionic loop.

Theorem: Asymptotically for high energies, the spectral action is

$$S_{CC} = \tag{109}$$
$$\int_M [\tfrac{2\Lambda_c}{16\pi G} - \tfrac{1}{16\pi G}R + a(5\,R^2 - 8\,\mathrm{Ricci}^2 - 7\,\mathrm{Riemann}^2)]\,dV\, + O(\Lambda^{-2}),$$

where the cosmological constant is $\Lambda_c = \tfrac{6f_0}{f_2}\Lambda^2$, Newton's constant is $G = \tfrac{3\pi}{f_2}\Lambda^{-2}$ and $a = \tfrac{f_4}{5760\pi^2}$. On the right-hand side of the theorem we have omitted surface terms, that is terms that do not contribute to the Euler–Lagrange equations. The Chamseddine–Connes action is universal in the sense that the 'cut off' function f only enters through its first three 'moments', $f_0 := \int_0^\infty u f(u)\,du$, $f_2 := \int_0^\infty f(u)\,du$ and $f_4 = f(0)$.

If we take for f a differentiable approximation of the characteristic function of the unit interval, $f_0 = 1/2$, $f_2 = f_4 = 1$, then the spectral action just counts the number of eigenvalues of the Dirac operator whose absolute values are below the 'cut off' Λ. In four dimensions, the minimax example is the flat 4-torus with all circumferences measuring 2π. Denote by $\psi_B(x)$, $B = 1, 2, 3, 4$, the four components of the spinor. The Dirac operator is

$$\partial\!\!\!/ = \begin{pmatrix} 0 & 0 & -i\partial_0 + \partial_3 & \partial_1 - i\partial_2 \\ 0 & 0 & \partial_1 + i\partial_2 & -i\partial_0 - \partial_3 \\ -i\partial_0 - \partial_3 & -\partial_1 + i\partial_2 & 0 & 0 \\ -\partial_1 - i\partial_2 & -i\partial_0 + \partial_3 & 0 & 0 \end{pmatrix}. \tag{110}$$

After a Fourier transform

$$\psi_B(x) =: \sum_{j_0,\dots,j_3 \in \mathbb{Z}} \hat\psi_B(j_0, \dots, j_3) \exp(-ij_\mu x^\mu), \quad B = 1, 2, 3, 4 \tag{111}$$

the eigenvalue equation $\partial\!\!\!/\,\psi = \lambda\psi$ reads

$$\begin{pmatrix} 0 & 0 & -j_0 - ij_3 & -ij_1 - j_2 \\ 0 & 0 & -ij_1 + j_2 & -j_0 + ij_3 \\ -j_0 + ij_3 & ij_1 + j_2 & 0 & 0 \\ ij_1 - j_2 & -j_0 - ij_3 & 0 & 0 \end{pmatrix} \begin{pmatrix} \hat\psi_1 \\ \hat\psi_2 \\ \hat\psi_3 \\ \hat\psi_4 \end{pmatrix} = \lambda \begin{pmatrix} \hat\psi_1 \\ \hat\psi_2 \\ \hat\psi_3 \\ \hat\psi_4 \end{pmatrix}. \tag{112}$$

Its characteristic equation is $\left[\lambda^2 - (j_0^2 + j_1^2 + j_2^2 + j_3^2)\right]^2 = 0$ and for fixed j_μ, each eigenvalue $\lambda = \pm\sqrt{j_0^2 + j_1^2 + j_2^2 + j_3^2}$ has multiplicity two. Therefore asymptotically for large Λ there are $4B_4\Lambda^4$ eigenvalues (counted with their multiplicity) whose absolute values are smaller than Λ. $B_4 = \pi^2/2$ denotes the volume of the unit ball in $\mathbb{R}^4$. En passant, we check . Let us arrange the absolute values of the eigenvalues in an increasing sequence and number them by naturals n, taking due account of their multiplicities. For large n, we have

$$|\lambda_n| \approx \left(\frac{n}{2\pi^2}\right)^{1/4}. \tag{113}$$

The exponent is indeed the inverse dimension. To check the heat kernel expansion, we compute the right-hand side of equation (110):

$$S_{\mathrm{CC}} = \int_M \frac{\Lambda_c}{8\pi G}\, dV = (2\pi)^4 \, \tfrac{f_0}{4\pi^2} \Lambda^4 = 2\pi^2 \Lambda^4, \tag{114}$$

which agrees with the asymptotic count of eigenvalues, $4B_4\Lambda^4$. This example was the flat torus. Curvature will modify the spectrum and this modification can be used to measure the curvature = gravitational field, exactly as the Zeemann or Stark effect measures the electro–magnetic field by observing how it modifies the spectral lines of an atom.

In the spectral action, we find the Einstein–Hilbert action, which is linear in curvature. In addition, the spectral action contains terms quadratic in the curvature. These terms can safely be neglected in weak gravitational fields like in our solar system. In homogeneous, isotropic cosmologies, these terms are a surface term and do not modify Einstein's equation. Nevertheless the quadratic terms render the (Euclidean) Chamseddine–Connes action positive. Therefore this action has minima. For instance, the 4-sphere with a radius of the order of the Planck length $\sqrt{G}$ is a minimum, a 'ground state'. This minimum breaks the diffeomorphism group spontaneously [23] down to the isometry group $SO(5)$. The little group is the isometry group, consisting of those lifted automorphisms that commute with the Dirac operator $\partial\!\!\!/$. Let us anticipate that the spontaneous symmetry breaking via the Higgs mechanism will be a mirage of this gravitational break down. Physically this ground state seems to regularize the initial cosmological singularity with its ultra strong gravitational field in the same way in which quantum mechanics regularizes the Coulomb singularity of the hydrogen atom.

We close this subsection with a technical remark. We noticed that the matrix $e^{-1\,\mu}{}_a$ in equation (103) is symmetric. A general, not necessarily symmetric matrix $\hat{e}^{-1\,\mu}{}_a$ can be obtained from a general Lorentz transformation $\Lambda \in {}^M SO(4)$:

$$e^{-1\,\mu}{}_a \Lambda^a{}_b = \hat{e}^{-1\,\mu}{}_b, \tag{115}$$

which is nothing but the polar decomposition of the matrix $\hat{e}^{-1}$. These transformations are the gauge transformations of general relativity in Cartan's formulation. They are invisible in Einstein's formulation because of the complete (symmetric) gauge fixing coming from the initial coordinate system $\tilde{x}^{\tilde{\mu}}$.

5.2   Almost Commutative Geometry

We are eager to see the spectral action in a noncommutative example. Technically the simplest noncommutative examples are almost commutative. To construct the latter, we need a natural property of spectral triples, commutative or not: The tensor product of two even spectral triples is an even spectral triple. If both are commutative, i.e. describing two manifolds, then their tensor product simply describes the direct product of the two manifolds.

Let $(\mathcal{A}_i, \mathcal{H}_i, \mathcal{D}_i, J_i, \chi_i)$, $i = 1, 2$ be two even, real spectral triples of even dimensions d_1 and d_2. Their tensor product is the triple $(\mathcal{A}_t, \mathcal{H}_t, \mathcal{D}_t, J_t, \chi_t)$ of dimension $d_1 + d_2$ defined by

$$
\begin{aligned}
\mathcal{A}_t &= \mathcal{A}_1 \otimes \mathcal{A}_2, \quad \mathcal{H}_t = \mathcal{H}_1 \otimes \mathcal{H}_2, \\
\mathcal{D}_t &= \mathcal{D}_1 \otimes 1_2 + \chi_1 \otimes \mathcal{D}_2, \\
J_t &= J_1 \otimes J_2, \quad \chi_t = \chi_1 \otimes \chi_2.
\end{aligned}
$$

The other obvious choice for the Dirac operator, $\mathcal{D}_1 \otimes \chi_2 + 1_1 \otimes \mathcal{D}_2$, is unitarily equivalent to the first one. By definition, an almost commutative geometry is a tensor product of two spectral triples, the first triple is a 4-dimensional spacetime, the calibrating example,

$$
\left(C^\infty(M), \mathcal{L}^2(S), \partial\!\!\!/, C, \gamma_5 \right), \tag{116}
$$

and the second is 0-dimensional. In accordance with , a 0-dimensional spectral triple has a finite dimensional algebra and a finite dimensional Hilbert space. We will label the second triple by the subscript $\cdot_f$ (for finite) rather than by $\cdot_2$. The origin of the word almost commutative is clear: we have a tensor product of an infinite dimensional commutative algebra with a finite dimensional, possibly noncommutative algebra.

This tensor product is, in fact, already familiar to you from the quantum mechanics of spin, whose Hilbert space is the infinite dimensional Hilbert space of square integrable functions on configuration space tensorized with the 2-dimensional Hilbert space $\mathbb{C}^2$ on which acts the noncommutative algebra of spin observables. It is the algebra $\mathbb{H}$ of quaternions, 2×2 complex matrices of the form $\begin{pmatrix} x & -\bar{y} \\ y & \bar{x} \end{pmatrix}$ $x, y \in \mathbb{C}$. A basis of $\mathbb{H}$ is given by $\{1_2, i\sigma_1, i\sigma_2, i\sigma_3\}$, the identity matrix and the three Pauli matrices (82) times i. The group of unitaries of $\mathbb{H}$ is $SU(2)$, the spin cover of the rotation group, the group of automorphisms of $\mathbb{H}$ is $SU(2)/\mathbb{Z}_2$, the rotation group.

A commutative 0-dimensional or finite spectral triple is just a collection of points, for examples see [31]. The simplest example is the two-point space,

$$
\mathcal{A}_f = \mathbb{C}_L \oplus \mathbb{C}_R \ni (a_L, a_R), \quad \mathcal{H}_f = \mathbb{C}^4,
$$

$$
\rho_f(a_L, a_R) = \begin{pmatrix} a_L & 0 & 0 & 0 \\ 0 & a_R & 0 & 0 \\ 0 & 0 & \bar{a}_R & 0 \\ 0 & 0 & 0 & \bar{a}_R \end{pmatrix}, \quad \mathcal{D}_f = \begin{pmatrix} 0 & m & 0 & 0 \\ \bar{m} & 0 & 0 & 0 \\ 0 & 0 & 0 & \bar{m} \\ 0 & 0 & m & 0 \end{pmatrix}, \quad m \in \mathbb{C},
$$

$$
J_f = \begin{pmatrix} 0 & 1_2 \\ 1_2 & 0 \end{pmatrix} \circ c\, c, \quad \chi_f = \begin{pmatrix} -1 & 0 & 0 & 0 \\ 0 & 1 & 0 & 0 \\ 0 & 0 & -1 & 0 \\ 0 & 0 & 0 & 1 \end{pmatrix}. \tag{117}
$$

The algebra has two points = pure states, δ_L and δ_R, $\delta_L(a_L, a_R) = a_L$. By Connes' formula (86), the distance between the two points is $1/|m|$. On the other hand $\mathcal{D}_t = \partial\!\!\!/ \otimes 1_4 + \gamma_5 \otimes \mathcal{D}_f$ is precisely the free massive Euclidean Dirac operator. It describes one Dirac spinor of mass $|m|$ together with its antiparticle. The tensor product of the calibrating example and the two point space is the two-sheeted universe, two identical spacetimes at constant distance. It was the first example in noncommutative geometry to exhibit spontaneous symmetry breaking [32,33].

One of the major advantages of the algebraic description of space in terms of a spectral triple, commutative or not, is that continuous and discrete spaces are included in the same picture. We can view almost commutative geometries as Kaluza–Klein models [34] whose fifth dimension is discrete. Therefore we will also call the finite spectral triple 'internal space'. In noncommutative geometry, 1-forms are naturally defined on discrete spaces where they play the role of connections. In almost commutative geometry, these discrete, internal connections will turn out to be the Higgs scalars responsible for spontaneous symmetry breaking.

Almost commutative geometry is an ideal playground for the physicist with low culture in mathematics that I am. Indeed Connes' reconstruction theorem immediately reduces the infinite dimensional, commutative part to Riemannian geometry and we are left with the internal space, which is accessible to anybody mastering matrix multiplication. In particular, we can easily make precise the last three axioms of spectral triples: orientability, Poincaré duality and regularity. In the finite dimensional case – let us drop the $\cdot_f$ from now on – orientability means that the chirality can be written as a finite sum,

$$\chi = \sum_j \rho(a_j) J\rho(\tilde{a}_j)J^{-1}, \quad a_j, \tilde{a}_j \in \mathcal{A}. \tag{118}$$

The Poincaré duality says that the intersection form

$$\cap_{ij} := \mathrm{tr}\left[\chi\,\rho(p_i)\,J\rho(p_j)J^{-1}\right] \tag{119}$$

must be non-degenerate, where the p_j are a set of minimal projectors of $\mathcal{A}$. Finally, there is the regularity condition. In the calibrating example, it ensures that the algebra elements, the functions on spacetime M, are not only continuous but differentiable. This condition is of course empty for finite spectral triples.

Let us come back to our finite, commutative example. The two-point space is orientable, $\chi = \rho(-1, 1)J\rho(-1, 1)J^{-1}$. It also satisfies Poincaré duality, there are two minimal projectors, $p_1 = (1, 0)$, $p_2 = (0, 1)$, and the intersection form is

$$\cap = \begin{pmatrix} 0 & -1 \\ -1 & 2 \end{pmatrix}.$$

5.3   The Minimax Example

It is time for a noncommutative internal space, a mild variation of the two point space:

$$\mathcal{A} = \mathbb{H} \oplus \mathbb{C} \ni (a,b), \quad \mathcal{H} = \mathbb{C}^6, \quad \rho(a,b) = \begin{pmatrix} a & 0 & 0 & 0 \\ 0 & \bar{b} & 0 & 0 \\ 0 & 0 & b1_2 & 0 \\ 0 & 0 & 0 & b \end{pmatrix}, \quad (120)$$

$$\tilde{\mathcal{D}} = \begin{pmatrix} 0 & \mathcal{M} & 0 & 0 \\ \mathcal{M}^* & 0 & 0 & 0 \\ 0 & 0 & 0 & \bar{\mathcal{M}} \\ 0 & 0 & \bar{\mathcal{M}}^* & 0 \end{pmatrix}, \quad \mathcal{M} = \begin{pmatrix} 0 \\ m \end{pmatrix}, \quad m \in \mathbb{C}, \quad (121)$$

$$J = \begin{pmatrix} 0 & 1_3 \\ 1_3 & 0 \end{pmatrix} \circ c\,c, \quad \chi = \begin{pmatrix} -1_2 & 0 & 0 & 0 \\ 0 & 1 & 0 & 0 \\ 0 & 0 & -1_2 & 0 \\ 0 & 0 & 0 & 1 \end{pmatrix}. \quad (122)$$

The unit is $(1_2, 1)$ and the involution is $(a,b)^* = (a^*, \bar{b})$, where a^* is the Hermitean conjugate of the quaternion a. The Hilbert space now contains one massless, left-handed Weyl spinor and one Dirac spinor of mass $|m|$ and $\mathcal{M}$ is the fermionic mass matrix. We denote the canonical basis of $\mathbb{C}^6$ symbolically by $(\nu, e)_L, e_R, (\nu^c, e^c)_L, e_R^c$. The spectral triple still describes two points, $\delta_L(a,b) = \frac{1}{2}\mathrm{tr}\,a$ and $\delta_R(a,b) = b$ separated by a distance $1/|m|$. There are still two minimal projectors, $p_1 = (1_2, 0)$, $p_2 = (0,1)$ and the intersection form $\cap = \begin{pmatrix} 0 & -2 \\ -2 & 2 \end{pmatrix}$ is invertible.

Our next task is to lift the automorphisms to the Hilbert space and fluctuate the 'flat' metric $\tilde{\mathcal{D}}$. All automorphisms of the quaternions are inner, the complex numbers considered as 2-dimensional real algebra only have one non-trivial automorphism, the complex conjugation. It is disconnected from the identity and we may neglect it. Then

$$\mathrm{Aut}(\mathcal{A}) = SU(2)/\mathbb{Z}_2 \ni \sigma_{\pm u}, \quad \sigma_{\pm u}(a,b) = (uau^{-1}, b). \quad (123)$$

The receptacle group, subgroup of $U(6)$ is readily calculated,

$$\mathrm{Aut}_{\mathcal{H}}(\mathcal{A}) = U(2) \times U(1) \ni U = \begin{pmatrix} U_2 & 0 & 0 & 0 \\ 0 & U_1 & 0 & 0 \\ 0 & 0 & \bar{U}_2 & 0 \\ 0 & 0 & 0 & \bar{U}_1 \end{pmatrix},$$

$$U_2 \in U(2), \ U_1 \in U(1). \quad (124)$$

The covariance property is fulfilled, $i_U \rho(a,b) = \rho(i_{U_2} a, b)$ and the projection, $p(U) = \pm(\det U_2)^{-1/2} U_2$, has kernel $\mathbb{Z}_2$. The lift,

$$L(\pm u) = \rho(\pm u, 1)J\rho(\pm u, 1)J^{-1} = \begin{pmatrix} \pm u & 0 & 0 & 0 \\ 0 & 1 & 0 & 0 \\ 0 & 0 & \pm\bar{u} & 0 \\ 0 & 0 & 0 & 1 \end{pmatrix}, \tag{125}$$

is double-valued. The spin group is the image of the lift, $L(\mathrm{Aut}(\mathcal{A})) = SU(2)$, a proper subgroup of the receptacle $\mathrm{Aut}_{\mathcal{H}}(\mathcal{A}) = U(2) \times U(1)$. The fluctuated Dirac operator is

$$\mathcal{D} := L(\pm u)\tilde{\mathcal{D}}L(\pm u)^{-1} = \begin{pmatrix} 0 & \pm u\mathcal{M} & 0 & 0 \\ (\pm u\mathcal{M})^* & 0 & 0 & 0 \\ 0 & 0 & 0 & \pm u\mathcal{M} \\ 0 & 0 & (\pm u\mathcal{M})^* & 0 \end{pmatrix}. \tag{126}$$

An absolutely remarkable property of the fluctuated Dirac operator in internal space is that it can be written as the flat Dirac operator plus a 1-form:

$$\mathcal{D} = \tilde{\mathcal{D}} + \rho(\pm u, 1)\,[\mathcal{D}, \rho(\pm u^{-1}, 1)] + J\,\rho(\pm u, 1)\,[\mathcal{D}, \rho(\pm u^{-1}, 1)]\,J^{-1}. \tag{127}$$

The anti-Hermitean 1-form

$$(-i)\rho(\pm u, 1)\,[\mathcal{D}, \rho(\pm u^{-1}, 1)] = (-i)\begin{pmatrix} 0 & h & 0 & 0 \\ h^* & 0 & 0 & 0 \\ 0 & 0 & 0 & 0 \\ 0 & 0 & 0 & 0 \end{pmatrix},$$

$$h := \pm u\mathcal{M} - \mathcal{M} \tag{128}$$

is the internal connection. The fluctuated Dirac operator is the covariant one with respect to this connection. Of course, this connection is flat, its field strength = curvature 2-form vanishes, a constraint that is relaxed by the equivalence principle. The result can be stated without going into the details of the reconstruction of 2-forms from the spectral triple: h becomes a general complex doublet, not necessarily of the form $\pm u\mathcal{M} - \mathcal{M}$.

Now we are ready to tensorize the spectral triple of spacetime with the internal one and compute the spectral action. The algebra $\mathcal{A}_t = C^\infty(M) \otimes \mathcal{A}$ describes a two-sheeted universe. Let us call again its sheets 'left' and 'right'. The Hilbert space $\mathcal{H}_t = L^2(S) \otimes \mathcal{H}$ describes the neutrino and the electron as genuine fields, that is spacetime dependent. The Dirac operator $\tilde{\mathcal{D}}_t = \partial\!\!\!/ \otimes 1_6 + \gamma_5 \otimes \tilde{\mathcal{D}}$ is the flat, free, massive Dirac operator and it is impatient to fluctuate.

The automorphism group close to the identity,

$$\mathrm{Aut}(\mathcal{A}_t) = [\mathrm{Diff}(M) \ltimes {}^M SU(2)/\mathbb{Z}_2] \times \mathrm{Diff}(M) \ni ((\sigma_L, \sigma_{\pm u}), \sigma_R), \tag{129}$$

now contains two independent coordinate transformations σ_L and σ_R on each sheet and a *gauged*, that is spacetime dependent, internal transformation $\sigma_{\pm u}$. The gauge transformations are inner, they act by conjugation $i_{\pm u}$. The receptacle group is

$$\mathrm{Aut}_{\mathcal{H}_t}(\mathcal{A}_t) = \mathrm{Diff}(M) \ltimes {}^M(Spin(4) \times U(2) \times U(1)). \tag{130}$$

It only contains one coordinate transformation, a point on the left sheet travels together with its right shadow. Indeed the covariance property forbids to lift an automorphism with $\sigma_L \neq \sigma_R$. Since the mass term multiplies left- and right-handed electron fields, the covariance property saves the locality of field theory, which postulates that only fields at the same spacetime point can be multiplied. We have seen examples where the receptacle has more elements than the automorphism group, e.g. six-dimensional spacetime or the present internal space. Now we have an example of automorphisms that do not fit into the receptacle. In any case the spin group is the image of the combined, now 4-valued lift $L_t(\sigma, \sigma_{\pm u})$,

$$L_t(\mathrm{Aut}(\mathcal{A}_t)) = \mathrm{Diff}(M) \ltimes {}^M(Spin(4) \times SU(2)). \tag{131}$$

The fluctuating Dirac operator is

$$\mathcal{D}_t = L_t(\sigma, \sigma_{\pm u}) \tilde{D}_t L_t(\sigma, \sigma_{\pm u})^{-1} = \begin{pmatrix} \partial\!\!\!/_L & \gamma_5 \varphi & 0 & 0 \\ \gamma_5 \varphi^* & \partial\!\!\!/_R & 0 & 0 \\ 0 & 0 & C \partial\!\!\!/_L C^{-1} & \gamma_5 \bar{\varphi} \\ 0 & 0 & \gamma_5 \bar{\varphi}^* & C \partial\!\!\!/_R C^{-1} \end{pmatrix}, \tag{132}$$

with

$$e^{-1} = \sqrt{\mathcal{J}\mathcal{J}^T}, \qquad \partial\!\!\!/_L = i e^{-1}{}^\mu{}_a \gamma^a [\partial_\mu + s(\omega(e)_\mu) + A_\mu], \tag{133}$$

$$A_\mu = - \pm u\, \partial_\mu(\pm u^{-1}), \quad \partial\!\!\!/_R = i e^{-1}{}^\mu{}_a \gamma^a [\partial_\mu + s(\omega(e)_\mu)], \tag{134}$$

$$\varphi = \pm u\mathcal{M}. \tag{135}$$

Note that the sign ambiguity in $\pm u$ drops out from the $su(2)$-valued 1-form $A = A_\mu dx^\mu$ on spacetime. This is not the case for the ambiguity in the 'Higgs' doublet φ yet, but this ambiguity does drop out from the spectral action. The variable φ is the homogeneous variable corresponding to the affine variable $h = \varphi - \mathcal{M}$ in the connection 1-form on internal space. The fluctuating Dirac operator $\mathcal{D}_t$ is still flat. This constraint has now three parts, $e^{-1} = \sqrt{\mathcal{J}(\sigma)\mathcal{J}(\sigma)^T}$, $A = -ud(u^{-1})$, and $\varphi = \pm u\mathcal{M}$. According to the equivalence principle, we will take e to be any symmetric, invertible matrix depending differentiably on spacetime, A to be any $su(2)$-valued 1-form on spacetime and φ any complex doublet depending differentiably on spacetime. This defines the new kinematics. The dynamics of the spinors = matter is given by the fluctuating Dirac operator $\mathcal{D}_t$, which is covariant with respect to i.e. minimally coupled to gravity, the gauge bosons and the Higgs boson. This dynamics is equivalently given by the Dirac action $(\psi, \mathcal{D}_t \psi)$ and this action delivers the awkward Yukawa couplings for free. The Higgs boson φ enjoys two geometric interpretations, first as connection in the discrete direction. The second derives from Connes' distance formula: $1/|\varphi(x)|$ is the – now x-dependent – distance between the two sheets. The calculation behind the second interpretation makes explicit use of the Kaluza–Klein nature of almost commutative geometries [35].

As in pure gravity, the dynamics of the new kinematics derives from the Chamseddine–Connes action,

$$S_{CC}[e, A, \varphi] = \text{tr } f(\mathcal{D}_t^2/\Lambda^2)$$
$$= \int_M [\ \tfrac{2\Lambda_c}{16\pi G} - \tfrac{1}{16\pi G} R + a(5\,R^2 - 8\,\text{Ricci}^2 - 7\,\text{Riemann}^2)$$
$$\tfrac{1}{2g_2^2} \text{tr } F_{\mu\nu}^* F^{\mu\nu} + \tfrac{1}{2}(D_\mu\varphi)^* D^\mu\varphi$$
$$\lambda|\varphi|^4 - \tfrac{1}{2}\mu^2|\varphi|^2 + \tfrac{1}{12}|\varphi|^2 R\ ]\, dV + O(\Lambda^{-2}), \qquad (136)$$

where the coupling constants are

$$\Lambda_c = \frac{6f_0}{f_2}\Lambda^2, \quad G = \frac{\pi}{2f_2}\Lambda^{-2}, \quad a = \frac{f_4}{960\pi^2},$$
$$g_2^2 = \frac{6\pi^2}{f_4}, \quad \lambda = \frac{\pi^2}{2f_4}, \quad \mu^2 = \frac{2f_2}{f_4}\Lambda^2. \qquad (137)$$

Note the presence of the conformal coupling of the scalar to the curvature scalar, $+\tfrac{1}{12}|\varphi|^2 R$. From the fluctuation of the Dirac operator, we have derived the scalar representation, a complex doublet φ. Geometrically, it is a connection on the finite space and as such unified with the Yang–Mills bosons, which are connections on spacetime. As a consequence, the Higgs self coupling λ is related to the gauge coupling g_2 in the spectral action, $g_2^2 = 12\lambda$. Furthermore the spectral action contains a negative mass square term for the Higgs $-\tfrac{1}{2}\mu^2|\varphi|^2$ implying a nontrivial ground state or vacuum expectation value $|\varphi| = v = \mu(4\lambda)^{-1/2}$ in flat spacetime. Reshifting to the inhomogeneous scalar variable $h = \varphi - v$, which vanishes in the ground state, modifies the cosmological constant by $V(v)$ and Newton's constant from the term $\tfrac{1}{12}v^2 R$:

$$\Lambda_c = 6\left(3\tfrac{f_0}{f_2} - \tfrac{f_2}{f_4}\right)\Lambda^2, \quad G = \frac{3\pi}{2f_2}\Lambda^{-2}. \qquad (138)$$

Now the cosmological constant can have either sign, in particular it can be zero. This is welcome because experimentally the cosmological constant is very close to zero, $\Lambda_c < 10^{-119}/G$. On the other hand, in spacetimes of large curvature, like for example the ground state, the positive conformal coupling of the scalar to the curvature dominates the negative mass square term $-\tfrac{1}{2}\mu^2|\varphi|^2$. Therefore the vacuum expectation value of the Higgs vanishes, the gauge symmetry is unbroken and all particles are massless. It is only after the big bang, when spacetime loses its strong curvature that the gauge symmetry breaks down spontaneously and particles acquire masses.

The computation of the spectral action is long, let us set some waypoints. The square of the fluctuating Dirac operator is $\mathcal{D}_t^2 = -\Delta + E$, where Δ is the covariant Laplacian, in coordinates:

$$\Delta = g^{\mu\tilde\nu}\left[\left(\frac{\partial}{\partial x^\mu}1_4 \otimes 1_\mathcal{H} + \tfrac{1}{4}\omega_{ab\mu}\gamma^{ab} \otimes 1_\mathcal{H} + 1_4 \otimes [\rho(A_\mu) + J\rho(A_\mu)J^{-1}]\right)\delta^\nu_{\tilde\nu}\right.$$
$$-\Gamma^\nu_{\tilde\nu\mu}1_4 \otimes 1_\mathcal{H}\right]$$
$$\times \left[\frac{\partial}{\partial x^\nu}1_4 \otimes 1_\mathcal{H} + \tfrac{1}{4}\omega_{ab\nu}\gamma^{ab} \otimes 1_\mathcal{H} + 1_4 \otimes [\rho(A_\nu) + J\rho(A_\nu)J^{-1}]\right], \qquad (139)$$

and where E, for endomorphism, is a zero order operator, that is a matrix of size $4 \dim \mathcal{H}$ whose entries are functions constructed from the bosonic fields and their first and second derivatives,

$$E = \tfrac{1}{2} \left[\gamma^\mu \gamma^\nu \otimes 1_{\mathcal{H}} \right] \mathbb{R}_{\mu\nu} \tag{140}$$

$$+ \begin{pmatrix} 1_4 \otimes \varphi\varphi^* & -i\gamma_5\gamma^\mu \otimes D_\mu\varphi & 0 & 0 \\ -i\gamma_5\gamma^\mu \otimes (D_\mu\varphi)^* & 1_4 \otimes \varphi^*\varphi & 0 & 0 \\ 0 & 0 & 1_4 \otimes \overline{\varphi\varphi^*} & -i\gamma_5\gamma^\mu \otimes \overline{D_\mu\varphi} \\ 0 & 0 & -i\gamma_5\gamma^\mu \otimes \overline{(D_\mu\varphi)^*} & 1_4 \otimes \overline{\varphi^*\varphi} \end{pmatrix}.$$

$\mathbb{R}$ is the total curvature, a 2-form with values in the (Lorentz $\oplus$ internal) Lie algebra represented on (spinors $\otimes \mathcal{H}$). It contains the curvature 2-form $R = d\omega + \omega^2$ and the field strength 2-form $F = dA + A^2$, in components

$$\mathbb{R}_{\mu\nu} = \tfrac{1}{4} R_{ab\mu\nu} \gamma^a \gamma^b \otimes 1_{\mathcal{H}} + 1_4 \otimes [\rho(F_{\mu\nu}) + J\rho(F_{\mu\nu})J^{-1}]. \tag{141}$$

The first term in equation (141) produces the curvature scalar, which we also (!) denote by R,

$$\tfrac{1}{2} \left[e^{-1\,\mu}_{\quad c} e^{-1\,\nu}_{\quad d} \gamma^c \gamma^d \right] \tfrac{1}{4} R_{ab\mu\nu} \gamma^a \gamma^b = \tfrac{1}{4} R 1_4. \tag{142}$$

We have also used the possibly dangerous notation $\gamma^\mu = e^{-1\,\mu}_{\quad a}\gamma^a$. Finally D is the covariant derivative appropriate for the representation of the scalars. The above formula for the square of the Dirac operator is also known as Lichérowicz formula. The Lichérowicz formula with arbitrary torsion can be found in [36].

Let $f : \mathbb{R}_+ \to \mathbb{R}_+$ be a positive, smooth function with finite moments,

$$f_0 = \int_0^\infty u f(u)\, du, \quad f_2 = \int_0^\infty f(u)\, du, \quad f_4 = f(0), \tag{143}$$

$$f_6 = -f'(0), \qquad f_8 = f''(0), \qquad \dots \tag{144}$$

Asymptotically, for large Λ, the distribution function of the spectrum is given in terms of the heat kernel expansion [37]:

$$S = \operatorname{tr} f(\mathcal{D}_t^2/\Lambda^2) = \frac{1}{16\pi^2} \int_M \left[\Lambda^4 f_0 a_0 + \Lambda^2 f_2 a_2 + f_4 a_4 + \Lambda^{-2} f_6 a_6 + \dots \right] dV, \tag{145}$$

where the a_j are the coefficients of the heat kernel expansion of the Dirac operator squared [30],

$$a_0 = \operatorname{tr}(1_4 \otimes 1_{\mathcal{H}}), \tag{146}$$

$$a_2 = \tfrac{1}{6} R \operatorname{tr}(1_4 \otimes 1_{\mathcal{H}}) - \operatorname{tr} E, \tag{147}$$

$$a_4 = \tfrac{1}{72} R^2 \operatorname{tr}(1_4 \otimes 1_{\mathcal{H}}) - \tfrac{1}{180} R_{\mu\nu} R^{\mu\nu} \operatorname{tr}(1_4 \otimes 1_{\mathcal{H}}) + \tfrac{1}{180} R_{\mu\nu\rho\sigma} R^{\mu\nu\rho\sigma} \operatorname{tr}(1_4 \otimes 1_{\mathcal{H}})$$
$$+ \tfrac{1}{12} \operatorname{tr}(\mathbb{R}_{\mu\nu} \mathbb{R}^{\mu\nu}) - \tfrac{1}{6} R \operatorname{tr} E + \tfrac{1}{2} \operatorname{tr} E^2 + \text{surface terms}. \tag{148}$$

As already noted, for large Λ the positive function f is universal, only the first three moments, f_0, f_2 and f_4 appear with non-negative powers of Λ. For the minimax model, we get (more details can be found in [38]):

$$a_0 \qquad\qquad = 4 \dim \mathcal{H} = 4 \times 6, \tag{149}$$

$$\mathrm{tr}\, E \qquad\qquad = \dim \mathcal{H}\, R + 16|\varphi|^2, \tag{150}$$

$$a_2 \qquad\qquad = \tfrac{2}{3} \dim \mathcal{H}\, R - \dim \mathcal{H}\, R - 16|\varphi|^2$$
$$= -\tfrac{1}{3} \dim \mathcal{H}\, R - 16|\varphi|^2, \tag{151}$$

$$\mathrm{tr}\,\left(\tfrac{1}{2}[\gamma^a,\gamma^b]\tfrac{1}{2}[\gamma^c,\gamma^d]\right) = 4\left[\delta^{ad}\delta^{bc} - \delta^{ac}\delta^{bd}\right], \tag{152}$$

$$\mathrm{tr}\,\{\mathbb{R}_{\mu\nu}\mathbb{R}^{\mu\nu}\} \qquad = -\tfrac{1}{2}\dim\mathcal{H}\,R_{\mu\nu\rho\sigma}R^{\mu\nu\rho\sigma}$$
$$-4\,\mathrm{tr}\,\{[\rho(F_{\mu\nu}) + J\rho(F_{\mu\nu})J^{-1}]^*$$
$$\times[\rho(F^{\mu\nu}) + J\rho(F^{\mu\nu})J^{-1}]\}$$
$$= -\tfrac{1}{2}\dim\mathcal{H}\,R_{\mu\nu\rho\sigma}R^{\mu\nu\rho\sigma}$$
$$-8\,\mathrm{tr}\,\{\rho(F_{\mu\nu})^*\rho(F^{\mu\nu})\}, \tag{153}$$

$$\mathrm{tr}\, E^2 \qquad\qquad = \tfrac{1}{4}\dim\mathcal{H}\,R^2 + 4\,\mathrm{tr}\,\{\rho(F_{\mu\nu})^*\rho(F^{\mu\nu})\}$$
$$+16|\varphi|^4 + 16(D_\mu\varphi)^*(D^\mu\varphi) + 8|\varphi|^2 R, \tag{154}$$

Finally we have up to surface terms,

$$a_4 = \tfrac{1}{360}\dim\mathcal{H}\,(5\,R^2 - 8\,\mathrm{Ricci}^2 - 7\,\mathrm{Riemann}^2) + \tfrac{4}{3}\mathrm{tr}\,\rho(F_{\mu\nu})^*\rho(F^{\mu\nu})$$
$$+8|\varphi|^4 + 8(D_\mu\varphi)^*(D^\mu\varphi) + \tfrac{4}{3}|\varphi|^2 R. \tag{155}$$

We arrive at the spectral action with its conventional normalization, equation (136), after a finite renormalization $|\varphi|^2 \to \frac{\pi^2}{f_4}|\varphi|^2$.

Our first timid excursion into gravity on a noncommutative geometry produced a rather unexpected discovery. We stumbled over a , which is precisely the electro–weak model for one family of leptons but with the $U(1)$ of hypercharge amputated. The sceptical reader suspecting a sleight of hand is encouraged to try and find a simpler, noncommutative finite spectral triple.

5.4   A Central Extension

We will see in the next section the technical reason for the absence of $U(1)$s as automorphisms: all automorphisms of finite spectral triples connected to the identity are inner, i.e. conjugation by unitaries. But conjugation by central unitaries is trivial. This explains that in the minimax example, $\mathcal{A} = \mathbb{H}\oplus\mathbb{C}$, the component of the automorphism group connected to the identity was $SU(2)/\mathbb{Z}_2 \ni (\pm u, 1)$. It is the domain of definition of the lift, equation (125),

$$L(\pm u, 1) = \rho(\pm u, 1)J\rho(\pm u, 1)J^{-1} = \begin{pmatrix} \pm u & 0 & 0 & 0 \\ 0 & 1 & 0 & 0 \\ 0 & 0 & \pm\bar{u} & 0 \\ 0 & 0 & 0 & 1 \end{pmatrix}. \tag{156}$$

It is tempting to centrally extend the lift to all unitaries of the algebra:

$$\mathbb{L}(w, v) = \rho(w, v)J\rho(w, v)J^{-1} = \begin{pmatrix} \bar{v}w & 0 & 0 & 0 \\ 0 & \bar{v}^2 & 0 & 0 \\ 0 & 0 & v\bar{w} & 0 \\ 0 & 0 & 0 & v^2 \end{pmatrix},$$

$$(w, v) \in SU(2) \times U(1). \tag{157}$$

An immediate consequence of this extension is encouraging: the extended lift is single-valued and after tensorization with the one from Riemannian geometry, the multi-valuedness will remain two.

Then redoing the fluctuation of the Dirac operator and recomputing the spectral action yields gravity coupled to the complete electro–weak model of the electron and its neutrino with a weak mixing angle of $\sin^2 \theta_w = 1/4$.

6 Connes' Do-It-Yourself Kit

Our first example of gravity on an almost commutative space leaves us wondering what other examples will look like. To play on the Yang–Mills–Higgs machine, one must know the classification of all real, compact Lie groups and their unitary representations. To play on the new machine, we must know all finite spectral triples. The first good news is that the list of algebras and their representations is infinitely shorter than the one for groups. The other good news is that the rules of Connes' machine are not made up opportunistically to suit the phenomenology of electro–weak and strong forces as in the case of the Yang–Mills–Higgs machine. On the contrary, as developed in the last section, these rules derive naturally from geometry.

6.1 Input

Our first input item is a finite dimensional, real, associative involution algebra with unit and that admits a finite dimensional faithful representation. Any such algebra is a direct sum of simple algebras with the same properties. Every such simple algebra is an algebra of $n \times n$ matrices with real, complex or quaternionic entries, $\mathcal{A} = M_n(\mathbb{R})$, $M_n(\mathbb{C})$ or $M_n(\mathbb{H})$. Their unitary groups $U(\mathcal{A}) := \{u \in \mathcal{A}, uu^* = u^*u = 1\}$ are $O(n)$, $U(n)$ and $USp(n)$. Note that $USp(1) = SU(2)$. The centre Z of an algebra $\mathcal{A}$ is the set of elements $z \in \mathcal{A}$ that commute with all elements $a \in \mathcal{A}$. The central unitaries form an abelian subgroup of $U(\mathcal{A})$. Let us denote this subgroup by $U^c(\mathcal{A}) := U(\mathcal{A}) \cap Z$. We have $U^c(M_n(\mathbb{R})) = \mathbb{Z}_2 \ni \pm 1_n$, $U^c(M_n(\mathbb{C})) = U(1) \ni \exp(i\theta)1_n$, $\theta \in [0, 2\pi)$, $U^c(M_n(\mathbb{H})) = \mathbb{Z}_2 \ni \pm 1_{2n}$. All automorphisms of the real, complex and quaternionic matrix algebras are inner with one exception, $M_n(\mathbb{C})$ has one outer automorphism, complex conjugation, which is disconnected from the identity automorphism. An inner automorphism σ is of the form $\sigma(a) = uau^{-1}$ for some $u \in U(\mathcal{A})$ and for all $a \in \mathcal{A}$. We will denote this inner automorphism by $\sigma = i_u$ and we will write $\mathrm{Int}(\mathcal{A})$ for the group of inner automorphisms. Of course a commutative algebra, e.g. $\mathcal{A} = \mathbb{C}$, has no inner automorphism. We have $\mathrm{Int}(\mathcal{A}) = U(\mathcal{A})/U^c(\mathcal{A})$, in particular $\mathrm{Int}(M_n(\mathbb{R})) = O(n)/\mathbb{Z}_2$, $n = 2, 3, ...$, $\mathrm{Int}(M_n(\mathbb{C})) = U(n)/U(1) = SU(n)/\mathbb{Z}_n$, $n = 2, 3, ...$, $\mathrm{Int}(M_n(\mathbb{H})) = USp(n)/\mathbb{Z}_2$, $n = 1, 2, ...$ Note the apparent injustice: the commutative algebra $\mathcal{C}^\infty(M)$ has the nonAbelian automorphism group $\mathrm{Diff}(M)$ while the noncommutative algebra $M_2(\mathbb{R})$ has the Abelian automorphism group $O(2)/\mathbb{Z}_2$. All exceptional groups are missing from our list of groups. Indeed they are automorphism groups of non-associative algebras, e.g. G_2 is the automorphism group of the octonions.

The second input item is a faithful representation ρ of the algebra $\mathcal{A}$ on a finite dimensional, complex Hilbert space $\mathcal{H}$. Any such representation is a direct sum of irreducible representations. $M_n(\mathbb{R})$ has only one irreducible representation, the fundamental one on $\mathbb{R}^n$, $M_n(\mathbb{C})$ has two, the fundamental one and its complex conjugate. Both are defined on $\mathcal{H} = \mathbb{C}^n \ni \psi$ by $\rho(a)\psi = a\psi$ and by $\rho(a)\psi = \bar{a}\psi$. $M_n(\mathbb{H})$ has only one irreducible representation, the fundamental one defined on $\mathbb{C}^{2n}$. For example, while $U(1)$ has an infinite number of inequivalent irreducible representations, characterized by an integer 'charge', its algebra $\mathbb{C}$ has only two with charge plus and minus one. While $SU(2)$ has an infinite number of inequivalent irreducible representations characterized by its spin, $0, \frac{1}{2}, 1, ...$, its algebra $\mathbb{H}$ has only one, spin $\frac{1}{2}$. The main reason behind this multitude of group representation is that the tensor product of two representations of one group is another representation of this group, characterized by the sum of charges for $U(1)$ and by the sum of spins for $SU(2)$. The same is not true for two representations of one associative algebra whose tensor product fails to be linear. (Attention, the tensor product of two representations of two algebras does define a representation of the tensor product of the two algebras. We have used this tensor product of Hilbert spaces to define almost commutative geometries.)

The third input item is the finite Dirac operator $\mathcal{D}$ or equivalently the fermionic mass matrix, a matrix of size $\dim\mathcal{H}_L \times \dim\mathcal{H}_R$.

These three items can however not be chosen freely, they must still satisfy all axioms of the spectral triple [39]. I do hope you have convinced yourself of the nontriviality of this requirement for the case of the minimax example.

The minimax example has taught us something else. If we want abelian gauge fields from the fluctuating metric, we must centrally extend the spin lift, an operation, that at the same time may reduce the multivaluedness of the original lift. Central extensions are by no means unique, its choice is our last input item [40].

To simplify notations, we concentrate on complex matrix algebras $M_n(\mathbb{C})$ in the following part. Indeed the others, $M_n(\mathbb{R})$ and $M_n(\mathbb{H})$, do not have central unitaries close to the identity. We have already seen that it is important to separate the commutative and noncommutative parts of the algebra:

$$\mathcal{A} = \mathbb{C}^M \oplus \bigoplus_{k=1}^{N} M_{n_k}(\mathbb{C}) \ni a = (b_1, ...b_M, c_1, ..., c_N), \quad n_k \geq 2. \tag{158}$$

Its group of unitaries is

$$U(\mathcal{A}) = U(1)^M \times \underset{k=1}{\overset{N}{\times}} U(n_k) \ni u = (v_1, ..., v_M, w_1, ..., w_N) \tag{159}$$

and its group of central unitaries

$$U^c(\mathcal{A}) = U(1)^{M+N} \ni u_c = (v_{c1}, ..., v_{cM}, w_{c1}1_{n_1}, ..., w_{cN}1_{n_N}). \qquad (160)$$

All automorphisms connected to the identity are inner, there are outer automorphisms, the complex conjugation and, if there are identical summands in $\mathcal{A}$, their permutations. In compliance with the minimax principle, we disregard the discrete automorphisms. Multiplying a unitary u with a central unitary u_c of course does not affect its inner automorphism $i_{u_c u} = i_u$. This ambiguity distinguishes between 'harmless' central unitaries $v_{c1}, ..., v_{cM}$ and the others, $w_{c1}, ..., w_{cN}$, in the sense that

$$\mathrm{Int}(\mathcal{A}) = U^n(\mathcal{A})/U^{nc}(\mathcal{A}), \qquad (161)$$

where we have defined the group of noncommutative unitaries

$$U^n(\mathcal{A}) := \mathop{\times}_{k=1}^{N} U(n_k) \ni w \qquad (162)$$

and $U^{nc}(\mathcal{A}) := U^n(\mathcal{A}) \cap U^c(\mathcal{A}) \ni w_c$. The map

$$\begin{aligned} i : U^n(\mathcal{A}) &\longrightarrow \mathrm{Int}(\mathcal{A}) \\ w &\longmapsto i_w \end{aligned} \qquad (163)$$

has kernel $\mathrm{Ker}\, i = U^{nc}(\mathcal{A})$.

The lift of an inner automorphism to the Hilbert space has a simple closed form [19], $L = \hat{L} \circ i^{-1}$ with

$$\hat{L}(w) = \rho(1, w)J\rho(1, w)J^{-1}. \qquad (164)$$

It satisfies $p(\hat{L}(w)) = i(w)$. If the kernel of i is contained in the kernel of $\hat{L}$, then the lift is well defined, as e.g. for $\mathcal{A} = \mathbb{H}$, $U^{nc}(\mathbb{H}) = \mathbb{Z}_2$.

$$\begin{array}{c} \mathrm{Aut}_{\mathcal{H}}(\mathcal{A}) \\ p\Big\downarrow \ L \quad \hat{L} \quad \ell \\ \mathrm{Int}(\mathcal{A}) \xleftarrow{\ i\ } U^n(\mathcal{A}) \underset{\mathrm{det}}{\leftrightarrows} U^{nc}(\mathcal{A}) \end{array} \qquad (165)$$

For more complicated real or quaternionic algebras, $U^{nc}(\mathcal{A})$ is finite and the lift L is multi-valued with a finite number of values. For noncommutative, complex algebras, their continuous family of central unitaries cannot be eliminated except for very special representations and we face a continuous infinity of values. The solution of this problem follows an old strategy: 'If you can't beat them, adjoin them'. Who is them? The harmful central unitaries $w_c \in U^{nc}(\mathcal{A})$ and adjoining means central extending. The central extension (157), only concerned a discrete group and a harmless $U(1)$. Nevertheless it generalizes naturally to the present setting:

$$\mathbb{L} : \mathrm{Int}(\mathcal{A}) \times U^{nc}(\mathcal{A}) \longrightarrow \mathrm{Aut}_{\mathcal{H}}(\mathcal{A})$$
$$(w_\sigma, w_c) \longmapsto (\hat{L} \circ i^{-1})(w_\sigma)\,\ell(w_c) \tag{166}$$

with

$$\ell(w_c) := \rho\Bigg(\prod_{j_1=1}^{N} (w_{cj_1})^{q_{1,j_1}}, ..., \prod_{j_M=1}^{N} (w_{cj_M})^{q_{M,j_M}}, \tag{167}$$

$$\prod_{j_{M+1}=1}^{N} (w_{cj_{M+1}})^{q_{M+1,j_{M+1}}} 1_{n_1}, ..., \prod_{j_{M+N}=1}^{N} (w_{cj_{M+N}})^{q_{M+N,j_{M+N}}} 1_{n_N}\Bigg) J\rho(...) J^{-1}$$

with the $(M + N) \times N$ matrix of charges q_{kj}. The extension satisfies indeed $p(\ell(w_c)) = 1 \in \mathrm{Int}(\mathcal{A})$ for all $w_c \in U^{nc}(\mathcal{A})$.

Having adjoined the harmful, continuous central unitaries, we may now stream line our notations and write the group of inner automorphisms as

$$\mathrm{Int}(\mathcal{A}) = \left(\underset{k=1}{\overset{N}{\times}} SU(n_k)\right) / \Gamma \ni [w_\sigma] = [(w_{\sigma 1}, ..., w_{\sigma N})] \bmod \gamma, \tag{168}$$

where Γ is the discrete group

$$\Gamma = \underset{k=1}{\overset{N}{\times}} \mathbb{Z}_{n_k} \ni (z_1 1_{n_1}, ..., z_N 1_{n_N}),$$
$$z_k = \exp[-m_k 2\pi i / n_k], \quad m_k = 0, ..., n_k - 1 \tag{169}$$

and the quotient is factor by factor. This way to write inner automorphisms is convenient for complex matrices, but not available for real and quaternionic matrices. Equation (161) remains the general characterization of inner automorphisms.

The lift $L(w_\sigma) = (\hat{L} \circ i^{-1})(w_\sigma)$, $w_\sigma = w \bmod U^{nc}(\mathcal{A})$, is multi-valued with, depending on the representation, up to $|\Gamma| = \prod_{j=1}^{N} n_j$ values. More precisely the multi-valuedness of L is indexed by the elements of the kernel of the projection p restricted to the image $L(\mathrm{Int}(\mathcal{A}))$. Depending on the choice of the charge matrix q, the central extension ℓ may reduce this multi-valuedness. Extending harmless central unitaries is useless for any reduction. With the multi-valued group homomorphism

$$(h_\sigma, h_c) : U^n(\mathcal{A}) \longrightarrow \mathrm{Int}(\mathcal{A}) \times U^{nc}(\mathcal{A})$$
$$(w_j) \longmapsto ((w_{\sigma j}, w_{cj})) = ((w_j (\det w_j)^{-1/n_j}, (\det w_j)^{1/n_j})), \tag{170}$$

we can write the two lifts L and ℓ together in closed form $\mathbb{L} : U^n(\mathcal{A}) \to \mathrm{Aut}_{\mathcal{H}}(\mathcal{A})$:

$$\mathbb{L}(w) = L(h_\sigma(w))\, \ell(h_c(w))$$

$$= \rho\left(\prod_{j_1=1}^{N} (\det w_{j_1})^{\tilde{q}_{1,j_1}}, \dots, \prod_{j_M=1}^{N} (\det w_{j_M})^{\tilde{q}_{M,j_M}}, \right.$$

$$\left. w_1 \prod_{j_{M+1}=1}^{N} (\det w_{j_{M+1}})^{\tilde{q}_{M+1,j_{M+1}}}, \dots, w_N \prod_{j_{N+M}=1}^{N} (\det w_{j_{N+M}})^{\tilde{q}_{N+M,j_{N+M}}} \right)$$

$$\times J\rho(\dots)J^{-1}. \tag{171}$$

We have set

$$\tilde{q} := \left(q - \begin{pmatrix} 0_{M\times N} \\ 1_{N\times N} \end{pmatrix} \right) \begin{pmatrix} n_1 & & \\ & \ddots & \\ & & n_N \end{pmatrix}^{-1}. \tag{172}$$

Due to the phase ambiguities in the roots of the determinants, the extended lift $\mathbb{L}$ is multi-valued in general. It is single-valued if the matrix $\tilde{q}$ has integer entries, e.g. $q = \begin{pmatrix} 0 \\ 1_N \end{pmatrix}$, then $\tilde{q} = 0$ and $\mathbb{L}(w) = \hat{L}(w)$. On the other hand, $q = 0$ gives $\mathbb{L}(w) = \hat{L}(i^{-1}(h_\sigma(w)))$, not always well defined as already noted. Unlike the extension (157), and unlike the map i, the extended lift $\mathbb{L}$ is not necessarily even. We do impose this symmetry $\mathbb{L}(-w) = \mathbb{L}(w)$, which translates into conditions on the charges, conditions that depend on the details of the representation ρ.

Let us note that the lift $\mathbb{L}$ is simply a representation up to a phase and as such it is not the most general lift. We could have added harmless central unitaries if any present, and, if the representation ρ is reducible, we could have chosen different charge matrices in different irreducible components. If you are not happy with central extensions, then this is a sign of good taste. Indeed commutative algebras like the calibrating example have no inner automorphisms and a huge centre. Truly noncommutative algebras have few outer automorphism and a small centre. We believe that almost commutative geometries with their central extensions are only low energy approximations of a truly noncommutative geometry where central extensions are not an issue.

6.2   Output

From the input data of a finite spectral triple, the central charges and the three moments of the spectral function, noncommutative geometry produces a coupled to gravity. Its entire Higgs sector is computed from the input data, Fig. 6. The Higgs representation derives from the fluctuating metric and the Higgs potential from the spectral action.

To see how the Higgs representation derives in general from the fluctuating Dirac operator $\mathcal{D}$, we must write it as 'flat' Dirac operator $\tilde{\mathcal{D}}$ plus internal 1-form H like we have done in equation (127) for the minimax example without

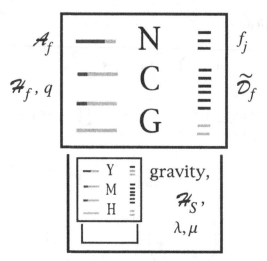

Fig. 6. Connes' slot machine

extension. Take the extended lift $\mathbb{L}(w) = \rho(w)J\rho(w)J^{-1}$ with the unitary

$$w = \prod_{j_1=1}^{N} (\det w_{j_1})^{\tilde{q}_{1j_1}}, ..., \prod_{j_M=1}^{N} (\det w_{j_M})^{\tilde{q}_{Mj_M}}, \tag{173}$$

$$w_1 \prod_{j_{M+1}=1}^{N} (\det w_{j_{M+1}})^{\tilde{q}_{M+1,j_{M+1}}}, ..., w_N \prod_{j_{N+M}=1}^{N} (\det w_{j_{N+M}})^{\tilde{q}_{N+M,j_{N+M}}}.$$

Then

$$\begin{aligned}
\mathcal{D} &= \mathbb{L}\tilde{\mathcal{D}}\mathbb{L}^{-1} \\
&= \left(\rho(w)\, J\rho(w)J^{-1}\right) \tilde{\mathcal{D}} \left(\rho(w)\, J\rho(w)J^{-1}\right)^{-1} \\
&= \rho(w)\, J\rho(w)J^{-1}\tilde{\mathcal{D}}\, \rho(w^{-1})\, J\rho(w^{-1})J^{-1} \\
&= \rho(w)\, J\rho(w)J^{-1}(\rho(w^{-1})\tilde{\mathcal{D}} + [\tilde{\mathcal{D}}, \rho(w^{-1})])J\rho(w^{-1})J^{-1} \\
&= J\rho(w)J^{-1}\tilde{\mathcal{D}}J\rho(w^{-1})J^{-1} + \rho(w)[\tilde{\mathcal{D}}, \rho(w^{-1})] \\
&= J\rho(w)\tilde{\mathcal{D}}\rho(w^{-1})J^{-1} + \rho(w)[\tilde{\mathcal{D}}, \rho(w^{-1})] \\
&= J(\rho(w)[\tilde{\mathcal{D}}, \rho(w^{-1})] + \tilde{\mathcal{D}})J^{-1} + \rho(w)[\tilde{\mathcal{D}}, \rho(w^{-1})] \\
&= \tilde{\mathcal{D}} + H + JHJ^{-1}, \tag{174}
\end{aligned}$$

with the internal 1-form, the Higgs scalar, $H = \rho(w)[\tilde{\mathcal{D}}, \rho(w^{-1})]$. In the chain (174) we have used successively the following three axioms of spectral triples, $[\rho(a), J\rho(\tilde{a})J^{-1}] = 0$, the first order condition $[[\tilde{\mathcal{D}}, \rho(a)], J\rho(\tilde{a})J^{-1}] = 0$ and $[\tilde{\mathcal{D}}, J] = 0$. Note that the unitaries, whose representation commutes with the internal Dirac operator, drop out from the Higgs, it transforms as a singlet under their subgroup.

The constraints from the axioms of noncommutative geometry are so tight that only very few s can be derived from noncommutative geometry as pseudo forces. No left-right symmetric model can [41], no Grand Unified Theory can [42],

Yang-Mills-Higgs

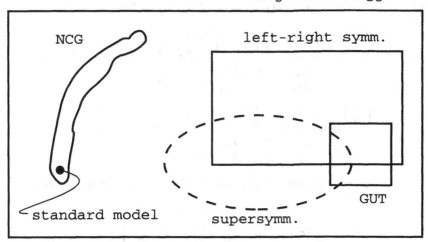

Fig. 7. Pseudo forces from noncommutative geometry

for instance the $SU(5)$ model needs 10-dimensional fermion representations, $SO(10)$ 16-dimensional ones, E_6 is not the group of an associative algebra. Moreover the last two models are left-right symmetric. Much effort has gone into the construction of a supersymmetric model from noncommutative geometry, in vain [43]. The standard model on the other hand fits perfectly into Connes' picture, Fig. 7.

6.3 The Standard Model

The first noncommutative formulation of the standard model was published by Connes & Lott [33] in 1990. Since then it has evolved into its present form [18–20,28] and triggered quite an amount of literature [44].

Spectral Triple. The internal algebra $\mathcal{A}$ is chosen as to reproduce $SU(2) \times U(1) \times SU(3)$ as subgroup of $U(\mathcal{A})$,

$$\mathcal{A} = \mathbb{H} \oplus \mathbb{C} \oplus M_3(\mathbb{C}) \ni (a, b, c). \tag{175}$$

The internal Hilbert space is copied from the Particle Physics Booklet [13],

$$\mathcal{H}_L = (\mathbb{C}^2 \otimes \mathbb{C}^N \otimes \mathbb{C}^3) \oplus (\mathbb{C}^2 \otimes \mathbb{C}^N \otimes \mathbb{C}), \tag{176}$$

$$\mathcal{H}_R = (\mathbb{C} \otimes \mathbb{C}^N \otimes \mathbb{C}^3) \oplus (\mathbb{C} \otimes \mathbb{C}^N \otimes \mathbb{C}^3) \oplus (\mathbb{C} \otimes \mathbb{C}^N \otimes \mathbb{C}). \tag{177}$$

In each summand, the first factor denotes weak isospin doublets or singlets, the second denotes N generations, $N = 3$, and the third denotes colour triplets or singlets. Let us choose the following basis of the internal Hilbert space, counting fermions and antifermions (indicated by the superscript $\cdot^c$ for 'charge conjugated') independently, $\mathcal{H} = \mathcal{H}_L \oplus \mathcal{H}_R \oplus \mathcal{H}_L^c \oplus \mathcal{H}_R^c = \mathbb{C}^{90}$:

$$\begin{pmatrix} u \\ d \end{pmatrix}_L, \begin{pmatrix} c \\ s \end{pmatrix}_L, \begin{pmatrix} t \\ b \end{pmatrix}_L, \begin{pmatrix} \nu_e \\ e \end{pmatrix}_L, \begin{pmatrix} \nu_\mu \\ \mu \end{pmatrix}_L, \begin{pmatrix} \nu_\tau \\ \tau \end{pmatrix}_L;$$

$$\begin{matrix} u_R, & c_R, & t_R, \\ d_R, & s_R, & b_R, \end{matrix} \quad e_R, \quad \mu_R, \quad \tau_R;$$

$$\begin{pmatrix} u \\ d \end{pmatrix}_L^c, \begin{pmatrix} c \\ s \end{pmatrix}_L^c, \begin{pmatrix} t \\ b \end{pmatrix}_L^c, \begin{pmatrix} \nu_e \\ e \end{pmatrix}_L^c, \begin{pmatrix} \nu_\mu \\ \mu \end{pmatrix}_L^c, \begin{pmatrix} \nu_\tau \\ \tau \end{pmatrix}_L^c;$$

$$\begin{matrix} u_R^c, & c_R^c, & t_R^c, \\ d_R^c, & s_R^c, & b_R^c, \end{matrix} \quad e_R^c, \quad \mu_R^c, \quad \tau_R^c.$$

This is the current eigenstate basis, the representation ρ acting on $\mathcal{H}$ by

$$\rho(a,b,c) := \begin{pmatrix} \rho_L & 0 & 0 & 0 \\ 0 & \rho_R & 0 & 0 \\ 0 & 0 & \bar\rho_L^c & 0 \\ 0 & 0 & 0 & \bar\rho_R^c \end{pmatrix} \tag{178}$$

with

$$\rho_L(a) := \begin{pmatrix} a \otimes 1_N \otimes 1_3 & 0 \\ 0 & a \otimes 1_N \end{pmatrix}, \quad \rho_R(b) := \begin{pmatrix} b1_N \otimes 1_3 & 0 & 0 \\ 0 & \bar b 1_N \otimes 1_3 & 0 \\ 0 & 0 & \bar b 1_N \end{pmatrix},$$

$$\rho_L^c(b,c) := \begin{pmatrix} 1_2 \otimes 1_N \otimes c & 0 \\ 0 & \bar b 1_2 \otimes 1_N \end{pmatrix}, \quad \rho_R^c(b,c) := \begin{pmatrix} 1_N \otimes c & 0 & 0 \\ 0 & 1_N \otimes c & 0 \\ 0 & 0 & \bar b 1_N \end{pmatrix}.$$

The apparent asymmetry between particles and antiparticles – the former are subject to weak, the latter to strong interactions – will disappear after application of the lift $\mathbb{L}$ with

$$J = \begin{pmatrix} 0 & 1_{15N} \\ 1_{15N} & 0 \end{pmatrix} \circ \text{ complex conjugation.} \tag{179}$$

For the sake of completeness, we record the chirality as matrix

$$\chi = \begin{pmatrix} -1_{8N} & 0 & 0 & 0 \\ 0 & 1_{7N} & 0 & 0 \\ 0 & 0 & -1_{8N} & 0 \\ 0 & 0 & 0 & 1_{7N} \end{pmatrix}. \tag{180}$$

The internal Dirac operator

$$\tilde{\mathcal{D}} = \begin{pmatrix} 0 & \mathcal{M} & 0 & 0 \\ \mathcal{M}^* & 0 & 0 & 0 \\ 0 & 0 & 0 & \bar{\mathcal{M}} \\ 0 & 0 & \bar{\mathcal{M}}^* & 0 \end{pmatrix} \tag{181}$$

is made of the fermionic mass matrix of the standard model,

$$
\mathcal{M} = \left(\begin{array}{cc} \left(\begin{smallmatrix} 1 & 0 \\ 0 & 0 \end{smallmatrix} \right) \otimes M_u \otimes 1_3 + \left(\begin{smallmatrix} 0 & 0 \\ 0 & 1 \end{smallmatrix} \right) \otimes M_d \otimes 1_3 & 0 \\ 0 & \left(\begin{smallmatrix} 0 \\ 1 \end{smallmatrix} \right) \otimes M_e \end{array} \right), \quad (182)
$$

with

$$
M_u := \begin{pmatrix} m_u & 0 & 0 \\ 0 & m_c & 0 \\ 0 & 0 & m_t \end{pmatrix}, \quad M_d := C_{\mathrm{KM}} \begin{pmatrix} m_d & 0 & 0 \\ 0 & m_s & 0 \\ 0 & 0 & m_b \end{pmatrix}, \quad (183)
$$

$$
M_e := \begin{pmatrix} m_e & 0 & 0 \\ 0 & m_\mu & 0 \\ 0 & 0 & m_\tau \end{pmatrix}. \quad (184)
$$

From the booklet we know that all indicated fermion masses are different from each other and that the Cabibbo–Kobayashi–Maskawa matrix C_{KM} is non-degenerate in the sense that no quark is simultaneously mass and weak current eigenstate.

We must acknowledge the fact – and this is far from trivial – that the finite spectral triple of the standard model satisfies all of Connes' axioms:
- It is orientable, $\chi = \rho(-1_2, 1, 1_3) J \rho(-1_2, 1, 1_3) J^{-1}$.
- Poincaré duality holds. The standard model has three minimal projectors,

$$
p_1 = (1_2, 0, 0), \quad p_2 = (0, 1, 0), \quad p_3 = \left(0, 0, \begin{pmatrix} 1 & 0 & 0 \\ 0 & 0 & 0 \\ 0 & 0 & 0 \end{pmatrix} \right) \quad (185)
$$

and the intersection form

$$
\cap = -2N \begin{pmatrix} 0 & 1 & 1 \\ 1 & -1 & -1 \\ 1 & -1 & 0 \end{pmatrix}, \quad (186)
$$

is non-degenerate. We note that Majorana masses are forbidden because of the axiom $\tilde{D}\chi = -\chi\tilde{D}$. On the other hand if we wanted to give Dirac masses to all three neutrinos we would have to add three right-handed neutrinos to the standard model. Then the intersection form,

$$
\cap = -2N \begin{pmatrix} 0 & 1 & 1 \\ 1 & -2 & -1 \\ 1 & -1 & 0 \end{pmatrix}, \quad (187)
$$

would become degenerate and Poincaré duality would fail.
- The first order axiom is satisfied precisely because of the first two of the six ad hoc properties of the standard model recalled in Sect. 3.3, colour couples vectorially and commutes with the fermionic mass matrix, $[\mathcal{D}, \rho(1_2, 1, c)] = 0$. As

an immediate consequence the Higgs scalar = internal 1-form will be a colour singlet and the gluons will remain massless, the third ad hoc property of the standard model in its conventional formulation.

• There seems to be some arbitrariness in the choice of the representation under $\mathbb{C} \ni b$. In fact this is not true, any choice different from the one in equations (179,179) is either incompatible with the axioms of spectral triples or it leads to charged massless particles incompatible with the Lorentz force or to a symmetry breaking with equal top and bottom masses. Therefore, the only flexibility in the fermionic charges is from the choice of the central charges [40].

Central Charges. The standard model has the following groups,

$$
\begin{aligned}
U(\mathcal{A}) &= SU(2) \times U(1) \times U(3) \ni u = (u_0, v, w), & (188)\\
U^c(\mathcal{A}) &= \mathbb{Z}_2 \times U(1) \times U(1) \ni u_c = (u_{c0}, v_c, w_c 1_3),\\
U^n(\mathcal{A}) &= SU(2) \times U(3) \ni (u_0, w),\\
U^{nc}(\mathcal{A}) &= \mathbb{Z}_2 \times U(1) \ni (u_{c0}, w_c 1_3),\\
\mathrm{Int}(\mathcal{A}) &= [SU(2) \times SU(3)]/\Gamma \ni u_\sigma = (u_{\sigma 0}, w_\sigma),\\
\Gamma &= \mathbb{Z}_2 \times \mathbb{Z}_3 \ni \gamma = (\exp[-m_0 2\pi i/2], \exp[-m_2 2\pi i/3]),
\end{aligned}
$$

with $m_0 = 0, 1$ and $m_2 = 0, 1, 2$. Let us compute the receptacle of the lifted automorphisms,

$$
\begin{aligned}
&\mathrm{Aut}_{\mathcal{H}}(\mathcal{A}) & (189)\\
&= [U(2)_L \times U(3)_c \times U(N)_{qL} \times U(N)_{\ell L} \times U(N)_{uR} \times U(N)_{dR}]/[U(1) \times U(1)]\\
&\quad \times U(N)_{eR}.
\end{aligned}
$$

The subscripts indicate on which multiplet the $U(N)$s act. The kernel of the projection down to the automorphism group $\mathrm{Aut}(\mathcal{A})$ is

$$
\begin{aligned}
\ker p &= [U(1) \times U(1) \times U(N)_{qL} \times U(N)_{\ell L} \times U(N)_{uR} \times U(N)_{dR}]/[U(1) \times U(1)]\\
&\quad \times U(N)_{eR}, & (190)
\end{aligned}
$$

and its restrictions to the images of the lifts are

$$
\ker p \cap L(\mathrm{Int}(\mathcal{A})) = \mathbb{Z}_2 \times \mathbb{Z}_3, \quad \ker p \cap \mathbb{L}(U^n(\mathcal{A})) = \mathbb{Z}_2 \times U(1). \quad (191)
$$

The kernel of i is $\mathbb{Z}_2 \times U(1)$ in sharp contrast to the kernel of $\hat{L}$, which is trivial. The isospin $SU(2)_L$ and the colour $SU(3)_c$ are the image of the lift $\hat{L}$. If $q \neq 0$, the image of ℓ consists of one $U(1) \ni w_c = \exp[i\theta]$ contained in the five flavour $U(N)$s. Its embedding depends on q:

$$
\begin{aligned}
\mathbb{L}(1_2, 1, w_c 1_3) &= \ell(w_c) & (192)\\
&= \mathrm{diag}\,(u_{qL} 1_2 \otimes 1_N \otimes 1_3, u_{\ell L} 1_2 \otimes 1_N, u_{uR} 1_N \otimes 1_3, u_{dR} 1_N \otimes 1_3, u_{eR} 1_N;\\
&\qquad \bar{u}_{qL} 1_2 \otimes 1_N \otimes 1_3, \bar{u}_{\ell L} 1_2 \otimes 1_N, \bar{u}_{uR} 1_N \otimes 1_3, \bar{u}_{dR} 1_N \otimes 1_3, \bar{u}_{eR} 1_N)
\end{aligned}
$$

with $u_j = \exp[iy_j\theta]$ and

$$y_{qL} = q_2, \quad y_{\ell L} = -q_1, \quad y_{uR} = q_1 + q_2, \quad y_{dR} = -q_1 + q_2, \quad y_{eR} = -2q_1. \quad (193)$$

Independently of the embedding, we have indeed *derived* the three fermionic conditions of the hypercharge fine tuning (57). In other words, in noncommutative geometry the massless electro–weak gauge boson necessarily couples vectorially.

Our goal is now to find the minimal extension ℓ that renders the extended lift symmetric, $\mathbb{L}(-u_0, -w) = \mathbb{L}(u_0, w)$, and that renders $\mathbb{L}(1_2, w)$ single-valued. The first requirement means $\{\ \tilde{q}_1 = 1$ and $\tilde{q}_2 = 0\ \}$ modulo 2, with

$$\begin{pmatrix} \tilde{q}_1 \\ \tilde{q}_2 \end{pmatrix} = \frac{1}{3}\left(\begin{pmatrix} q_1 \\ q_2 \end{pmatrix} - \begin{pmatrix} 0 \\ 1 \end{pmatrix} \right). \quad (194)$$

The second requirement means that $\tilde{q}$ has integer coefficients.

The first extension which comes to mind has $q = 0$, $\tilde{q} = \begin{pmatrix} 0 \\ -1/3 \end{pmatrix}$. With respect to the interpretation (168) of the inner automorphisms, one might object that this is not an extension at all. With respect to the *generic* characterization (161), it certainly is a non-trivial extension. Anyhow it fails both tests. The most general extension that passes both tests has the form

$$\tilde{q} = \begin{pmatrix} 2z_1 + 1 \\ 2z_2 \end{pmatrix}, \quad q = \begin{pmatrix} 6z_1 + 3 \\ 6z_2 + 1 \end{pmatrix}, \quad z_1, z_2 \in \mathbb{Z}. \quad (195)$$

Consequently, $y_{\ell_L} = -q_1$ cannot vanish, the neutrino comes out electrically neutral in compliance with the Lorentz force. As common practise, we normalize the hypercharges to $y_{\ell_L} = -1/2$ and compute the last remaining hypercharge y_{q_L},

$$y_{q_L} = \frac{q_2}{2q_1} = \frac{\frac{1}{6} + z_2}{1 + 2z_1}. \quad (196)$$

We can change the sign of y_{q_L} by permuting u with d^c and d with u^c. Therefore it is sufficient to take $z_1 = 0, 1, 2, \ldots$ The minimal such extension, $z_1 = z_2 = 0$, recovers nature's choice $y_{q_L} = \frac{1}{6}$. Its lift,

$$\mathbb{L}(u_0, w) = \rho(u_0, \det w, w) J \rho(u_0, \det w, w) J^{-1}, \quad (197)$$

is the anomaly free fermionic representation of the standard model considered as $SU(2) \times U(3)$. The double-valuedness of $\mathbb{L}$ comes from the discrete group $\mathbb{Z}_2$ of central quaternionic unitaries $(\pm 1_2, 1_3) \in \mathbb{Z}_2 \subset \Gamma \subset U^{nc}(\mathcal{A})$. On the other hand, O'Raifeartaigh's [5] $\mathbb{Z}_2$ in the group of the standard model (45), $\pm(1_2, 1_3) \in \mathbb{Z}_2 \subset U^{nc}(\mathcal{A})$, is not a subgroup of Γ. It reflects the symmetry of $\mathbb{L}$.

Fluctuating Metric. The stage is set now for fluctuating the metric by means of the extended lift. This algorithm answers en passant a long standing question in Yang–Mills theories: To gauge or not to gauge? Given a fermionic Lagrangian, e.g. the one of the standard model, our first reflex is to compute its symmetry group. In noncommutative geometry, this group is simply the internal receptacle (190). The painful question in Yang–Mills theory is what subgroup of this symmetry group should be gauged? For us, this question is answered by the choices of the spectral triple and of the spin lift. Indeed the image of the extended lift is the gauge group. The fluctuating metric promotes its generators to gauge bosons, the $W^\pm$, the Z, the photon and the gluons. At the same time, the Higgs representation is derived, equation (174):

$$H = \rho(u_0, \det w, w)[\tilde{\mathcal{D}}, \rho(u_0, \det w, w)^{-1}] = \begin{pmatrix} 0 & \hat{H} & 0 & 0 \\ \hat{H}^* & 0 & 0 & 0 \\ 0 & 0 & 0 & 0 \\ 0 & 0 & 0 & 0 \end{pmatrix} \tag{198}$$

with

$$\hat{H} = \begin{pmatrix} \begin{pmatrix} h_1 M_u & -\bar{h}_2 M_d \\ h_2 M_u & \bar{h}_1 M_d \end{pmatrix} \otimes 1_3 & 0 \\ 0 & \begin{pmatrix} -\bar{h}_2 M_e \\ \bar{h}_1 M_e \end{pmatrix} \end{pmatrix} \tag{199}$$

and

$$\begin{pmatrix} h_1 & -\bar{h}_2 \\ h_2 & \bar{h}_1 \end{pmatrix} = \pm u_0 \begin{pmatrix} \det w & 0 \\ 0 & \det \bar{w} \end{pmatrix} - 1_2. \tag{200}$$

The Higgs is characterized by one complex doublet, $(h_1, h_2)^T$. Again it will be convenient to pass to the homogeneous Higgs variable,

$$\mathcal{D} = \mathbb{L}\tilde{\mathcal{D}}\mathbb{L}^{-1} = \tilde{\mathcal{D}} + H + JHJ^{-1}$$

$$= \Phi + J\Phi J^{-1} = \begin{pmatrix} 0 & \hat{\Phi} & 0 & 0 \\ \hat{\Phi}^* & 0 & 0 & 0 \\ 0 & 0 & 0 & \hat{\Phi} \\ 0 & 0 & \hat{\Phi}^* & 0 \end{pmatrix} \tag{201}$$

with

$$\hat{\Phi} = \begin{pmatrix} \begin{pmatrix} \varphi_1 M_u & -\bar{\varphi}_2 M_d \\ \varphi_2 M_u & \bar{\varphi}_1 M_d \end{pmatrix} \otimes 1_3 & 0 \\ 0 & \begin{pmatrix} -\bar{\varphi}_2 M_e \\ \bar{\varphi}_1 M_e \end{pmatrix} \end{pmatrix} = \rho_L(\phi)\mathcal{M} \tag{202}$$

and

$$\phi = \begin{pmatrix} \varphi_1 & -\bar{\varphi}_2 \\ \varphi_2 & \bar{\varphi}_1 \end{pmatrix} = \pm u_0 \begin{pmatrix} \det w & 0 \\ 0 & \det \bar{w} \end{pmatrix}. \tag{203}$$

In order to satisfy the first order condition, the representation of $M_3(\mathbb{C}) \ni c$ had to commute with the Dirac operator. Therefore the Higgs is a colour singlet and the gluons will remain massless. The first two of the six intriguing properties of the standard model listed in Sect. 3.3 have a geometric *raison d'être*, the first order condition. In turn, they imply the third property: we have just shown that the Higgs $\varphi = (\varphi_1, \varphi_2)^T$ is a colour singlet. At the same time the fifth property follows from the fourth: the Higgs of the standard model is an isospin doublet because of the parity violating couplings of the quaternions $\mathbb{H}$. Furthermore, this Higgs has hypercharge $y_\varphi = -\frac{1}{2}$ and the last fine tuning of the sixth property (57) also derives from Connes' algorithm: the Higgs has a component with vanishing electric charge, the physical Higgs, and the photon will remain massless.

In conclusion, in Connes version of the standard model there is only one intriguing input property, the fourth: explicit parity violation in the algebra representation $\mathcal{H}_L \oplus \mathcal{H}_R$, the five others are mathematical consequences.

Spectral Action. Computing the spectral action $S_{CC} = f(\mathcal{D}_t^2/\Lambda^2)$ in the standard model is not more difficult than in the minimax example, only the matrices are a little bigger,

$$\mathcal{D}_t = \mathbb{L}_t \tilde{\mathcal{D}}_t \mathbb{L}_t^{-1} = \begin{pmatrix} \partial\!\!\!/_L & \gamma_5\hat{\Phi} & 0 & 0 \\ \gamma_5\hat{\Phi}^* & \partial\!\!\!/_R & 0 & 0 \\ 0 & 0 & C\partial\!\!\!/_L C^{-1} & \gamma_5\bar{\hat{\Phi}} \\ 0 & 0 & \gamma_5\bar{\hat{\Phi}}^* & C\partial\!\!\!/_R C^{-1} \end{pmatrix}. \quad (204)$$

The trace of the powers of $\hat{\Phi}$ are computed from the identities $\hat{\Phi} = \rho_L(\phi)\mathcal{M}$ and $\phi^*\phi = \phi\phi^* = (|\varphi_1|^2 + |\varphi_2|^2)1_2 = |\varphi|^2 1_2$ by using that ρ_L as a representation respects multiplication and involution.

The spectral action produces the complete action of the standard model coupled to gravity with the following relations for coupling constants:

$$g_3^2 = g_2^2 = \tfrac{9}{N}\lambda. \quad (205)$$

Our choice of central charges, $\tilde{q} = (1,0)^T$, entails a further relation, $g_1^2 = \tfrac{3}{5}g_2^2$, i.e. $\sin^2\theta_w = 3/8$. However only products of the Abelian gauge coupling g_1 and the hypercharges y_j appear in the Lagrangian. By rescaling the central charges, we can rescale the hypercharges and consequently the Abelian coupling g_1. It seems quite moral that noncommutative geometry has nothing to say about Abelian gauge couplings.

Experiment tells us that the weak and strong couplings are unequal, equation (49) at energies corresponding to the Z mass, $g_2 = 0.6518 \pm 0.0003$, $g_3 = 1.218 \pm 0.01$. Experiment also tells us that the coupling constants are not constant, but that they evolve with energy. This evolution can be understood theoretically in terms of renormalization: one can get rid of short distance divergencies in perturbative quantum field theory by allowing energy depending gauge, Higgs, and Yukawa couplings where the theoretical evolution depends on the particle content of the model. In the standard model, g_2 and g_3 come together with increasing energy, see Fig. 8. They would become equal at astronomical energies, $\Lambda = 10^{17}$ GeV, if one believed that between presently explored energies, 10^2 GeV,

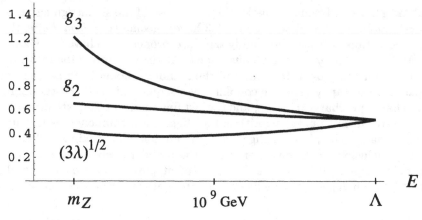

Fig. 8. Running coupling constants

and the 'unification scale' Λ, no new particles exist. This hypothesis has become popular under the name 'big desert' since Grand Unified Theories. It was believed that new gauge bosons, 'lepto-quarks' with masses of order Λ existed. The lepto-quarks together with the $W^{\pm}$, the Z, the photon and the gluons generate the simple group $SU(5)$, with only one gauge coupling, $g_5^2 := g_3^2 = g_2^2 = \frac{5}{3}g_1^2$ at Λ. In the minimal $SU(5)$ model, these lepto-quarks would mediate proton decay with a half life that today is excluded experimentally.

If we believe in the big desert, we can imagine that – while almost commutative at present energies – our geometry becomes truly noncommutative at time scales of $\hbar/\Lambda \sim 10^{-41}$ s. Since in such a geometry smaller time intervals cannot be resolved, we expect the coupling constants to become energy independent at the corresponding energy scale Λ. We remark that the first motivation for noncommutative geometry in spacetime goes back to Heisenberg and was precisely the regularization of short distance divergencies in quantum field theory, see e.g. [45]. The big desert is an opportunistic hypothesis and remains so in the context of noncommutative geometry. But in this context, it has at least the merit of being consistent with three other physical ideas:

Planck time: There is an old hand waving argument combining of phase space with the Schwarzschild horizon to find an uncertainty relation in spacetime with a scale Λ smaller than the Planck energy $(\hbar c^5/G)^{1/2} \sim 10^{19}$ GeV: To measure a position with a precision Δx we need, following Heisenberg, at least a momentum $\hbar/\Delta x$ or, by special relativity, an energy $\hbar c/\Delta x$. According to general relativity, such an energy creates an horizon of size $G\hbar c^{-3}/\Delta x$. If this horizon exceeds Δx all information on the position is lost. We can only resolve positions with Δx larger than the Planck length, $\Delta x > (\hbar G/c^3)^{1/2} \sim 10^{-35}$ m. Or we can only resolve time with Δt larger than the Planck time, $\Delta t > (\hbar G/c^5)^{1/2} \sim 10^{-43}$ s. This is compatible with the above time uncertainty of $\hbar/\Lambda \sim 10^{-41}$ s.

Stability: We want the Higgs self coupling λ to remain positive [46] during its perturbative evolution for all energies up to Λ. A negative Higgs self coupling would mean that no ground state exists, the Higgs potential is unstable. This requirement is met for the self coupling given by the constraint (205) at energy Λ, see Fig. 8.

Triviality: We want the Higgs self coupling λ to remain perturbatively small [46] during its evolution for all energies up to Λ because its evolution is computed from a perturbative expansion. This requirement as well is met for the self coupling given by the constraint (205), see Fig. 8. If the top mass was larger than 231 GeV or if there were $N = 8$ or more generations this criterion would fail.

Since the big desert gives a minimal and consistent picture we are curious to know its numerical implication. If we accept the constraint (205) with $g_2 = 0.5170$ at the energy $\Lambda = 0.968 \; 10^{17}$ GeV and evolve it down to lower energies using the perturbative renormalization flow of the standard model, see Fig. 8, we retrieve the experimental nonAbelian gauge couplings g_2 and g_3 at the Z mass by construction of Λ. For the Higgs coupling, we obtain

$$\lambda = 0.06050 \pm 0.0037 \quad \text{at} \quad E = m_Z. \tag{206}$$

The indicated error comes from the experimental error in the top mass, $m_t = 174.3 \pm 5.1$ GeV, which affects the evolution of the Higgs coupling. From the Higgs coupling at low energies we compute the Higgs mass,

$$m_H = 4\sqrt{2}\,\frac{\sqrt{\lambda}}{g_2}\,m_W = 171.6 \pm 5 \text{ GeV}. \tag{207}$$

For details of this calculation see [47].

6.4   Beyond the Standard Model

A social reason, that made the Yang–Mills–Higgs machine popular, is that it is an inexhaustible source of employment. Even after the standard model, physicists continue to play on the machine and try out extensions of the standard model by adding new particles, 'let the desert bloom'. These particles can be gauge bosons coupling only to right-handed fermions in order to restore left-right symmetry. The added particles can be lepto-quarks for grand unification or supersymmetric particles. These models are carefully tuned not to upset the phenomenological success of the standard model. This means in practice to choose Higgs representations and potentials that give masses to the added particles, large enough to make them undetectable in present day experiments, but not too large so that experimentalists can propose bigger machines to test these models. Independently there are always short lived deviations from the standard model predictions in new experiments. They never miss to trigger new, short lived models with new particles to fit the 'anomalies'. For instance, the literature contains hundreds of superstring inspired s, each of them with hundreds of parameters, coins, waiting for the standard model to fail.

Of course, we are trying the same game in Connes' do–it–yourself kit. So far, we have not been able to find one single consistent extension of the standard model [41–43,48]. The reason is clear, we have no handle on the Higgs representation and potential, which are on the output side, and, in general, we meet two problems: light physical scalars and degenerate fermion masses in irreducible multiplets. The extended standard model with arbitrary numbers of quark generations, $N_q \geq 0$, of lepton generations, $N_\ell \geq 1$, and of colours N_c, somehow manages to avoid both problems and we are trying to prove that it is unique as such. The minimax model has $N_q = 0$, $N_\ell = 1$, $N_c = 0$. The standard model has $N_q = N_\ell =: N$ and $N_c = 3$ to avoid Yang–Mills anomalies [12]. It also has $N = 3$ generations. So far, the only realistic extension of the standard model that we know of in noncommutative geometry, is the addition of right-handed neutrinos and of Dirac masses in one or two generations. These might be necessary to account for observed neutrino oscillations [13].

7   Outlook and Conclusion

Noncommutative geometry reconciles Riemannian geometry and uncertainty and we expect it to reconcile general relativity with quantum field theory. We also expect it to improve our still incomplete understanding of quantum field theory. On the perturbative level such an improvement is happening right now: Connes, Moscovici, and Kreimer discovered a subtle link between a noncommutative generalization of the index theorem and perturbative quantum field theory. This link is a Hopf algebra relevant to both theories [49].

In general, Hopf algebras play the same role in noncommutative geometry as Lie groups play in Riemannian geometry and we expect new examples of noncommutative geometry from its merging with the theory of Hopf algebras. Reference [50] contains a simple example where quantum group techniques can be applied to noncommutative particle models.

The running of coupling constants from perturbative quantum field theory must be taken into account in order to perform the high precision test of the standard model at present day energies. We have invoked an extrapolation of this running to astronomical energies to make the constraint $g_2 = g_3$ from the spectral action compatible with experiment. This extrapolation is still based on quantum loops in flat Minkowski space. While acceptable at energies below the scale Λ where gravity and the noncommutativity of space seem negligible, this approximation is unsatisfactory from a conceptual point of view and one would like to see quantum fields constructed on a noncommutative space. At the end of the nineties first examples of quantum fields on the (flat) noncommutative torus or its non-compact version, the Moyal plane, were published [51]. These examples came straight from the spectral action. The noncommutative torus is motivated from quantum mechanical phase space and was the first example of a noncommutative spectral triple [52]. Bellissard [53] has shown that the noncommutative torus is relevant in solid state physics: one can understand the quantum Hall effect by taking the Brillouin zone to be noncommutative. Only recently other examples of noncommutative spaces like noncommutative spheres where uncovered [54]. Since 1999, quantum fields on the noncommutative torus are being studied extensively including the fields of the standard model [55]. So far, its internal part is not treated as a noncommutative geometry and Higgs bosons and potentials are added opportunistically. This problem is avoided naturally by considering the tensor product of the noncommutative torus with a finite spectral triple, but I am sure that the axioms of noncommutative geometry can be rediscovered by playing long enough with model building.

In quantum mechanics and in general relativity, time and space play radically different roles. Spatial position is an observable in quantum mechanics, time is not. In general relativity, spacial position loses all meaning and only proper time can be measured. Distances are then measured by a particular observer as (his proper) time of flight of photons going back and forth multiplied by the speed of light, which is supposed to be universal. This definition of distances is operational thanks to the high precision of present day atomic clocks, for example in the GPS. The 'Riemannian' definition of the meter, the forty millionth part of a complete geodesic on earth, had to be abandoned in favour of a quantum mechanical definition of the second via the spectrum of an atom. Connes' definition of geometry via the spectrum of the Dirac operator is the precise counter part of today's experimental situation. Note that the meter stick is an extended (rigid ?) object. On the other hand an atomic clock is a pointlike object and experiment tells us that the atom is sensitive to the potentials at the location of the clock, the potentials of all forces, gravitational, electro–magnetic, ... The special role of time remains to be understood in noncommutative geometry [56] as well as the notion of spectral triples with Lorentzian signature and their 1+3 split [57].

Let us come back to our initial claim: Connes derives the standard model of electro–magnetic, weak, and strong forces from noncommutative geometry and, at the same time, unifies them with gravity. If we say that the Balmer–Rydberg formula is derived from quantum mechanics, then this claim has three levels:

Explain the nature of the variables: The choice of the discrete variables n_j, contains already a – at the time revolutionary – piece of physics, energy quantization. Where does it come from?

Explain the ansatz: Why should one take the power law (11)?

Explain the experimental fit: The ansatz comes with discrete parameters, the 'bills' q_j, and continuous parameters, the 'coins' g_j, which are determined by an experimental fit. Where do the fitted values, 'the winner', come from?

How about deriving gravity from Riemannian geometry? Riemannian geometry has only one possible variable, the metric g. The minimax principle dictates the Lagrangian ansatz:

$$S[g] = \int_M [\Lambda^c - \tfrac{1}{16\pi G} R^q] \, dV. \tag{208}$$

Experiment rules on the parameters: $q = 1$, $G = 6.670 \cdot 10^{-11}$ m^{3}s^{-2}kg, Newton's constant, and $\Lambda^c \sim 0$. Riemannian geometry remains silent on the third

Table 3. Deriving some YMH forces from gravity

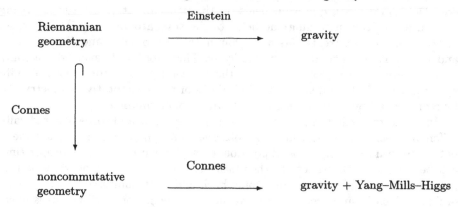

level. Nevertheless, there is general agreement, gravity derives from Riemannian geometry.

Noncommutative geometry has only one possible variable, the Dirac operator, which in the commutative case coincides with the metric. Its fluctuations explain the variables of the additional forces, gauge and Higgs bosons. The minimax principle dictates the Lagrangian ansatz: the spectral action. It reproduces the Einstein–Hilbert action and the ansatz of Yang, Mills and Higgs, see Table 3. On the third level, noncommutative geometry is not silent, it produces lots of constraints, all compatible with the experimental fit. And their exploration is not finished yet.

I hope to have convinced one or the other reader that noncommutative geometry contains elegant solutions of long standing problems in fundamental physics and that it proposes concrete strategies to tackle the remaining ones. I would like to conclude our outlook with a sentence by Planck who tells us how important the opinion of our young, unbiased colleagues is. Planck said, a new theory is accepted, not because the others are convinced, because they die.

Acknowledgements

It is a pleasure to thank Eike Bick and Frank Steffen for the organization of a splendid School. I thank the participants for their unbiased criticism and Kurusch Ebrahimi-Fard, Volker Schatz, and Frank Steffen for a careful reading of the manuscript.

Appendix

A.1   Groups

Groups are an extremely powerful tool in physics. Most symmetry transformations form a group. Invariance under continuous transformation groups entails conserved quantities, like energy, angular momentum or electric charge.

A group G is a set equipped with an associative, not necessarily commutative (or 'Abelian') multiplication law that has a neutral element 1. Every group element g is supposed to have an inverse g^{-1}.

We denote by $\mathbb{Z}_n$ the *cyclic group* of n elements. You can either think of $\mathbb{Z}_n$ as the set $\{0, 1, ..., n-1\}$ with multiplication law being addition modulo n and neutral element 0. Or equivalently, you can take the set $\{1, \exp(2\pi i/n), \exp(4\pi i/n), ..., \exp((n-1)2\pi i/n)\}$ with multiplication and neutral element 1. $\mathbb{Z}_n$ is an Abelian subgroup of the permutation group on n objects.

Other immediate examples are matrix groups: The *general linear groups* $GL(n, \mathbb{C})$ and $GL(n, \mathbb{R})$ are the sets of complex (real), invertible $n \times n$ matrices. The multiplication law is matrix multiplication and the neutral element is the $n \times n$ unit matrix 1_n. There are many important subgroups of the general linear groups: $SL(n, \cdot)$, $\cdot = \mathbb{R}$ or $\mathbb{C}$, consist only of matrices with unit determinant. S stands for special and will always indicate that we add the condition of unit determinant. The *orthogonal group* $O(n)$ is the group of real $n \times n$ matrices g satisfying $gg^T = 1_n$. The *special orthogonal group* $SO(n)$ describes the rotations in the Euclidean space $\mathbb{R}^n$. The *Lorentz group* $O(1, 3)$ is the set of real 4×4 matrices g satisfying $g\eta g^T = \eta$, with $\eta = \mathrm{diag}\{1, -1, -1, -1\}$. The *unitary group* $U(n)$ is the set of complex $n \times n$ matrices g satisfying $gg^* = 1_n$. The *unitary symplectic group* $USp(n)$ is the group of complex $2n \times 2n$ matrices g satisfying $gg^* = 1_{2n}$ and $g\mathcal{I}g^T = \mathcal{I}$ with

$$
\mathcal{I} := \begin{pmatrix} \begin{pmatrix} 0 & 1 \\ -1 & 0 \end{pmatrix} & \cdots & 0 \\ \vdots & \ddots & \vdots \\ 0 & \cdots & \begin{pmatrix} 0 & 1 \\ -1 & 0 \end{pmatrix} \end{pmatrix} . \tag{A.1}
$$

The **center** $Z(G)$ of a group G consists of those elements in G that commute with all elements in G, $Z(G) = \{z \in G, zg = gz \text{ for all } g \in G\}$. For example, $Z(U(n)) = U(1) \ni \exp(i\theta) 1_n$, $Z(SU(n)) = \mathbb{Z}_n \ni \exp(2\pi ik/n) 1_n$.

All matrix groups are subsets of $\mathbb{R}^{2n^2}$ and therefore we can talk about **compactness** of these groups. Recall that a subset of $\mathbb{R}^N$ is compact if and only if it is closed and bounded. For instance, $U(1)$ is a circle in $\mathbb{R}^2$ and therefore compact. The Lorentz group on the other hand is unbounded because of the boosts.

The matrix groups are *Lie groups* which means that they contain infinitesimal elements X close to the neutral element: $\exp X = 1 + X + O(X^2) \in G$. For instance,

$$
X = \begin{pmatrix} 0 & \epsilon & 0 \\ -\epsilon & 0 & 0 \\ 0 & 0 & 0 \end{pmatrix}, \quad \epsilon \text{ small}, \tag{A.2}
$$

describes an infinitesimal rotation around the z-axis by an infinitesimal angle ϵ. Indeed

$$\exp X = \begin{pmatrix} \cos\epsilon & \sin\epsilon & 0 \\ -\sin\epsilon & \cos\epsilon & 0 \\ 0 & 0 & 1 \end{pmatrix} \in SO(3), \quad 0 \le \epsilon < 2\pi, \qquad (A.3)$$

is a rotation around the z-axis by an arbitrary angle ϵ. The infinitesimal transformations X of a Lie group G form its **Lie algebra** $\mathfrak{g}$. It is closed under the commutator $[X, Y] = XY - YX$. For the above matrix groups the Lie algebras are denoted by lower case letters. For example, the Lie algebra of the special unitary group $SU(n)$ is written as $su(n)$. It is the set of complex $n \times n$ matrices X satisfying $X + X^* = 0$ and $\operatorname{tr} X = 0$. Indeed, $1_n = (1_n + X + ...)(1_n + X + ...)^* = 1_n + X + X^* + O(X^2)$ and $1 = \det \exp X = \exp \operatorname{tr} X$. Attention, although defined in terms of complex matrices, $su(n)$ is a **real** vector space. Indeed, if a matrix X is anti-Hermitean, $X + X^* = 0$, then in general, its complex scalar multiple iX is no longer anti-Hermitean.

However, in real vector spaces, eigenvectors do not always exist and we will have to **complexify** the real vector space $\mathfrak{g}$: Take a basis of $\mathfrak{g}$. Then $\mathfrak{g}$ consists of linear combinations of these basis vectors with real coefficients. The **complexification** $\mathfrak{g}^{\mathbb{C}}$ of $\mathfrak{g}$ consits of linear combinations with complex coefficients.

The *translation group* of $\mathbb{R}^n$ is $\mathbb{R}^n$ itself. The multiplication law now is vector addition and the neutral element is the zero vector. As the vector addition is commutative, the translation group is Abelian.

The *diffeomorphism group* $\mathrm{Diff}(M)$ of an open subset M of $\mathbb{R}^n$ (or of a manifold) is the set of differentiable maps σ from M into itself that are invertible (for the composition $\circ$) and such that its inverse is differentiable. (Attention, the last condition is not automatic, as you see by taking $M = \mathbb{R} \ni x$ and $\sigma(x) = x^3$.) By virtue of the chain rule we can take the composition as multiplication law. The neutral element is the identity map on M, $\sigma = 1_M$ with $1_M(x) = x$ for all $x \in M$.

A.2   Group Representations

We said that $SO(3)$ is the rotation group. This needs a little explanation. A rotation is given by an axis, that is a unit eigenvector with unit eigenvalue, and an angle. Two rotations can be carried out one after the other, we say 'composed'. Note that the order is important, we say that the 3-dimensional rotation group is nonAbelian. If we say that the rotations form a group, we mean that the composition of two rotations is a third rotation. However, it is not easy to compute the multiplication law, i.e., compute the axis and angle of the third rotation as a function of the axes and angles of the two initial rotations. The equivalent 'representation' of the rotation group as 3×3 matrices is much more convenient because the multiplication law is simply matrix multiplication. There are several 'representations' of the 3-dimensional rotation group in terms of matrices of different sizes, say $N \times N$. It is sometimes useful to know all these representations. The $N \times N$ matrices are linear maps, 'endomorphisms', of the N-dimensional vector space $\mathbb{R}^N$ into itself. Let us denote by $\text{End}(\mathbb{R}^N)$ the set of all these matrices. By definition, a representation of the group G on the vector space $\mathbb{R}^N$ is a map $\rho : G \to \text{End}(\mathbb{R}^N)$ reproducing the multiplication law as matrix multiplication or in nobler terms as composition of endomorphisms. This means $\rho(g_1 g_2) = \rho(g_1) \rho(g_2)$ and $\rho(1) = 1_N$. The representation is called **faithful** if the map ρ is injective. By the minimax principle we are interested in the faithful representations of lowest dimension. Although not always unique, physicists call them fundamental representations. The fundamental representation of the 3-dimensional rotation group is defined on the vector space $\mathbb{R}^3$. Two N-dimensional representations ρ_1 and ρ_2 of a group G are **equivalent** if there is an invertible $N \times N$ matrix C such that $\rho_2(g) = C\rho_1(g)C^{-1}$ for all $g \in G$. C is interpreted as describing a change of basis in $\mathbb{R}^N$. A representation is called **irreducible** if its vector space has no proper invariant subspace, i.e. a subspace $W \subset \mathbb{R}^N$, with $W \neq \mathbb{R}^N, \{0\}$ and $\rho(g)W \subset W$ for all $g \in G$.

Representations can be defined in the same manner on complex vector spaces, $\mathbb{C}^N$. Then every $\rho(g)$ is a complex, invertible matrix. It is often useful, e.g. in quantum mechanics, to represent a group on a Hilbert space, we put a scalar product on the vector space, e.g. the standard scalar product on $\mathbb{C}^N \ni v, w$, $(v, w) := v^* w$. A **unitary** representation is a representation whose matrices $\rho(g)$ all respect the scalar product, which means that they are all unitary. In quantum mechanics, unitary representations are important because they preserve probability. For example, take the **adjoint representation** of $SU(n) \ni g$. Its Hilbert space is the complexification of its Lie algebra $su(n)^{\mathbb{C}} \ni X, Y$ with scalar product $(X, Y) := \text{tr}(X^* Y)$. The representation is defined by conjugation, $\rho(g)X := gXg^{-1}$, and it is unitary, $(\rho(g)X, \rho(g)Y) = (X, Y)$. In Yang–Mills theories, the gauge bosons live in the adjoint representation. In the Abelian case, $G = U(1)$, this representation is 1-dimensional, there is one gauge boson, the photon, $A \in u(1)^{\mathbb{C}} = \mathbb{C}$. The photon has no electric charge, which means that it transforms trivially, $\rho(g)A = A$ for all $g \in U(1)$.

Unitary equivalence of representations is defined by change of orthonormal bases. Then C is a unitary matrix. A key theorem for particle physics states that all irreducible unitary representations of any compact group are finite dimensional. If we accept the definition of elementary particles as orthonormal basis vectors of unitary representations, then we understand why Yang and Mills only take compact groups. They only want a finite number of elementary particles. Unitary equivalence expresses the quantum mechanical superposition principle observed for instance in the $K^0 - \bar{K}^0$ system. The unitary matrix C is sometimes referred to as mixing matrix.

Bound states of elementary particles are described by tensor products: the tensor product of two unitary representations ρ_1 and ρ_2 of one group defined on two Hilbert spaces $\mathcal{H}_1$ and $\mathcal{H}_2$ is the unitary representation $\rho_1 \otimes \rho_2$ defined on $\mathcal{H}_1 \otimes \mathcal{H}_2 \ni \psi_1 \otimes \psi_2$ by $(\rho_1 \otimes \rho_2)(g)(\psi_1 \otimes \psi_2) := \rho_1(g)\psi_1 \otimes \rho_2(g)\psi_2$. In the case of electro–magnetism, $G = U(1) \ni \exp(i\theta)$ we know that all irreducible unitary representations are 1-dimensional, $\mathcal{H} = \mathbb{C} \ni \psi$ and characterized by the electric charge q, $\rho(\exp(i\theta))\psi = \exp(iq\psi)\psi$. Under tensorization the electric charges are added. For $G = SU(2)$, the irreducible unitary representations are characterized by the spin, $\ell = 0, \frac{1}{2}, 1, \ldots$ The addition of spin from quantum mechanics is precisely tensorization of these representations.

Let ρ be a representation of a Lie group G on a vector space and let $\mathfrak{g}$ be the Lie algebra of G. We denote by $\tilde{\rho}$ the Lie algebra representation of the group representation ρ. It is defined on the same vector space by $\rho(\exp X) = \exp(\tilde{\rho}(X))$. The $\tilde{\rho}(X)$s are not necessarily invertible endomorphisms. They satisfy $\tilde{\rho}([X, Y]) = [\tilde{\rho}(X), \tilde{\rho}(Y)] := \tilde{\rho}(X)\tilde{\rho}(Y) - \tilde{\rho}(Y)\tilde{\rho}(X)$.

An **affine** representation is the same construction as above, but we allow the $\rho(g)$s to be invertible affine maps, i.e. linear maps plus constants.

A.3   Semi-Direct Product and Poincaré Group

The direct product $G \times H$ of two groups G and H is again a group with multiplication law: $(g_1, h_1)(g_2, h_2) := (g_1 g_2, h_1 h_2)$. In the direct product, all elements of the first factor commute with all elements of the second factor: $(g, 1^H)(1^G, h) = (1^G, h)(g, 1^H)$. We write 1^H for the neutral element of H. Warning, you sometimes see the misleading notation $G \otimes H$ for the direct product.

To be able to define the semi-direct product $G \ltimes H$ we must have an action of G on H, that is a map $\rho : G \to \text{Diff}(H)$ satisfying $\rho_g(h_1 h_2) = \rho_g(h_1)\,\rho_g(h_2)$, $\rho_g(1^H) = 1^H$, $\rho_{g_1 g_2} = \rho_{g_1} \circ \rho_{g_2}$ and $\rho_{1^G} = 1_H$. If H is a vector space carrying a representation or an affine representation ρ of the group G, we can view ρ as an action by considering H as translation group. Indeed, invertible linear maps and affine maps are diffeomorphisms on H. As a set, the semi-direct product $G \ltimes H$ is the direct product, but the multiplication law is modified by help of the action:

$$(g_1, h_1)(g_2, h_2) := (g_1 g_2, h_1\, \rho_{g_1}(h_2)). \tag{A.4}$$

We retrieve the direct product if the action is trivial, $\rho_g = 1_H$ for all $g \in G$. Our first example is the invariance group of electro–magnetism coupled to gravity $\text{Diff}(M) \ltimes {}^M U(1)$. A diffeomorphism $\sigma(x)$ acts on a gauge function $g(x)$ by $\rho_\sigma(g) := g \circ \sigma^{-1}$ or more explicitly $(\rho_\sigma(g))(x) := g(\sigma^{-1}(x))$. Other examples come with other gauge groups like $SU(n)$ or spin groups.

Our second example is the Poincaré group, $O(1, 3) \ltimes \mathbb{R}^4$, which is the isometry group of Minkowski space. The semi-direct product is important because Lorentz transformations do not commute with translations. Since we are talking about the Poincaré group, let us mention the theorem behind the definition of particles as orthonormal basis vectors of unitary representations: The irreducible, unitary representations of the Poincaré group are characterized by mass and spin. For fixed mass $M \geq 0$ and spin ℓ, an orthonormal basis is labelled by the momentum $\boldsymbol{p}$ with $E^2/c^2 - \boldsymbol{p}^2 = c^2 M^2$, $\psi = \exp(i(Et - \boldsymbol{p} \cdot \boldsymbol{x})/\hbar)$ and the z-component m of the spin with $|m| \leq \ell$, $\psi = Y_{\ell,m}(\theta, \varphi)$.

A.4   Algebras

Observables can be added, multiplied and multiplied by scalars. They form naturally an associative algebra $\mathcal{A}$, i.e. a vector space equipped with an associative product and neutral elements 0 and 1. Note that the multiplication does not always admit inverses, a^{-1}, e.g. the neutral element of addition, 0, is not invertible. In quantum mechanics, observables are self adjoint. Therefore, we need an involution $\cdot^*$ in our algebra. This is an anti-linear map from the algebra into itself, $(\lambda a + b)^* = \bar{\lambda} a^* + b^*$, $\lambda \in \mathbb{C}$, $a, b \in \mathcal{A}$, that reverses the product, $(ab)^* = b^* a^*$, respects the unit, $1^* = 1$, and is such that $a^{**} = a$. The set of $n \times n$ matrices with complex coefficients, $M_n(\mathbb{C})$, is an example of such an algebra, and more generally, the set of endomorphisms or operators on a given Hilbert space $\mathcal{H}$. The multiplication is matrix multiplication or more generally composition of operators, the involution is Hermitean conjugation or more generally the adjoint of operators.

A representation ρ of an abstract algebra $\mathcal{A}$ on a Hilbert space $\mathcal{H}$ is a way to write $\mathcal{A}$ concretely as operators as in the last example, $\rho : \mathcal{A} \to \text{End}(\mathcal{H})$. In the group case, the representation had to reproduce the multiplication law. Now it has to reproduce, the linear structure: $\rho(\lambda a + b) = \lambda \rho(a) + \rho(b)$, $\rho(0) = 0$, the multiplication: $\rho(ab) = \rho(a)\rho(b)$, $\rho(1) = 1$, and the involution: $\rho(a^*) = \rho(a)^*$. Therefore the tensor product of two representations ρ_1 and ρ_2 of $\mathcal{A}$ on Hilbert spaces $\mathcal{H}_1 \ni \psi_1$ and $\mathcal{H}_2 \ni \psi_2$ is not a representation: $((\rho_1 \otimes \rho_2)(\lambda a))(\psi_1 \otimes \psi_2) = (\rho_1(\lambda a)\psi_1) \otimes (\rho_2(\lambda a)\psi_2) = \lambda^2 (\rho_1 \otimes \rho_2)(a)(\psi_1 \otimes \psi_2)$.

The group of unitaries $U(\mathcal{A}) := \{u \in \mathcal{A}, uu^* = u^*u = 1\}$ is a subset of the algebra $\mathcal{A}$. Every algebra representation induces a unitary representation of its group of unitaries. On the other hand, only few unitary representations of the group of unitaries extend to an algebra representation. These representations describe elementary particles. Composite particles are described by tensor products, which are not algebra representations.

An anti-linear operator J on a Hilbert space $\mathcal{H} \ni \psi, \tilde{\psi}$ is a map from $\mathcal{H}$ into itself satisfying $J(\lambda \psi + \tilde{\psi}) = \bar{\lambda} J(\psi) + J(\tilde{\psi})$. An anti-linear operator J is anti-unitary if it is invertible and preserves the scalar product, $(J\psi, J\tilde{\psi}) = (\tilde{\psi}, \psi)$. For example, on $\mathcal{H} = \mathbb{C}^n \ni \psi$ we can define an anti-unitary operator J in the following way. The image of the column vector ψ under J is obtained by taking the complex conjugate of ψ and then multiplying it with a unitary $n \times n$ matrix U, $J\psi = U\bar{\psi}$ or $J = U \circ$ complex conjugation. In fact, on a finite dimensional Hilbert space, every anti-unitary operator is of this form.

References

1. A. Connes, A. Lichnérowicz and M. P. Schützenberger, *Triangle de Pensées*, O. Jacob (2000), English version: *Triangle of Thoughts*, AMS (2001)
2. G. Amelino-Camelia, *Are we at the dawn of quantum gravity phenomenology?*, Lectures given at 35th Winter School of Theoretical Physics: From Cosmology to Quantum Gravity, Polanica, Poland, 1999, gr-qc/9910089
3. S. Weinberg, *Gravitation and Cosmology*, Wiley (1972)
   R. Wald, *General Relativity*, The University of Chicago Press (1984)
4. J. D. Bjørken and S. D. Drell, *Relativistic Quantum Mechanics*, McGraw–Hill (1964)
5. L. O'Raifeartaigh, *Group Structure of Gauge Theories*, Cambridge University Press (1986)

6. M. Göckeler and T. Schücker, *Differential Geometry, Gauge Theories, and Gravity*, Cambridge University Press (1987)
7. R. Gilmore, *Lie Groups, Lie Algebras and some of their Applications*, Wiley (1974)
   H. Bacry, *Lectures Notes in Group Theory and Particle Theory*, Gordon and Breach (1977)
8. N. Jacobson, *Basic Algebra I, II*, Freeman (1974,1980)
9. J. Madore, *An Introduction to Noncommutative Differential Geometry and its Physical Applications*, Cambridge University Press (1995)
   G. Landi, *An Introduction to Noncommutative Spaces and their Geometry*, hep-th/9701078, Springer (1997)
10. J. M. Gracia-Bondía, J. C. Várilly and H. Figueroa, *Elements of Noncommutative Geometry*, Birkhäuser (2000)
11. J. W. van Holten, *Aspects of BRST quantization*, hep-th/0201124, in this volume
12. J. Zinn-Justin, *Chiral anomalies and topology*, hep-th/0201220, in this volume
13. The Particle Data Group, *Particle Physics Booklet* and http://pdg.lbl.gov
14. G. 't Hooft, *Renormalizable Lagrangians for Massive Yang–Mills Fields*, Nucl. Phys. B35 (1971) 167
    G. 't Hooft and M. Veltman, *Regularization and Renormalization of Gauge Fields*, Nucl. Phys. B44 (1972) 189
    G. 't Hooft and M. Veltman, *Combinatorics of Gauge Fields*, Nucl. Phys. B50 (1972) 318
    B. W. Lee and J. Zinn-Justin, *Spontaneously broken gauge symmetries I, II, III and IV*, Phys. Rev. D5 (1972) 3121, 3137, 3155; Phys. Rev. D7 (1973) 1049
15. S. Glashow, *Partial-symmetries of weak interactions*, Nucl. Phys. 22 (1961) 579
    A. Salam in Elementary Particle Physics: Relativistic Groups and Analyticity, Nobel Symposium no. 8, page 367, eds.: N. Svartholm, Almqvist and Wiksell, Stockholm 1968
    S. Weinberg, *A model of leptons*, Phys. Rev. Lett. 19 (1967) 1264
16. J. Iliopoulos, *An introduction to gauge theories*, Yellow Report, CERN (1976)
17. G. Esposito-Farèse, *Théorie de Kaluza–Klein et gravitation quantique*, Thèse de Doctorat, Université d'Aix-Marseille II, 1989
18. A. Connes, *Noncommutative Geometry*, Academic Press (1994)
19. A. Connes, *Noncommutative Geometry and Reality*, J. Math. Phys. 36 (1995) 6194
20. A. Connes, *Gravity coupled with matter and the foundation of noncommutative geometry*, hep-th/9603053, Comm. Math. Phys. 155 (1996) 109
21. H. Rauch, A. Zeilinger, G. Badurek, A. Wilfing, W. Bauspiess and U. Bonse, *Verification of coherent spinor rotations of fermions*, Phys. Lett. 54A (1975) 425
22. E. Cartan, *Leçons sur la théorie des spineurs*, Hermann (1938)
23. A. Connes, *Brisure de symétrie spontanée et géométrie du point de vue spectral*, Séminaire Bourbaki, 48ème année, 816 (1996) 313
    A. Connes, *Noncommutative differential geometry and the structure of space time*, Operator Algebras and Quantum Field Theory, eds.: S. Doplicher et al., International Press, 1997
24. T. Schücker, *Spin group and almost commutative geometry*, hep-th/0007047
25. J.-P. Bourguignon and P. Gauduchon, *Spineurs, opérateurs de Dirac et variations de métriques*, Comm. Math. Phys. 144 (1992) 581
26. U. Bonse and T. Wroblewski, *Measurement of neutron quantum interference in noninertial frames*, Phys. Rev. Lett. 1 (1983) 1401
27. R. Colella, A. W. Overhauser and S. A. Warner, *Observation of gravitationally induced quantum interference*, Phys. Rev. Lett. 34 (1975) 1472

28. A. Chamseddine and A. Connes, *The spectral action principle*, hep-th/9606001, Comm. Math. Phys.186 (1997) 731

29. G. Landi and C. Rovelli, *Gravity from Dirac eigenvalues*, gr-qc/9708041, Mod. Phys. Lett. A13 (1998) 479

30. P. B. Gilkey, *Invariance Theory, the Heat Equation, and the Atiyah–Singer Index Theorem*, Publish or Perish (1984)
    S. A. Fulling, *Aspects of Quantum Field Theory in Curved Space-Time*, Cambridge University Press (1989)

31. B. Iochum, T. Krajewski and P. Martinetti, *Distances in finite spaces from non-commutative geometry*, hep-th/9912217, J. Geom. Phys. 37 (2001) 100

32. M. Dubois-Violette, R. Kerner and J. Madore, *Gauge bosons in a noncommutative geometry*, Phys. Lett. 217B (1989) 485

33. A. Connes, *Essay on physics and noncommutative geometry*, in *The Interface of Mathematics and Particle Physics*, eds.: D. G. Quillen et al., Clarendon Press (1990)
    A. Connes and J. Lott, *Particle models and noncommutative geometry*, Nucl. Phys. B 18B (1990) 29
    A. Connes and J. Lott, *The metric aspect of noncommutative geometry*, in the proceedings of the 1991 Cargèse Summer Conference, eds.: J. Fröhlich et al., Plenum Press (1992)

34. J. Madore, *Modification of Kaluza Klein theory*, Phys. Rev. D 41 (1990) 3709

35. P. Martinetti and R. Wulkenhaar, *Discrete Kaluza–Klein from Scalar Fluctuations in Noncommutative Geometry*, hep-th/0104108, J. Math. Phys. 43 (2002) 182

36. T. Ackermann and J. Tolksdorf, *A generalized Lichnerowicz formula, the Wodzicki residue and gravity*, hep-th/9503152, J. Geom. Phys. 19 (1996) 143
    T. Ackermann and J. Tolksdorf, *The generalized Lichnerowicz formula and analysis of Dirac operators*, hep-th/9503153, J. reine angew. Math. 471 (1996) 23

37. R. Estrada, J. M. Gracia-Bondía and J. C. Várilly, *On summability of distributions and spectral geometry*, funct-an/9702001, Comm. Math. Phys. 191 (1998) 219

38. B. Iochum, D. Kastler and T. Schücker, *On the universal Chamseddine–Connes action: details of the action computation*, hep-th/9607158, J. Math. Phys. 38 (1997) 4929
    L. Carminati, B. Iochum, D. Kastler and T. Schücker, *On Connes' new principle of general relativity: can spinors hear the forces of space-time?*, hep-th/9612228, Operator Algebras and Quantum Field Theory, eds.: S. Doplicher et al., International Press, 1997

39. M. Paschke and A. Sitarz, *Discrete spectral triples and their symmetries*, q-alg/9612029, J. Math. Phys. 39 (1998) 6191
    T. Krajewski, *Classification of finite spectral triples*, hep-th/9701081, J. Geom. Phys. 28 (1998) 1

40. S. Lazzarini and T. Schücker, *A farewell to unimodularity*, hep-th/0104038, Phys. Lett. B 510 (2001) 277

41. B. Iochum and T. Schücker, *A left-right symmetric model à la Connes–Lott*, hep-th/9401048, Lett. Math. Phys. 32 (1994) 153
    F. Girelli, *Left-right symmetric models in noncommutative geometry?* hep-th/0011123, Lett. Math. Phys. 57 (2001) 7

42. F. Lizzi, G. Mangano, G. Miele and G. Sparano, *Constraints on unified gauge theories from noncommutative geometry*, hep-th/9603095, Mod. Phys. Lett. A11 (1996) 2561

43. W. Kalau and M. Walze, *Supersymmetry and noncommutative geometry*, hep-th/9604146, J. Geom. Phys. 22 (1997) 77

44. D. Kastler, *Introduction to noncommutative geometry and Yang–Mills model building*, Differential geometric methods in theoretical physics, Rapallo (1990), 25

— , *A detailed account of Alain Connes' version of the standard model in noncommutative geometry, I, II and III*, Rev. Math. Phys. 5 (1993) 477, Rev. Math. Phys. 8 (1996) 103

D. Kastler and T. Schücker, *Remarks on Alain Connes' approach to the standard model in non-commutative geometry*, Theor. Math. Phys. 92 (1992) 522, English version, 92 (1993) 1075, hep-th/0111234

— , *A detailed account of Alain Connes' version of the standard model in noncommutative geometry, IV*, Rev. Math. Phys. 8 (1996) 205

— , *The standard model à la Connes–Lott*, hep-th/9412185, J. Geom. Phys. 388 (1996) 1

J. C. Várilly and J. M. Gracia-Bondía, *Connes' noncommutative differential geometry and the standard model*, J. Geom. Phys. 12 (1993) 223

T. Schücker and J.-M. Zylinski, *Connes' model building kit*, hep-th/9312186, J. Geom. Phys. 16 (1994) 1

E. Alvarez, J. M. Gracia-Bondía and C. P. Martín, *Anomaly cancellation and the gauge group of the Standard Model in Non-Commutative Geometry*, hep-th/9506115, Phys. Lett. B364 (1995) 33

R. Asquith, *Non-commutative geometry and the strong force*, hep-th/9509163, Phys. Lett. B 366 (1996) 220

C. P. Martín, J. M. Gracia-Bondía and J. C. Várilly, *The standard model as a noncommutative geometry: the low mass regime*, hep-th/9605001, Phys. Rep. 294 (1998) 363

L. Carminati, B. Iochum and T. Schücker, *The noncommutative constraints on the standard model à la Connes*, hep-th/9604169, J. Math. Phys. 38 (1997) 1269

R. Brout, *Notes on Connes' construction of the standard model*, hep-th/9706200, Nucl. Phys. Proc. Suppl. 65 (1998) 3

J. C. Várilly, *Introduction to noncommutative geometry*, physics/9709045, EMS Summer School on Noncommutative Geometry and Applications, Portugal, september 1997, ed.: P. Almeida

T. Schücker, *Geometries and forces*, hep-th/9712095, EMS Summer School on Noncommutative Geometry and Applications, Portugal, september 1997, ed.: P. Almeida

J. M. Gracia-Bondía, B. Iochum and T. Schücker, *The Standard Model in Non-commutative Geometry and Fermion Doubling*, hep-th/9709145, Phys. Lett. B 414 (1998) 123

D. Kastler, *Noncommutative geometry and basic physics*, Lect. Notes Phys. 543 (2000) 131

— , *Noncommutative geometry and fundamental physical interactions: the Lagrangian level*, J. Math. Phys. 41 (2000) 3867

K. Elsner, *Noncommutative geometry: calculation of the standard model Lagrangian*, hep-th/0108222, Mod. Phys. Lett. A16 (2001) 241

45. R. Jackiw, *Physical instances of noncommuting coordinates*, hep-th/0110057

46. N. Cabibbo, L. Maiani, G. Parisi and R. Petronzio, *Bounds on the fermions and Higgs boson masses in grand unified theories*, Nucl. Phys. B158 (1979) 295

47. L. Carminati, B. Iochum and T. Schücker, *Noncommutative Yang–Mills and noncommutative relativity: A bridge over troubled water*, hep-th/9706105, Eur. Phys. J. C8 (1999) 697

48. B. Iochum and T. Schücker, *Yang–Mills–Higgs versus Connes–Lott*, hep-th/9501142, Comm. Math. Phys. 178 (1996) 1

I. Pris and T. Schücker, *Non-commutative geometry beyond the standard model*, hep-th/9604115, J. Math. Phys. 38 (1997) 2255

I. Pris and T. Krajewski, *Towards a Z' gauge boson in noncommutative geometry*, hep-th/9607005, Lett. Math. Phys. 39 (1997) 187

M. Paschke, F. Scheck and A. Sitarz, *Can (noncommutative) geometry accommodate leptoquarks?* hep-th/9709009, Phys . Rev. D59 (1999) 035003

T. Schücker and S. ZouZou, *Spectral action beyond the standard model*, hep-th/0109124

49. A. Connes and H. Moscovici, *Hopf algebra, cyclic cohomology and the transverse index theorem*, Comm. Math. Phys. 198 (1998) 199

D. Kreimer, *On the Hopf algebra structure of perturbative quantum field theories*, q-alg/9707029, Adv. Theor. Math. Phys. 2 (1998) 303

A. Connes and D. Kreimer, *Renormalization in quantum field theory and the Riemann-Hilbert problem. 1. The Hopf algebra structure of graphs and the main theorem*, hep-th/9912092, Comm. Math. Phys. 210 (2000) 249

A. Connes and D. Kreimer, *Renormalization in quantum field theory and the Riemann–Hilbert problem. 2. The beta function, diffeomorphisms and the renormalization group*, hep-th/0003188, Comm. Math. Phys. 216 (2001) 215

for a recent review, see J. C. Várilly, *Hopf algebras in noncommutative geometry*, hep-th/010977

50. S. Majid and T. Schücker, $\mathbb{Z}_2 \times \mathbb{Z}_2$ *Lattice as Connes–Lott-quantum group model*, hep-th/0101217, J. Geom. Phys. 43 (2002) 1

51. J. C. Várilly and J. M. Gracia-Bondía, *On the ultraviolet behaviour of quantum fields over noncommutative manifolds*, hep-th/9804001, Int. J. Mod. Phys. A14 (1999) 1305

T. Krajewski, *Géométrie non commutative et interactions fondamentales*, Thése de Doctorat, Université de Provence, 1998, math-ph/9903047

C. P. Martín and D. Sanchez-Ruiz, *The one-loop UV divergent structure of U(1) Yang-Mills theory on noncommutative* $\mathbb{R}^4$, hep-th/9903077, Phys. Rev. Lett. 83 (1999) 476

M. M. Sheikh-Jabbari, *Renormalizability of the supersymmetric Yang–Mills theories on the noncommutative torus*, hep-th/9903107, JHEP 9906 (1999) 15

T. Krajewski and R. Wulkenhaar, *Perturbative quantum gauge fields on the noncommutative torus*, hep-th/9903187, Int. J. Mod. Phys. A15 (2000) 1011

S. Cho, R. Hinterding, J. Madore and H. Steinacker, *Finite field theory on noncommutative geometries*, hep-th/9903239, Int. J. Mod. Phys. D9 (2000) 161

52. M. Rieffel, *Irrational Rotation C^*-Algebras*, Short Comm. I.C.M. 1978

A. Connes, C^* *algèbres et géométrie différentielle*, C.R. Acad. Sci. Paris, Ser. A-B (1980) 290, English version hep-th/0101093

A. Connes and M. Rieffel, *Yang–Mills for non-commutative two-tori*, Contemp. Math. 105 (1987) 191

53. J. Bellissard, $K-$*theory of C^*-algebras in solid state physics*, in: Statistical Mechanics and Field Theory: Mathematical Aspects, eds.: T. C. Dorlas et al., Springer (1986)

J. Bellissard, A. van Elst and H. Schulz-Baldes, *The noncommutative geometry of the quantum Hall effect*, J. Math. Phys. 35 (1994) 5373

54. A. Connes and G. Landi, *Noncommutative manifolds, the instanton algebra and isospectral deformations*, math.QA/0011194, Comm. Math. Phys. 216 (2001) 215

A. Connes and M. Dubois-Violette, *Noncommutative finite-dimensional manifolds I. Spherical manifolds and related examples*, math.QA/0107070

55. M. Chaichian, P. Prešnajder, M. M. Sheikh-Jabbari and A. Tureanu, *Noncommutative standard model: Model building*, hep-th/0107055

    X. Calmet, B. Jurčo, P. Schupp, J. Wess and M. Wohlgenannt, *The standard model on non-commutative space-time*, hep-ph/0111115

56. A. Connes and C. Rovelli, *Von Neumann algebra Automorphisms and time-thermodynamics relation in general covariant quantum theories*, gr-qc/9406019, Class. Quant. Grav. 11 (1994) 1899

    C. Rovelli, *Spectral noncommutative geometry and quantization: a simple example*, gr-qc/9904029, Phys. Rev. Lett. 83 (1999) 1079

    M. Reisenberger and C. Rovelli, *Spacetime states and covariant quantum theory*, gr-qc/0111016

57. W. Kalau, *Hamiltonian formalism in non-commutative geometry*, hep-th/9409193, J. Geom. Phys. 18 (1996) 349

    E. Hawkins, *Hamiltonian gravity and noncommutative geometry*, gr-qc/9605068, Comm. Math. Phys. 187 (1997) 471

    T. Kopf and M. Paschke, *A spectral quadruple for the De Sitter space*, math-ph/0012012

    A. Strohmaier, *On noncommutative and semi-Riemannian geometry*, math-ph/0110001

Index

Lecture Notes in Physics

For information about Vols. 1–612
please contact your bookseller or Springer
LNP Online archive: springerlink.com